ALLE ZEIT WACH 1842

Guido di Prisco
Bruno Maresca
Bruno Tota
(Eds.)

Biology of Antarctic Fish

With 105 Figures
and 30 Tables

Springer-Verlag
Berlin Heidelberg New York
London Paris Tokyo
Hong Kong Barcelona
Budapest

Professor Dr. *Guido di Prisco*
Institute of Protein Biochemistry and Enzymology (IBPE)
Italian National Research Council
Via Marconi 10, I-80125 Naples
Italy

Professor Dr. *Bruno Maresca*
International Institute of Genetics and Biophysics, CNR
I-80125 Naples
Italy

Professor Dr. *Bruno Tota*
University of Calabria
I-87030 Arcavacata di Rende (CS)
Italy

ISBN-13:978-3-642-76219-2 e-ISBN-13:978-3-642-76217-8
DOI: 10.1007/978-3-642-76217-8

Library of Congress Cataloging-in-Publication Data
Biology of antarctic fish / Guido de Prisco, Bruno Maresca, Bruno Tota, editors. p. cm.
Includes bibliographical references and index.

1. Marine fishes – Antarctic Ocean – Congresses.
I. Di Prisco, Guido, 1937- . II. Maresca, B. (Bruno), 1950- . III. Tota, Bruno.
QL637.2.B56 1991 597.092'4-dc20 91-35000 CIP

Softcover reprint of the hardcover 1st edition 1991

Production Editor: Martin Langner
Typesetting, International Typesetters, Inc. Makati, M. M., The Philippines
31/3145-543210 – Printed on acid-free paper

Preface

Antarctica originated as a fragment of the ancient supercontinent Gondwana, which drifted to its present polar position and was isolated climatically and biologically by the establishment of a circumpolar oceanic circulation. This gave rise to a well-defined, roughly circular ocean front: the Antarctic Convergence. During the past 25 million years, fish have remained in isolation south of the Convergence, where the progressive cooling of the environment became a key factor in influencing their evolution. At present, the year-round temperature of the coastal waters is very near −1.87°C, the equilibrium temperature of the ice-seawater mixture.

The unique characteristics produced in antarctic fish in the course of their evolution during the long isolation south of the Convergence encourage the investigation of this ecosystem. Populations of complex organisms such as fish contain a hierarchy of levels of organization: individuals, organ systems, cells, organelles, membranes, molecules (in order of decreasing complexity). Indeed, the temperature of the environment affects each of the levels of organization of fish life. Thus, the biology of fish of the Antarctic Ocean, where the extremely low temperature is a feature of the utmost importance, can be best explored when multidisciplinary approaches and efforts are combined in a general framework.

The need to produce a clear, up-to-date picture of the present state of knowledge in a large number of fields of the biology of fish belonging to this unique environment, gave rise to a strong motivation to organize a meeting along these lines.

The Second International Conference: Biology of Antarctic Fishes ("Ravello 2") took place in Ravello (Italy) at the end of May 1990. It was sponsored by the Italian National Research Council and by the National Programme for Antarctic Research. "Ravello 2" was a follow-up of the first Conference ("Ravello 1", 1986), which dealt with a much broader range of topics inherent to marine biology of antarctic organisms.

The Conference lectures were focused on ecology, evolution, life history, physiology, biochemistry and molecular biology of antarctic fish, and are all contained in this volume. We are possibly dealing with the most comprehensive and complete contribution so far available in the field.

We express our gratitude to all the speakers, thanks to whom the Conference has been so stimulating and productive. We also wish to express our sincere hope to reconvene in "Ravello 3" in the near future.

Naples/Italy, October 1990

Guido di Prisco
Bruno Maresca
Bruno Tota

The speakers at the Second International Conference: Biology of Antarctic Fishes ("Ravello 2"). From left to right: *front row,* Eastman, Detrich, Targett, Duhamel, Radtke; *second row,* Hemmingsen, Ekau, Hubold, Wells, Sidell, North, El-Sayed, Somero, Clarke, Johnston, DeVries, Macdonald, Kellermann, accompanying person; *at back,* White, Tota, di Prisco, Maresca.

Contents

Part 1 Ecology, Evolution and Life History

Part 2 Physiology, Biochemistry and Molecular Biology

Part I

Ecology, Evolution and Life History

Ecology of Notothenioid Fish in the Weddell Sea

G. Hubold

1 Introduction

Low water temperature, and the seasonality and scarcity of food are key factors in the life of polar ectotherms (Clarke 1983). Life in cold water requires special adaptations against freezing in teleost fish (DeVries 1971), and the need to accumulate energy reserves during summer may govern the seasonal activity pattern. In addition to these typical high-latitude features, environmental factors such as water-mass distribution and current patterns also affect the life of fish in the Antarctic. Hydrography becomes a critical factor when pelagic stages are included in the life history (Harden Jones 1968; Cushing 1974; Sinclair 1988).

For evolutionary adaptation, stability or predictability of the environment is important. Antarctic fish occur over a latitudinal range of more than 20° in the Atlantic sector. While temperature and production decrease towards the south, stability of temperature and predictability of hydrographic features increase in the permanent pack-ice zone surrounding the continent (Hempel 1990). Environmental conditions similar to those presently found may have prevailed over the past 20 million years, although several major advances and retreats of the continental ice sheet will have had effects on the composition of the fish fauna (DeWitt 1970; Hubold and Ekau 1987; Clarke and Crame 1989). High latitude species may therefore by confronted with *very severe conditions*, but may also have had the possibility to develop *very specialized adaptations* to match these conditions in the course of evolution. Fish species from high antarctic latitudes (e.g., McMurdo Sound) were found to have more specialized biochemical adaptations than subantarctic species (DeVries and Eastman 1981). It is therefore of interest to investigate a wider spectrum of biological characteristics of high-latitude antarctic fish and relate these to the key factors of the particular environment, such as the southern Weddell Sea.

Institute for Polar Ecology, University of Kiel, Olshausenstr. 40–60 D-2300 Kiel 1(W), FRG

2 Data Base

Due to very difficult access to the permanent pack-ice zone, the antarctic continental shelves are poorly studied. In the Weddell Sea, early taxonomic data on fish are available from the "Scotia" expediton of 1901-1903 to Coats Land (Norman 1938). First hydrographic investigations date back to the "Deutschland" expedition in 1911–1912 (Brennecke 1921). Hydrographic investigations were intensified in the 1960s and 1970s (Gill 1973; Carmack and Foster 1977), and a review was published by Hellmer and Bersch (1985).

Since 1979/80, systematic biological and hydrographical investigations have been carried out on six expeditions to the shelf area between Atka Bay (8° W) and the Filchner-Ronne ice shelf (60° W) within the frame of a multidisciplinary German and European research project. More than 200 000 fish were collected by bongo nets, a rectangular midwater trawl (RMT) 8+1, pelagic krill trawls, Agassiz trawls, benthopelagic and demersal trawls. Details of the cruises and station work are provided in the respective expedition reports (Schneppenheim 1982; Hubold and Drescher 1982; Drescher et al. 1983; Hempel 1983, 1985; Schnack-Schiel 1987; Arntz et al. 1990). The expedition work included studies of oceanography, plankton, benthos, mammals and birds in addition to the fish investigations. By now, this data set is probably the most comprehensive ecological inventory of a high antarctic shelf sea, and a number of results on the Weddell Sea ecosystem have been published (see review by Hempel 1990).

3 The Weddell Sea Environment

3.1 Abiotic Factors

The Weddell Sea is the largest of several deep embayments of the antarctic continent. On the eastern side, within the current system of the east west drift (EWD), a coastal current conveys surface water masses from lower to higher latitudes as far south as 78° S (Fig. 1). Current velocity can be in the order of 30–40 cm s^{-1} (Carmack and Foster 1977). The current is also present near the bottom where a diverse and abundant epifauna, typically composed of filter feeders, makes use of the particles advected in the current (Voß 1988; Gutt 1988). Large organisms such as sponges provide structured microhabitats for fish on the sea floor of the eastern Weddell Sea shelf (Ekau and Gutt 1990).

Based on year-round mooring observations, a counter-current to the Weddell Sea coastal current was recently described on the outer continental slope in depths between 1000 and 2000 m (Fahrbach et al. 1987; Kottmeier and Fahrbach 1989). Coastal current and counter current are stable hydrographic structures providing predictable conditions for the drift patterns of plankton organisms and pelagic stages of fish in the Weddell Sea.

On a medium scale, the distribution of the principal water masses in the EWD and coastal current leads to differentiated temperature conditions on the shelf

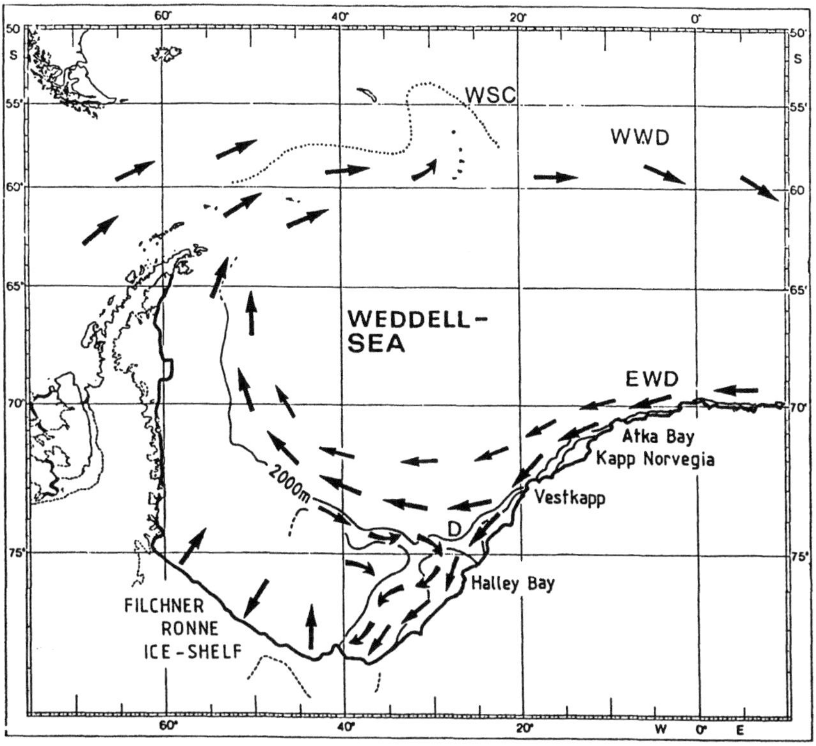

Fig. 1. Topography of the Weddell Sea. Currents (*arrows*) after Gill 1973; Carmack and Foster 1977; Robin et al. 1983; see text)

(Fig. 2). Water temperature in the eastern shelf water (ESW) is constantly below –1.8 °C. Due to advection of warm meltwater from the north and east, temperatures can reach positive values during summer in a surface layer of 50–100 m thickness (Gammelsrod and Slotsvik 1981; Rohardt et al. 1990). The stable summer surface layer is another predictable feature of the Weddell Sea hydrography.

At a depth of > 500 m, oceanic warm deep water (WDW) approaches the shelf and fills the deep trenches of the eastern shelf, providing constantly temperatures between 0 and + 0.8 °C in near-bottom layers. The warm deep water is not present on the southern shelf or in the Filchner Depression, where cold antarctic bottom water (ABW) is formed and spreads one the sea floor.

Both the surface summer water and the WDW are inhabited by notothenioid fish. Total temperature range experienced by the fish on the southern Weddell Sea shelf, however, is no more than –2.2 to +1 °C.

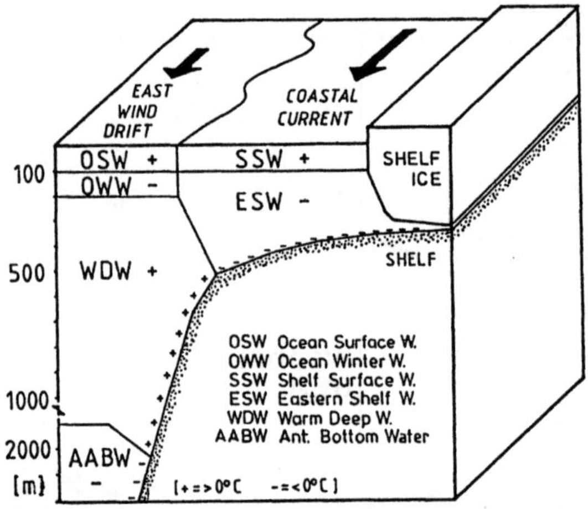

Fig. 2. Principal water masses found in the southeastern Weddell Sea (After Gammelsröd and Slotsvik 1981; Wegner 1982; Fahrbach et al. 1987; Rohardt et al. 1990; Wegner, unpubl.; Hellmer, unpubl.)

3.2 Plankton Biomass and Production

Primary production in the Atlantic sector decreases from 50–100 g c m^{-2} $year^{-1}$ in the Bransfield Strait and Antarctic Peninsula area (von Bodungen et al. 1986) to 20–40 g C m^{-2} $year^{-1}$ in the eastern Weddell Sea (von Bodungen et al. 1988). In the Bransfield Strait, zooplankton biomass (wet weight) can locally exceed 250 g m^{-2} in krill swarms (Siegel 1986). Mean zooplankton biomass is about 30 g wet wt. m^{-2} in the northern Weddell Sea, but only 10 g m^{-2} in the south. Principal species in the southern Weddell Sea are calanoid copepods of 1–5 mm size and the ice krill *E. crystallorophias* of > 15 mm. Gelatinous plankton (tunicates, coelenterates and chaetognaths) accounts for ca. 20% of zooplankton biomass in the 1–40-mm-size range (Boysen-Ennen et al., in prep.). Based on the biomass estimates, zooplankton production can be assessed by using P:B ratios (Voronina et al. 1980; Allen 1971) as approximately 170 and 40 g m^{-2} $year^{-1}$, in the northern and southern shelf areas, respectively. Due to its species composition of mainly crustaceous organisms, most of the pelagic production can be utilized as food by fish.

3.3 Benthos

Benthos biomass on eastern artarctic continental shelves is about 500 g WW m^{-2} (Permitin 1970). Most of this biomass is due to organisms not suitable as fish food (e.g., 74% sponges; Belyaev and Ushakov 1957, in Arnaud 1977). Similar benthos communities were described in the Weddell Sea (Voß 1988). Low P:B ratios of antarctic benthos limit the availability of demersal food for antarctic-shelf fish even more (Dearborn 1965; Picken 1979; Luxemore 1985; Clarke 1985).

Low food level, together with a relatively high share of pelagic production advected in the coastal current, is therefore a key environmental factor for the life of fish on the southern Weddell Sea shelf.

4 The Weddell Sea Fish Fauna

4.1 Fish Biomass

As a result of the low primary and secondary production, mean fish biomass in the eastern and southern Weddell Sea is one order of magnitude lower than that of the subantarctic islands (e.g., South Georgia; Fig. 3). Biomass estimates based on demersal trawling (swept-area method) yielded 0.3–1.6 t km^{-2} on the eastern shelf, and 0.1–0.4 t km^{-2} in the southernmost parts (Ekau 1988, 1990). Pelagic biomass is <0.1 t km^{-2} on the eastern shelf and in the order of 0.8 t km^{-2} in the south (after Hubold and Ekau 1987). Thus, from northeast to southwest, total biomass remains fairly constant, but there is a shift from demersal to pelagic species. This shift reflects the amount of advected zooplankton in the southern Weddell Sea (Hubold et al. 1988). The dominant demersal fish species is the channichthyid *Chionodraco*

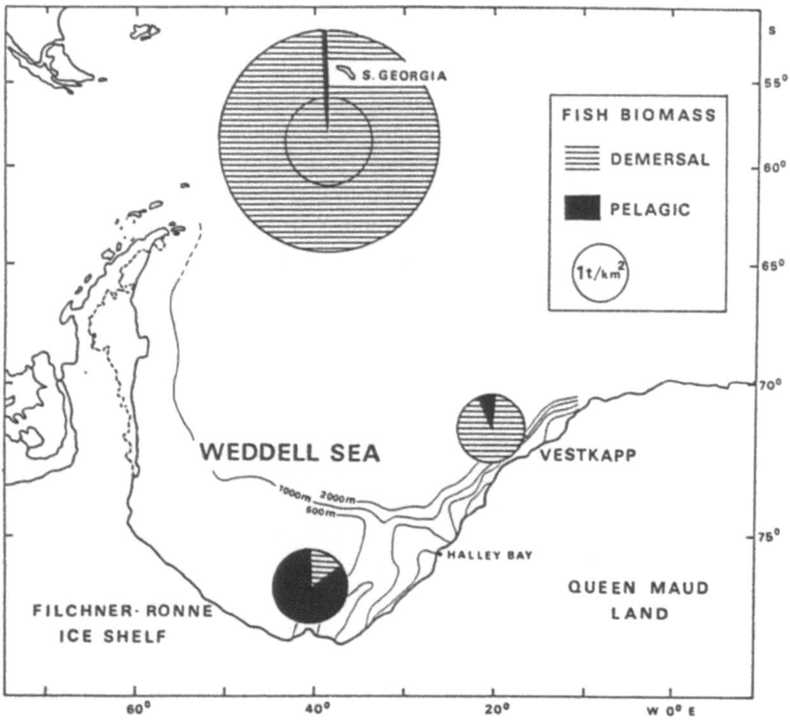

Fig. 3. Fish biomass at three localities in the Atlantic sector. South Georgia (pristine stock 14 t km^{-2}, Everson 1977b). Southern Weddell Sea (Hubold and Ekau 1987; Ekau 1988)

myersi (49% by numbers; Ekau 1990). Pelagic fish biomass is almost entirely due to one species, *Pleuragramma antarcticum.*

4.2 Species Composition and Diversity

In spite of low fish biomass and the dominance of single species, the fish fauna of the eastern and southern Weddell Sea is diverse on the species level. It consists of more than 75 species from 8 orders, 14 families, and 43 genera. Excluding oceanic (mesopelagic) species, the Weddell Sea fish fauna is richer in species (69) and genera (37) than the subantarctic island shelves (Elephant Island, 44 and 31, respectively; Nast et al. 1987). Notothenioids contribute 70% of species and dominate even more than in the Antarctic Ocean as a whole (57%; Andriashev 1987).

Species composition in the Weddell Sea differs considerably from that of the subantarctic islands and Antarctic Peninsula, where the genera *Notothenia* and *Champsocephalus* dominate. Of the coastal species in the Weddell Sea and on the Elephant Island shelf, only 28 are common to both areas (Nast et al. 1987; Hubold, in prep.). Some of the subantarctic species occur occasionally in the southern Weddell Sea, where they live in the deeper layers in the WDW (Ekau 1990).

The family Nototheniidae is the most diverse with 33% of the species and 94% of the specimens. Artedidraconidae represent 28% of the species, Bathydraconidae 22%, and Channichthyidae 17% of the species.

Common demersal species are *Chionodraco myersi* (20% of specimens, 50% by biomass; Ekau 1990), *T. scotti, T. lepidorhinus, T. eulepidotus,* and the small *Dolloidraco longedorsalis* (10% of specimens each). By biomass, other important species are *Chionodraco hamatus, Pagetopsis maculatus, Cryodraco antarcticus, Cygnodraco mawsoni* and *P. antarcticum* (Ekau 1990).

The Weddell Sea fish fauna is similar to that of the Ross Sea (Iwami and Abe 1981) and Prydz Bay (Williams, pers. comm.), when trawl catches are considered. Species composition of trap and line catches is different from trawls, but again, trap catches are comparable between high-latitude shelf seas. *P. hansoni* and *P. bernacchii* are typical species collected in traps, where they represent 66–88% of the catch in the Weddell Sea and in East Antarctica (Naito 1982; Ekau 1988). They represent, however, less than 0.4% in the demersal fish community in the Weddell Sea as described from bottom trawl catches.

The diversity of the demersal fish communities on the spatial scales of bottom trawl hauls (10^4–10^5 m^2) is high (Shannon-Weaver Diversity Index H' = 1.8; evenness = 0.62; species richness Margaleff index, 2.7). Diversity is higher than in bottom trawl hauls from the North Sea (H = 1.05) or southeast Greenland (H' = 1.2; data from S. Ehrich, Hamburg, pers. comm.; Hubold, in prep.) or on subantarctic island shelves (Tiedtke and Kock 1989; Targett 1981). Within the Weddell Sea, there are only small differences in diversity from northeast to southwest (Schwarzbach 1988).

In the water column, diversity is low (H = 0.7; species richness = 1; evenness = 0.5). The main structuring element in the pelagic realm is the transition from oceanic to neritic fauna, which coincides with the hydrographic frontal zone over

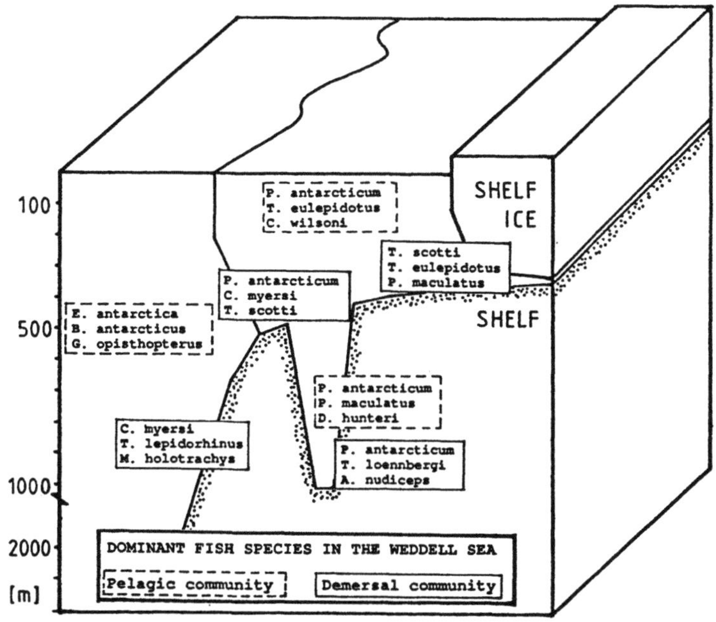

Fig. 4. Dominant species in fish assemblages defined by cluster analysis

the slope of the continental shelf. The oceanic fauna consists mainly of myctophids, bathylagids, and paralepidids.

Typical species assemblages can be defined by multivariate analysis. The clusters correspond primarily to depth strata and secondarily to latitudinal differences. These communities may be termed "shallow shelf", "deep shelf", "slope" and "Filchner Depression" (Schwarzbach 1988; Hubold, in prep.). In the water column, the oceanic-neritic separation is even stronger than the seasonal differences, which are marked in the pelagic communities on the shelf. A northeastern and a southern shelf assemblage can be distinguished in the pelagic realm.

Dominant species (by numbers) in the clusters are shown in Fig. 4. Each species assemblage on the eastern shelf contains a dominant channichthyid and one or two nototheniid species. On the southern shelf, the channichthyids are among the dominant species only in the water column. Non-notothenioid species appear in the warm deep water over the continental slope.

4.3 Feeding and Activity

Demersal fish in the Weddell Sea feed on polychaetes, amphipods, mysids, and deep living zooplankton (Kock et al. 1984; Schwarzbach 1988). Most of these prey organisms are actively moving epibentic or benthopelagic species, or planktonic species advected in the coastal current. This "active" prey movement enables the

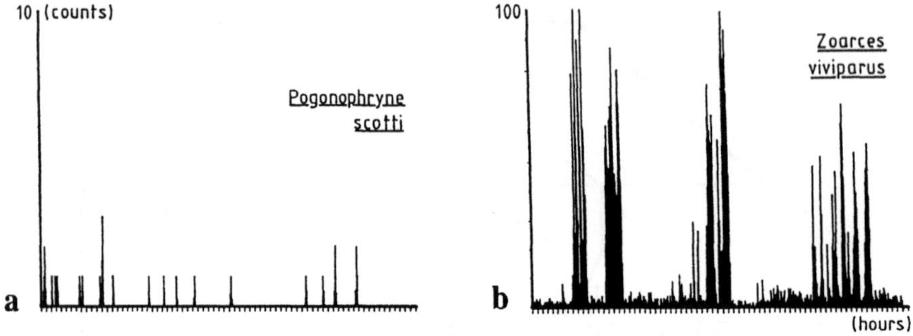

Fig. 5.a. Spontaneous activity of *Pogonophryne scotti* in captivity as measured by light-path interruptions. Temperature 0 °C; constant dim light. **b.** Spontaneous activity of boreal Zoarces viviparus under the same experimental conditions, at 4 °C

fish to adopt an economic "sit-and-wait" feeding strategy, somewhat analogous to that of filter-feeding epibenthos. For fishes of this feeding type, it is therefore of advantage to sit on elevated places, and such a "sit-and-wait" behaviour can be repeatedly observed in the underwater videos from the Weddell Sea shelf (J. Gutt, unpubl.). Some fish also ascend into the near-bottom water layer and stay in place against the current. These species show a more active life style and belong mostly to the Channichthyidae and Nototheniidae. The family Artedidraconidae contains some very sluggish species, and individuals of *A. longedorsalis* do not swim away even when touched with a stick. In a laboratory study, the slightly more active *Pogonophryne scotti* was observed to sit motionless for several hours (Fig. 5a). A boreal zoarcid under the same experimental conditions showed significantly higher activity (Fig. 5b).

4.4 Metabolism

Sluggish zoarcids have very low metabolic rates. Active notothenioid species such as *Pagothenia* spp. show higher resting metabolism (Wohlschlag 1964; Wells and Tetens 1984; Wells 1987). Scope for activity was therefore suggested to be one factor determining resting metabolic rates of antarctic notothenioid fish (Wells 1987).

Part of the elevated resting metabolism of the notothenioid species may be due to the maintenance of an efficient nervous and sensory system, and to the synthesis of sufficient levels of antifreeze proteins and connected physiological activity (DeVries and Somero 1970). Zoarcids do not possess the high molecular antifreeze proteins of notothenioids and are not an endemic antarctic fish family. The investigation of specialized, endemic, yet sluggish notothenioid species can yield basic estimates on the metabolic cost for those specialized adaptations which are not directly related to activity.

A truly sluggish notothenioid species is *Pogonophryne scotti*. Although the blood of *P. scotti* has not been tested for antifreeze capacity so far, *Pogonophryne*

Table 1. Routine oxygen consumption, freezing resistance and activity types of selected polar fish species[a]

Species	Freezing resistance	Resting VO_2 at $T < -1$ C ($mg\ O_2\ kg^{-1}\ h^{-1}$)	Source
Antarctic Species			
Nototheniidae:			
P. borchqrevinki		48	Wells (1987)
		56	After Wohlschlag in Wells (1978)
	−2.09		DeVries and Somero (1970)
T. bernacchii		51	After Wohlschlag (1964)
		42	After Wohlschlag in Wells (1987)
		56	Wells (unpubl. in Macdonald et al. (1987)
	−1.98		DeVries and Somero (1970)
T. hansoni		60	Wohlschlag 1964 in Macdonald et al. (1987)
		59	Wohlschlag 1964 in Wells (1987)
		77	Wells (unpubl. in Macdonald et al. (1987)
	−2.01		DeVries and Somero (1970)
T. loennbergi		37	Wohlschlag (1964) in Macdonald et al. (1987)
		35	Wells (unpubl. in Macdonald et al. 1987)
	−1.83		DeVries and Somero (1970)
T. eulepidotus (115 g)		63	Saint-Paul et al. (unpubl.)
T. centronotus		46	Macdonald et al. (1987)
Artedidraconidae:			
P. scotti		28	Saint-Paul et al. (1988)
Bathydraconidae:			
G. acuticeps		45	Wells (unpubl. in Macdonald et al. 1987)
Other taxa: Zoarcidae			
R. dearborni		15	After Wells (1986)
		19	After Wohlschlag (1964) in Macdonald et al. (1987)
		10	Wells (unpubl. in Macdonald et al. 1987)
	−1.54		DeVries and Somero (1970)
Arctic species			
Cottidae spp.		28	After Holeton (1974)
Zoarcidae spp.		15	After Holeton (1974)
Gadidae:			
B. saida		41	After Holeton (1974)
	−2.2		Denstad et al. (1987)
G. ogac		35	Mikhail and Welch (1989)
	−1.6 Winter		Scholander et al. (1957)

[a]O_2 measurements at < − 1°C, standardized for 100 g fish.

spp. survive cooling to the freezing point in the presence of ice crystals. Furthermore, these species are on the alert and capable of learning in captivity (Hubold, unpubl.). Resting metabolism of *P. scotti* falls between that of zoarcids and nototheniids (Saint Paul et al. 1988). In addition to the minimum metabolic rate of 10–20 mg $O_2kg^{-1}h^{-1}$ displayed by the zoarcid *Rhigophila (Lycodichthys) dearborni*, *P. scotti* respires approximately 15 mg $O_2kg^{-1}h^{-1}$ for metabolic purposes, which are not related to potentially high activity (Fig. 6). A compilation of respiration rates of high antarctic and arctic fish shows a good correlation with activity pattern and degree of adaptation (inferred from taxonomic category; Table 1; Fig. 6). Oxygen consumption of taxonomically or ecologically comparable arctic species is similar to that of antarctic species. The difference in resting oxygen consumption between a sluggish zoarcid and a sluggish cottid or notothenioid may be a rough measure for the cost of maintaining high levels of freezing resistance and nervous function in the specialized or endemic species. Additional levels of resting metabolism in the active species are surprizing and may provide "one strategy whereby a fish could increase its active rate" (Wells 1987).

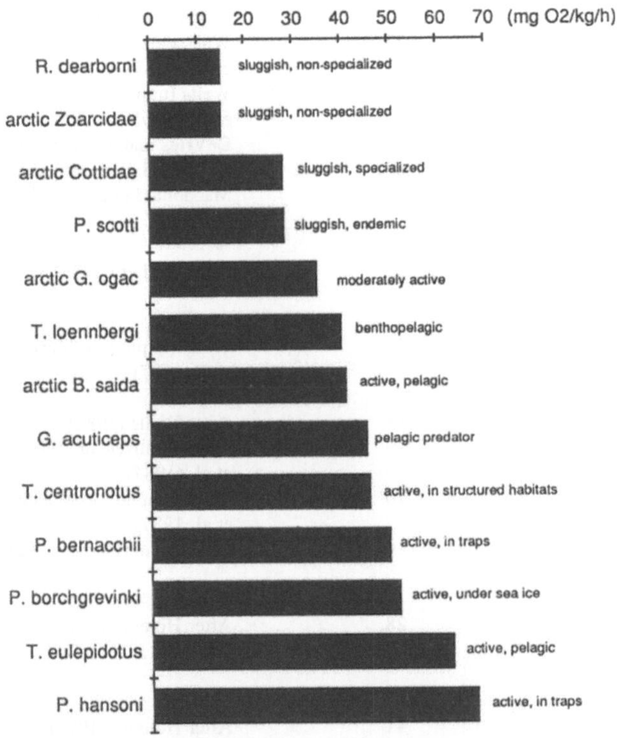

Fig. 6. Resting oxygen consumptionof antarctic and arctic fish measured at temperatures < 0 °C (data from Scholander 1957; Wohlschlag 1964; DeVries and Somero 1970; Holeton 1974; Wells 1986; Denstad et al. 1987; Macdonald et al. 1987; Wells 1987; Saint Paul et al. 1988 and unpubl.; Mikhail and Welch 1989; according to Table 1)

4.5 Size and Growth

Most antarctic fish are small. On an average, the Weddell Sea species are smaller than subantarctic taxa. Seventy-seven percent of the species are smaller than 35 cm, and 50% are smaller than 25 cm (L_{max}). Except for *Dissostichus mawsoni*, large notothenioids are lacking in the Weddell Sea.

Growth of antarctic fish is generally slow (Everson 1977a). Most species, however, do not grow significantly slower than ecologically similar boreal species (Olsen 1954; Everson 1977a). When growth parameters k and W_∞ of the von Bertalanffy growth equation are compared in the auximetric grid (Pauly 1979), *Dissostichus mawsoni* grows similar to boreal *Lophius piscatorius*, notothenioids grow like gadoids, and some channichthyids grow almost as fast as mackerels (Pauly 1979; Hubold, in prep.).

Growth of Weddell Sea nototheniids is lower than that of subantarctic species (Table 2; Fig. 7). Exceptionally slow growth is observed for *Trematomus* and *Pagothenia* spp., *Aethotaxis mitopteryx* and*Pleuragramma antarcticum*. In contrast, growth of a channichthyid (*Pagetopsis maculatus*) from the Weddell Sea is probably fast (Rake and von Dorrien, pers. comm.)

Slow growth of the Weddell Sea fish cannot be related to a single environmental factor. Slow-growing *T. scotti* and *P. antarcticum* live in cold shelf water, as does fast-growing *Pagetopsis maculatus*. Adult *A. mitopteryx* are found in warm deep water on the continental slope. *Pleuragramma and Aethotaxis* are plankton feeders, whereas *T. scotti* feeds exclusively on a variety of benthic animals. Two of the species may have to spend energy for a pelagic mode of life, and *P. antarcticum* is likely to conduct spawning migrations. The three slowest growing species belong to the more fertile species, as judged by the abundance of adults and larvae and direct fecundity investigations. Thus, reproductive effort may be one key factor affecting growth of Weddell Sea fishes.

4.6 Reproduction

Reproduction in the polar environment requires adaptations on various levels. Adult fish must allocate considerable energy for egg production and/or spawning behaviour (e.g., brood protection and spawning migrations) to provide early life-history stages with sufficient energy reserves to cope with the short seasonal production peaks. Time of spawning, egg incubation time, and larval hatch must be matched to assure the presence of the pelagic larvae during a convenient phase of the seasonal production cycle.

Reproduction parameters of approximately 15 notothenioid species from the Weddell Sea are known (Ekau 1990; Hubold, in prep.). In addition, data of few continental antarctic species from East Antarctica and the Ross Sea are available (Wohlschlag 1961; Dearborn 1965: Andriashev 1965; Butskaya and Faleeva 1987; Hureau 1970; Burchett et al. 1984; Faleeva and Gheracimchook, unpubl.). A compilation of existing data on antarctic fish fecundity has been prepared by Kock and Kellermann (1991). According to the relatively sparse data, age at first maturity of 5–9 years in high antarctic fish is **not** significantly retarded compared to subant-

Table 2. Growth parameters of antarctic fish species

No. species	P	W_{inf}(g)	L_{inf}(cm)	K	T_o	Source
01 *Dissostichus mawsoni*	3.66	75967	185	0.06	–4.7	Burchett et al. (1984)
02 *Notothenia rossii marmorata*	3.24	11683	89	0.15	–0.5	Shust and Pinskaya (1978); Freytag (1980); North and White (1987); Burchett (1983); Olsen (1954); Everson (1977); Scherbich and Slepokurov (1976);
03 *Notothenia rossii rossii*	3.16	12270	91	0.12	–1.0	Hureau (1970); Tankevic (1986);
04 *Pseudochaenichthys georgianus*	2.81	2330	58	0.28	1.3	Kock (1980); Kock et al. (1985); Chojnacki and Palczewski (1981)
05 *Chaenocephalus aceratus*	2.75	2690	69	0.21	0.5	Kock (1980); Gubsch (1980)
06 *Channichthys rhinoceratus*	2.61	1562	56	0.26	0.6	Hureau (1966)
07 *Notothenia gibberifrons*	2.58	2397	54	0.16	–0.3	Shust and Pinskaya (1978); Hoffmann (1982); Kock et al. (1985); Boronin and Altman (1979)
08 *Chionodraco rastrospinosus*	2.52	1724	51	0.19	–0.8	Rowedder (1984)
09 *Champsocephalus gunnari*	2.51	956	66	0.34	0.2	Kock (1981); Kock et al. (1985); Sosinski (1981, 1982)
10 *Notothenia squamifrons*	2.50	3188	55	0.1	0.03	Kock et al. (1985); Duhamel and Ozouf-Costaz (1985); Dumamel and Pletikosic (1983)
11 *Notothenia neglecta*	2.40	2236	49	0.11	–1.3	Everson (1970)
12 *Notothenia ramsayi*	2.35	769	40	0.29	0.2	Ekau (1982)
13 *Pagothenia hansoni*	2.27	1353	37	0.14	–1.4	Shust and Pinskaya (1978); Hureau (1970)
14 *Trematomus nicolai*	1.98	1191	38	0.08	0.1	Ekau (1988)
15 *Pagothenia hansoni*	1.87	924	37	0.08	0.3	Ekau (1988)
16 *Pagothenia bernacchii*	1.73	595	31	0.09	0.3	Ekau (1988)
17 *Trematomus centronotus*	1.71	862	34	0.06	0.03	Ekau (1988)
18 *Pagothenia bernacchii*	1.67	309	30	0.15	0.3	Pauly (1979) after Wohlschlag (1962); Hureau (1970)
19 *Trematomus eulepidotus*	1.66	380	27	0.12	–0.5	Ekau (1988)
20 *Trematomus centronotus*	1.65	450	28	0.1	–1.9	Zuur (1977)
21 *Trematomus lepidorhinus*	1.63	354	26	0.12	0.3	Ekau (1988)
22 *Trematomus loennbergi*	1.60	365	29	0.11	0.2	Ekau (1988)
23 *Pleuragramma antarcticum*	1.59	226	31	0.17	0.1	Gheracimchook (1987 and pers. comm.)
24 *Aethotaxis mitopteryx*	1.52	833	41	0.04	–3.6	Ekau (1988)

Table 2. (*continued*)

No. of species	P	W_{inf}(g)	L_{inf}(cm)	K	T_o	Source
25 *Pagothenia borchgrevinki*	1.50	224	25	0.14	0.03	Ekau (1988)
26 *Pagetopsis maculatus*	1.42	43	17	0.61	0.3	Rake and von Dorrien (unpubl.)
27 *Pleuragramma antarcticum*	1.08	171	31	0.07	1.2	Hubold and Tomo (1988)
28 *Harpagifer bispinis*	0.91	37	11	0.22	1.3	North and White (1987) after Daniels (1982); Tomo and Cordero-Funes (1972)
29 *Trematomus scotti*	0.71	37	14	0.14	1.0	Ekau (1988)

[a] P refers to "Parameter of growth Performance" (Pauly 1979). Mean values from published data. Von Bertalanffy parameters: W_{inf} = infinite weight; L_{inf} = infinite length; K = growth coefficient; T_o(years) = age at theoretical length 0. Numbers refer to Fig. 7.

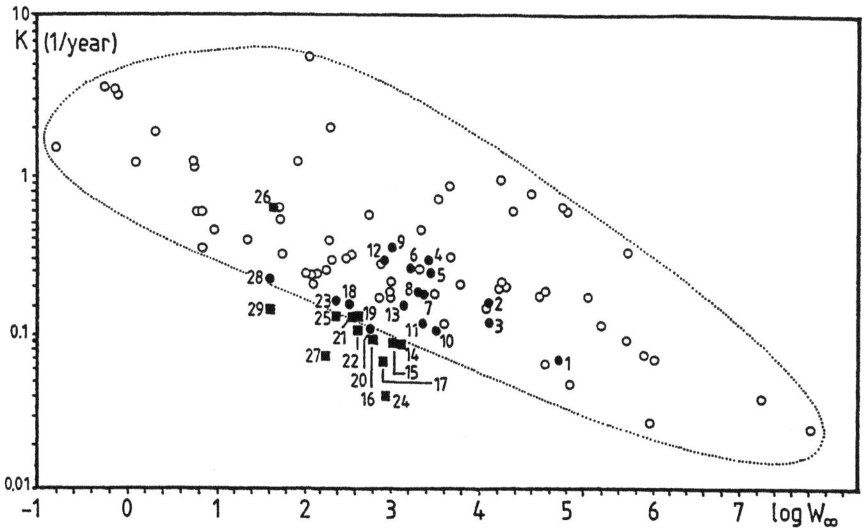

Fig. 7. Growth of antarctic fish species in the "Auximetric Grid" (Pauly 1979). *Numeration* of antarctic species according to Table 2. ■ Weddell Sea; ● Antarctic Ocean; ○ World Ocean; *K* growth coefficient of the von Bertalanffy growth equation

arctic species (Hubold, in prep.). Species with wide latitudinal distribution, however, tend to have later maturation in the south (Hureau 1970; Duhamel 1987).

Fecundity of Weddell Sea fish is relatively low, between 200 and 13 000 eggs per female. High fecundity is found in *P. antarcticum*; low fecundity and brood caring is typical for artedidraconids and some nototheniids. Channichthyids have a relatively low fecundity. Fecundity is positively related to size. Thus, low absolute fecundity of Weddell Sea fish is at least partly a result of the generally small species' sizes. Relative fecundity values between 10 and 200 (eggs/g) fall within the limits set by subantarctic species (Kock and Kellermann 1991).

Reproductive effort of Weddell Sea fish, as calculated by relative gonad volume (egg/g x egg volume) and converted to theoretical gonado-somatic-indices are between 13 and > 50% (Hubold, in prep.). This effort is as high as in subantarctic species (Kock and Kellermann 1991), and compares with that of boreal cyprinids and salmonids (Wootton 1990). Eggs are generally large in Weddell Sea fish, between 2 (*P. antarcticum*) and 4.4 mm (oocytes, *Cryodraco antarcticus*). Considering the energy limitation of the high latitudes, the relative effort invested in reproduction of southern species tends to be high.

4.7 Early Life History

Developmental time is positively related with egg size in fish (Blaxter 1969). In cold water, development of large fish eggs (e.g., salmonids) takes between 3 and 6 months. Developmental times of antarctic fish eggs of comparable size are in the same range and thus do not appear to be temperature compensated (Hubold, in prep.). Long developmental time increases mortality risk by predation, especially because antarctic benthos are not in hibernation during winter. This calls for brood-caring or protective activity, which may be an additional energy consumer in the energy budget of the fish. Especially for sluggish species, activity related to nesting and territorial behaviour (as observed, for example, in *Harpagifer antarcticus* and *Pogonophryne phyllopogon*, respectively) may represent considerable energy expenditure.

Brood caring is one possible strategy to ensure reproduction success. The few large larvae of brood-caring species directly enter the demersal food web and are probably not closely related with seasonal surface production. These species are less sensitive to changes of hydrographic conditions than pelagic species and make up the larger part of the diverse benthic fish community.

Of more than 75 fish species occuring in the Weddell Sea, planktonic stages of less than half (33) are known. Most of these only appear in the surface layer for a short time in summer. There is an established seasonal sequence of species in the summer plankton of the Weddell Sea (Hubold 1990), as well as in other antarctic areas, e.g., the Bransfield Strait (Kellermann 1989). In winter, pelagic stages of only seven species were found in the Weddell Sea.

The pelagic stages and species utilize the relatively high water column production and thus achieve higher stock sizes than the strictly benthic species. They are probably sensitive to environmental changes and represent a relatively young community which evolved in the interglacial periods (DeWitt 1970; Hubold and Ekau 1987).

Planktonic stages concentrate in the surface mixed layer, where feeding conditions seem to be near optimum. Depending on the area, 95–99% of *Pleuragramma* larvae had stomachs filled with 9–29 food particles each (von Dorrien 1989). During the planktonic phase, fast growth (e.g., 0.2 mm/day in *P. antarcticum*) can be achieved (Keller 1983; Hubold 1985).

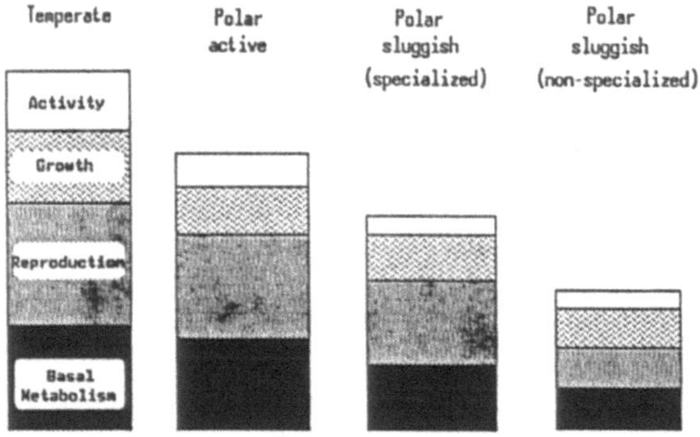

Fig. 8. Concepts of different types of metabolism in polar fish (after Clarke 1980). *Polar active* represents highly evolved active species of the Notothenioidei, e.g., *Trematomus* and *Pagothenia* spp. and channichthyids, as well as arctic gadoids. *Polar sluggish, specialized* represents the majority of endemic antarctic artedidraconids, bathydraconids and some nototheniids, as well as arctic cottids with relatively high basal metabolism, high reproductive effort and relatively low activity and growth. *Polar sluggish, nonspecialized* represents the polar type sensu Clarke. This type is found, e.g., in the family Zoarcidae (arbitraily energy units)

4.8 Energy Budget

As a result of ecological adaptation to the Weddell Sea environment, the energy budgets of single fish will not be simply characterized by an overall depression compared to boreal species, as suggested by Clarke (1980, 1983). Especially reproductive effort and basal metabolism are comparably high, whereas activity and growth may be reduced. In the diagram suggested by Clarke (1980) therefore the "cold adapted" species (sensu Wohlschlag 1964 and Dunbar 1968) and the "polar" species (sensu Clarke) may represent different ecotypes (Fig.8). "Cold adapted" may relate to the active polar species, e.g., from the endemic genera *Trematomus* and *Pagothenia*, and possibly to the predatory channichthyids. These species show relatively high metabolic activity, which is, however, still lower than in ecologically similar temperate species. The "polar" type (sensu Clarke) represents the sluggish non-specialized taxa, such as zoarcids, with a strongly depressed metabolism. The majority of the species, however, is probably of an intermediate, sluggish type, with relatively high resting metabolism (due to high levels of antifreeze and nervous function) and considerable reproductive effort, but with low expenditure for activity and growth.

Acknowledgements. The present review is partly based upon published and unpublished data from colleagues in Bremerhaven and Kiel who participated in the Weddell Sea project headed by Prof. G. Hempel. Werner Ekau kindly advised in the drafting of the manuscript. Special thanks are due to Andrew Clarke for his detailed and constructive critics, and to Andreas Kunzmann for thoroughly checking the final version.

References

Allen KR (1971) Relation between production and biomass. J Fish Res Bd Canada 28:1573–1581
Andriashev AP (1965) A general review of the antarctic fish fauna. In: Oye P van, Mieghem J van (eds) Biogeography and ecology in Antarctica. Junk, The Hague, Monographiae Biologicae 15:491–550
Andriashev AP (1987) A general review of the antarctic bottom fish fauna. In: Kullander SO, Fernholm B (eds), Proc V Congr Europ Ichthyol, Stockholm 1985, 357–372
Arnaud PM (1977) Adaptations within the antarctic marine benthic ecosystem. In: Llano GA (ed) Adaptations within antarctic ecosystems. Houston, 135–157
Arntz W, Ernst W, Hempel I (1990) The expedition ANTARKTIS VII/4 (EPOS leg 3) and VII/5 of RV "Polarsten" in 1989. Rep Polar Res 68:1–214
Blaxter JHS (1969) Eggs and larvae. In: Hoar WS, Randall DJ (eds), Fish physiology, vol 3: Reproduction and growth, bioluminescence, pigments, and poisons. Academic Press, New York, pp 177–252
Boronin AV, Altman YS (1979) Growth and natural mortality of green Notothenia. (in russ). Trudy Atlant Nauchno-issled Ryb Khoz Okeanogr 81:66–71
Boysen-Ennen E, Hagen W, Hubold G, Piatkowski U (in prep) Zooplankton biomass in the Weddell Sea, Antarctica
Brennecke W (1921) Die ozeanographischen Arbeiten der deutschen Antarktischen Expedition 1911–1912. Arch Deut Seewarte 39 (1):1–215
Burchett MS (1983) Food, feeding and behaviour of *Notothenia rossii* nearshore at South Georgia, Br Antarct Surv Bull 1, (1983):45–59
Burchett MS, DeVries A, Briggs AJ (1984) Age determination and growth of *Dissostichus mawsoni* (Norman, 1937) (Pisces, Nototheniidae) from McMurdo Sound (Antarctica). Cybium (8) 1:27–31
Butskaya NA, Faleeva TI (1987) Seasonal changes in the gonads and fecundity of antarctic fishes *Trematomus bernacchii, Trematomus hansoni* and *Pagothenia borchgrevinki* (Nototheniidae). Vopr Ikhtiol 1:114–123
Carmack EC, Foster TD (1977) Water masses and circulation in the Wenddell Sea. In: Dunbar MJ (ed) Polar Oceans. Proc SCOR/SCAR Polar Ocean Conf, Montreal 1974: Arctic Inst. North America, pp 151–165
Chojnacki J, Palczewski P (1981) Age and growth rate of Pseudochaenichthys georgianus Norman 1937 (Fan Channichthyidae) of the South Georgia region. Pol Polar Res 2/3–4:145–157
Clarke A (1980) A reappraisal of the concept of metabolic cold adaptation in polar marine invertebrates. Biol J. Lin Soc 14:77–92
Clarke A (1983) Life in cold water: the physiological ecology of polar marine ectotherms. Oceanogr Mar Biol Ann Rev 21:341–453
Clarke A (1985) Food webs and interactions: an overview of the antarctic ecosystem. In: Bonner WN, Walton WH (eds) Key environments: Antarctica. Pergamon Press, Oxford, pp 329–350
Clarke A, Crame JA (1989) The origin of the Southern ocean marine fauna. In:Crame JA (ed) Origins and Evolution of the antarctic biota. Geol Soc Spec Publ 47:253–268
Cushing DH (1974) The possible density-dependence of larval mortality and adult mortality in fishes. In: Blaxter JHS (ed) The early life history of fish. Springer, Berlin Heidelberg New York, pp 103–111
Daniels RA (1982) Feeding ecology of some fishes of the Antarctic Peninsula. Fish Bull 80 (3):575–588
Dearborn JH (1965) Food of Weddell seals at the McMurdo Sound, Antarctica. J Mammal 46(1):37–43
Denstad JP, Aunaas T, Borseth JF, Vollan Aarset A, Zachariassen KE (1987) Thermal hysteresis antifreezė agents in fishes from Spitsbergen waters. Polar Res 5:171–174
DeVries AL (1971) Glycoproteins as biological antifreeze agents in antarctic fishes. Science 171:1152–1155
DeVries AL, Somero GN (1970) The physiology and biochemistry of low temperature adaptation in antarctic marine organisms. SCAR Symp on antarctic ice and water masses, Tokyo 1970, pp 101–113
DeVries AL, Eastman JT (1981) Physiology and ecology of notothenioid fishes of the Ross Sea. J Roy Soc New Zealand, 11 (4):329–340
DeWitt HH (1970) The character of the midwater fish fauna of the Ross Sea, Antarctica. In: Holdgate MW (ed) Antarctic Ecology, vol 1. Academic Press, London, pp 305–314

Drescher HE, Hubold G,Piatkowski U, Plotz J, Voss J (1983) Das Biologische Programm der Ant-I-Expedition mit FS Polarstern. Ber, Polarforsch 12:1–34

Duhamel G (1987) Reproduction des Nototheniidae et Channichthydae des Iles Kerguelen. Colloque sur l'ecologie marine des ils subantarctiques et antarctiques. Museum Nat d'Histoire Naturelle. CNFRA 57:91–107

Duhamel G, Ozouf-Costaz C (1985) Age, growth and reproductive biology of *Notothenia squamifrons* Gunther 1880 from the Indian sector of the Southern Ocean. Polar Biol 4:143–153

Duhamel G, Pletikosic M (1983) Donnees biologiques sur les Nototheniidae des Iles Crozet. Cybium 7(3):43–57

Dunbar MJ (1968) Ecological development in polar regions. A study in evolution. Prentice Hall, Englewood Cliffs, New York, pp 1–119

Ekau W (1982) Biological investigations on *Notothenia ramsayi* Regan 1913 (Pisces, Notothenioidei) Arch Fischwiss 33(1/2):43–68

Ekau W (1988) Okomorphologie nototheniider Fische aus dem Weddellmeer, Antarktis. Ber Polarforsch 51:1–140

Ekau W (1990) Reproduction in high antarctic fish (Notothernioidei). Proc 6th CEI, Budapest (in press)

Ekau W, Gutt J (1990) Notothenioid fishes from the Weddell Sea and their habitat, observed by underwater photography and television. Proc 12th Symp on Polar Biology, Tokyo 1989 (in press)

Everson I (1970) The population dynamics and energy budget of *Notothenia neglecta* Nybelin at Signy Island, South Orkney Islands. Bull Br Ant Surv 23:25–50

Everson I (1977a) Antarctic marine secondary production and the phenomenon of cold adaptation. Phil Trans R Soc Lond B 279:55–66

Everson I (1977b) The living resources of the Southern Ocean. FAO, Rome, pp 1–156

Fahrbach E, Klindt H, Muus D, Rohardt G, Salameh P (1987) Physical oceanography. In: Schnack-Schiel S (ed) Die Winter-Expedition mit FS "Polarstern" in die Antarktis (ANT V/1–3). Ber Polarforsch 39:156–169

Freytag G (1980) Length, age, and growth of *Notothenia rossii* marmorata Fischer 1885 in the West Antarctic waters. Arch Fischwiss 30(1)39–66

Gammelsrod T, Slotsvik N (1981) Physical oceanography of the Weddell Sea. Some results from the German exped 1979–80. Polarforschung 51(1):101–111

Gill AE (1973) Circulation and bottom water production in the Weddell Sea. Deep Sea Res 20:111–140

Gheracimchook VV (1987) On the fecundity of antarctic sidestripe, *Pleuragramma antarcticum*. Vopr Ikhtiol 5:858–860

Gubsch G (1980) Untersuchungen zur Altersbestimmung und zum Wachstum beim Eisfisch *Chaenocephalus aceratus* Lonnberg. Fisch-Forsch 18(1):7–10

Gutt J (1988) Zur Verbreitung und Okologie der Seegurken (Holothuroidea, Echinodermata) im Weddellmeer (Antarktis). Ber Polarforsch 41:1–87

Harden Jones FR (1968) Fish migration. Arnold, London, pp 1–325

Hellmer H, Bersch M (1985) The Southern Ocean. Ber Polarforsch 26:1–115

Hempel G (1983) Die Expedition ANTARKTIS-I mit FS"Polarstern" 1982/83. Ber Polarforsch 14:1–141

Hempel G (1985) Die Expedition ANTARKTIS III mit FS "Polarstern" 1984/85. Ber Polarforsch 25:1–209

Hempel G (1990) The Weddell Sea: a high polar ecosystem. In: Sherman K, Alexander LM, Gold BD (eds) Large marine ecosystems: patterns, processes, and yields. AAAS Publ No 90–308:5–18

Hoffmann U (1982) Zur Altersbestimmung und zum Wachstum von *Notothenia gibberifrons* Lonnberg bei South Georgia. Fisch–Forsch 20(2):49–53

Holeton GF (1974) Metabolic cold adaptation of polar fish: fact or artefact? Physiol Zool 47:137–152

Hubold G (1985) On the early life history of the high antarctic silverfish Pleuragramma antarcticum. In: Siegfried WR, Condy PR, Laws RM (eds) Antarctic nutrient cycles and food webs. Proc 4th SCAR Symp on Antarctic Biology. Springer, Heidelberg New York Tokyo, pp 445–451

Hubold G (1990) Seasonal patterns of Ichthyoplankton distribution and abundance in the southern Weddell Sea. In: Kerry KR, Hempel G (eds) Antarctic ecosystems – ecological change and conservation. Springer Berlin Heidelberg New York, pp 149–158

Hubold G (in prep) Zur Okologie der Fische im Weddellmeer. Habilitationsschrift, Univ Kiel
Hubold G, Dresclher HE (1982) Filchner–Schelfeis–Expedition 1980/81 mit MS POLARSIRKEL: Liste der Zooplanktonfange und Lichtarkemessungen. Ber Polarforsch 4:1–30
Hubold G, Ekau W (1987) Midwater fish fauna of the Weddell Sea, Antarctica. Proc V Congr Europ Ichthyol, Stockholm 1985, pp 391–396
Hubold G, Tomo AP (1989) Age and growth of antarctic silverfish *Pleuragramma antarcticum* Boulenger, 1902, from the southern Weddell Sea and Antarctic Peninsula. Polar Biol 9:205–212
Hubold G, Hempel I, Meyer M (1988) Zooplankton communities in the southern Weddell Sea (Antarctica). Polar Biol 8:225–233
Hureau JC (1966) Biologie de *Chaenichthys rhinoceratus* Richardson, et probleme du sang incolore des Chaenichthyidae poissons des Mares Australes. Bull Soc Zool Fr 91:735–751
Hureau JC (1970) Biologie comparee de quelques Poissons antarctiques (Nototheniidae). Bull Int Oceanogr Monaco 68(1391):1–244
Iwami T, Abe T (1981) The collection of the fishes trawled in the Ross Sea. Namkyoku Shiro (Antarct Rec) 71:130–141
Keller R (1983) Contributions to the early life history of *Pleuragramma antarcticum* Boul. 1902 (Pisces, Nototheniidae) in the Weddell Sea. Meeresforschung 30:10–24
Kellermann A (1989) The larval fish community in the zone of seasonal pack-ice cover and its seasonal and annual variability. Arch Fischwiss 39 Beih 1:81–109
Kock KH (1980) Graphical analysis of length frequency distribution of *Champsocephalus gunnari* Lonnberg (Channichthyidae) from South Georgia. Cybium Ser 3 (8):33–42
Kock KH (1981) Fischereibiologische Untersuchungen an drei antarktischen Fischarten: *Champsocephalus gunnari* Lonnberg, 1905, Chaenocephalus aceratus (Lonnberg, 1906) und *Pseudochaenichthys georgianus* Norman, 1937 (Notothenioidei, Channichthyidae). Mitt Inst Seefisch Hamburg 32:1–226
Kock KH, Schneppenheim R, Siegel V (1984) A contribution to the fish fauna of the Weddell Sea. Arch Fischwiss 34(2/3):103–120
Kock KH, Duhamel G, Hureau JC (1985) Biology and status of exploited antarctic fish stocks: A review. Biomass Sci Ser 6:1–143
Kock KH, Kellermann A (1991) Reproduction in antarctic fish – a review. Ant Sci 3(2):125–150
Kottmeier C, Fahrbach E (1989) Wechselwirkung zwischen Wasser, Eis und Luft in der antarktischen Kustenzone. Promet 19(1/2):15–22
Luxmoore RA (1985) The energy budget of a population of the antarctic isopod Serolis polita. In:Siegfried WR, Condy PR, Laws RM (eds) Antarctic nutrient cycles and food webs. Springer Berlin Heidelberg New York Tokyo, pp 389–396
Macdonald JA, Montgomery JC, Wells RGM (1987) Comparative physiology of antarctic fishes. Adv Mar Biol 24:321–388
Mikhail MY, Welch HE (1989) Biology of Greenland cod, *Gadus ogac*, at Saqvaqjuac, northwest coast of Hudson Bay, Environ Biol Fishes 26:49–62
Naito Y (1982) Fish fauna in the northeastern parts of Lutzow-Holm Bay with some notes on the stomach contents. In: Proc fifth Symp on Antarctic Biology . Mem Natl Inst Polar Res, Spec Issue 23:63–71
Nast F, Kock KH, Sahrhage D, Stein M, Tiedtke JE (1987) Hydrography, krill and fish and their possible relationships around Elephant Island. In: Sahrhage D (ed) Antarctic Ocean and Resources Variability. Springer Berlin Heidelberg New York Tokyo,pp 183–198
Norman JR (1938) Coast fishes. Part III: The antarctic zone. Discovery Rep. Vol XVIII: 1–105
North AW, White MG (1987) Reproductive strategies of antarctic fish. Proc V Congr Europ Ichthyol Stockholm 1985, pp 381–390
Olsen S (1954) South Georgian Cod (*Notothenia rossii marmorata* Fischer). Norsk Hvalfangst-Tidende 43 (7):373–382
Pauly D (1979) Gill size and temperature as governing factors in fish growth. A generalization of Von Bertalanffy's growth formula. Ber Inst Meeresk Kiel 63:1–156
Permitin YY (1970) The consumption of krill by antarcltic fish. In: Holdgate MW (ed) Antarctic ecology. Academic Press London, pp 177–182

Picken GB (1979) Growth, production and biomass of the antarctic gastropod *Laevilacunaria antartica*

Robin GQ, Doake CSM, Kohnen H, Crabtree RD, Jordan SR, Moller D (1983) Regime of the Filchner Ronne ice shelves, Antartica. Nature (Lond) 302(5909):582–586

Rohardt G, Ruhland G, Schleif U (1990) The Halley Bay – Kapp Norvegia comparison. In: Arntz W, Ernst W, Hempel I (eds) The expedition ANTARKTKUS VII/4 (EPOS leg 3) and VII/5 of RV "Polarstern" in 1989. Rep Polarforsch 68:39–49

Rowedder U (1984) Nahrungsokologie antarktischer Fische. Ph.D. Thesis, Univ Kiel, pp 1–101

Saint-Paul U, Hubold G, Ekau W (1988) Acclimation effects on routine oxygen consumption of the antarctic fish *Pogonophryne scotti* (Artedidraconidae). Polar Biol 9:125–128

Scherbich LV, Slepkurov VA (1976) Natural mortality of the marbled notothenia. (in russ). Trudy Atlant Nauchno-Issled Ryb Khoz Okeanogr 60:76–84

Schnack-Schiel S (1987) Die expedition ANTARKTIS V 1–3 mit FS "Polarstern" 1986. Ber Polarforsch 39:1–260

Schneppenheim R (1982) Marin-biologisches Begleitprogramm zur Standorterkundung 1979/80 mit MS "Polarsirkel (Pre-Site Survey) Stationsliste der Mikronekton- und Zooplanktonfange sowie der Bodenfischerei. Ber Polarforsch 7:1–32

Scholander PF, VanDamm L, Kanwisher JW, Hammel HT, Gordon MS (1957) Supercooling and osmoregulation in arctic fish. J Cell Comp Physiol 49(1):5–24

Schwarzbach W (1988) Die Fischfauna der ostlichen und sudostlichen Weddellsee: geographische Verbreitung, Nahrung und trophische Stellung der Fischarten. Ber Polarforsch 55:1–94

Shust KV, Pinskaya IA (1978) Age and rate of growth of six species of nototheniid fishes (Fam. Nototheniidae). J Ichthyol 18(5):743–749

Siegel V (1986) Structure and composition of the antarctic krill stock in the Bransfield Strait (Antarctic Peninsula) during the Second International Biomass Experiment (SIBEX). Arch Fischwiss 37:51–72

Sinclair M (1988) Marine populations. An essay on population regulation and speciation. Books in recruitment fishery oceanography. Wash Sea Grant Prog, Univ Wash Press, Seattle, pp 1–252

Sosinski J (1981) Comparative biology of antarctic icefish Champsocephalus gunnari Lonnberg 1905 from the Antarctic. Skedia I Materialy Ser V No 48

Sosinski J (1982) Zosoby Antarktyki. Rep Morsk Inst Ryb 1980/81:77–80

Tankevich PB (1986) Morphometric characteristics of the Kerguelen marbled Notothenia *Notothenia rossi* rossi Richardson from the Indian Ocean sector of the southern ocean. Vopr. Ikhtiol 26(2):328–332

Targett TE (1981) Trophic ecology and structure of coastal antarctic fish communities. Mar Ecol Prog Ser 4:243–263

Tiedtke J, Kock KH (1989) Structure and composition of the demersal fish fauna around Elephant Island. Arch Fisch Wiss 39 Beih 1:143–169

Tomo AP, Cordero-Funes JR (1972) Edad y crecimiento en largo del *Harpagifer bispinis antarcticus*. Contrib Inst Ant Argentino 96:1–19

von Bodungen B, Smetacek VS, Tilzer MM, Zeitzchel B (1986) Primary production and sedimentation during spring in the Antarctic Peninsula region. Deep-Sea Res 33:177–194

von Bodungen B, Nöthig RM, Sui Q (1988) New production of phytoplankton and sedimentation during summer 1985 in the southeastern Weddell Sea. J Comp Biochem Physiol 90B:475–487

von Dorrien C (1989) Ichthyoplankton in Abhangigkeit von Hydrographie und Zooplankton im Weddellmeer. Diplomarbeit, Univ Kiel, pp 1–66

Voronina NM, Menshutkin KN, Tseytlin VB (1980) Production of the common species of antarctic copepods *Calanoides acutus*. Oceanol Acad Sci USSR 20(1):90–93

Voß J (1988) Zoogeographie und Gemeinschaftsanalyse des Makrozoobenthos des Weddellmeeres (Antarktis). Ber Polarforsch 45:1–144

Wegner G (1982) Two temperature sections through these southern Atlantic and southeastern Weddell Sea. Meeresforschung 29(4):239–243

Wells RMG (1986) Cutaneous oxygen uptake in the atlantic icequab, *Rhigophila dearborni* (Pisces: Zoarcidae). Polar Biol 5:175–179

Wells RMG, Tetens V (1984) Recovery from stress following capture and anaesthesia of antarctic fish: haematology and blood chemistry. J Fish Biol 24:567–576
Wohlschlag DE (1961) Growth of an antarctic fish at freezing temperatures. Copeia (1):11–18
Wohlschlag DE (1962) Antarctic fish growth and metabolic differences related to sex. Ecology 43(4):589–597
Wohlschlag DE (1963) An antarctic fish with unusual low metabolism. Ecology 44(3):557–564
Wohlschlag DE (1964) Respiratory metabolism and growth of some antarctic fishes. In: Carrick R, Holdgate MW, Prevost J (eds) Biologie antarctique (Sci Committee on Antarctic Research, Proc 1. Symp Antarct Biol, Paris). Hermann, Paris, pp 489–502
Wootton RJ (1990) Ecology of teleost fishes. Chapman and Hal Fish and Fisheries Ser 1, London, pp 1–404
Zuur R (1977) The age and growth of *Trematomus bernachii* and *T. centronotus*. B.Sc. Thesis, Univ Canterberry, pp 1–110

Morphological Adaptations and Mode of Life in High Antarctic Fish

W. Ekau

1 Introduction

The Antarctic Ocean is characterized by a circumpolar system of currents and hydrographic fronts. This causes a high transport of water masses between east and west, but reduces exchange mechanisms in the upper 1000-m water column in a north-south direction. Thus the Antarctic Convergence is a great barrier for pelagic organisms, except birds and mammals. For benthic animals, the deep-sea basins form a border in distribution. This hydrographic and topographic isolation of the Antarctic Continent resulted in a very high percentage of endemism in different taxonomic groups. Eighty-eight percent of the fish species and 76% of the genera are endemic to the Antarctic Ocean (Andriashev 1987). Values for other benthic groups range between 60 and 90% (Hempel 1985).

About 200 demersal fish species are described for the Southern Ocean , belonging mainly (about 110 species) to the 6 families of the perciform suborder Notothenioidei (Andriashev 1987). With the process of isolation of the Antarctic Continent from Gondwana and the formation of a circum-Antarctic Ocean since the last 30 million years, the environmental conditions changed to low water temperatures and a pronounced seasonally variable coverage by sea ice (Grobe 1987). The monophyletic notothenioids adapted well to these conditions and were able to occupy a large number of very different ecological niches. As a result of this capability to adapt, they contribute more than 90%, in numbers and biomass, of the demersal fish fauna in the Southern Ocean (DeWitt 1971; Ekau 1988,1990; Tiedtke and Kock 1989). Today the Notothenioidei comprises a group of fish with a large variety of morphological forms. The degree of variation in body shape and occupation of different ecological niches within closely related species is found elsewhere only in isolated environments such as the great African lakes, or similar habitats.

Several authors have discussed an evolution of the notothenioids from bottom-dwelling species to pelagic mode of life (Dewitt 1970; Eastman and DeVries 1982; Voskoboinikova 1982; Andersen 1984; Andriashev 1987; Miller 1987; Eastman and Grande 1989) as the general trend in demersal antarctic fish. During this evolution, the fish have had to compensate the absence of a swim bladder by incorporating lipids, reducing density and evolving hydrodynamic adaptations for

Alfred-Wegener-Institute for Polar and Marine Research, D-2850 Bremerhaven, FRG positive buoyancy, such as in the most highly evolved species *Pleuragramma antarcticum*, the antarctic silverfish (DeWitt 1970; Andriashev 1987). It is the dominant species in the pelagic of the permanent pack-ice zone (DeWitt 1970; Hubold and Ekau 1987). However, other species also show tendencies to pelagic life. Many of them have demersal eggs, but pelagic larval stages. Investigations on the ichthyoplankton of the high antarctic shelf areas show an overwhelming dominance of *P. antarcticum*, but also a high abundance of *Trematomus* and channichthyid larvae (DeWitt 1970; Williams 1985; Hubold and Ekau 1987). In the Weddell Sea, *Trematomus eulepidotus* is the second most abundant species in pelagic trawls (Hubold and Ekau 1987). *Aethotaxis mitopteryx, T. lepidorhinus, T. scotti, Pagothenia bernacchii*, and *P. hansoni* appear frequently in the catches. The use of pelagic organisms as a food source by these species, also as adults, was demonstrated by several authors (Targett 1981; Takahashi and Nemoto 1984; Schwarzbach 1988).

Different habitat, food, and life style of fish often correlates with special morphological adaptations (Aleyev 1969; Gosline 1971; Eastman and DeVries 1982; Ojeda 1986). Methods and criteria were established and discussed by other authors, but are related mainly to fresh water fish of rivers and lakes (e.g., Gatz 1979; Barel 1983, 1984; Mahon 1984; Watson and Balon 1984). While many publications dealing with ecomorphology are only descriptive, Mahon (1984) could separate some relevant characters to distinguish between species from fast and slow running waters by principal component analysis. Davis and Birdsong (1973) used cluster analysis to divide coral reef fish into "benthos foragers" and "water-column foragers".

In this work an attempt was made to classify the *Trematomus* and *Pagothenia* species from the Weddell Sea according to ecological aspects, and to rank them according to demersal to pelagic life-styles, using different statistical methods.

2 Methods

The material for this study was collected during two cruises with the German RV "Polarstern" to the Weddell Sea in January/February 1985 and October/November 1986. A total of 16 Bottom and 36 Agassiz trawls were carried out on the eastern and southern continental shelf (Fig. 1) between 200- and 1200-m water depth. Detailed information on trawling stations and catches are given in Ekau (1988).

A total of 676 specimens of 9 nototheniid species were investigated for morphological studies. For a detailed description of the body shape of the fish, measurements of 38 morphometric characters were carried out on freshly defrosted specimens to length classes of up to 1 mm. Standard length of the fish investigated ranged from 40–290 mm. A detailed list of the length ranges in each species is given by Ekau (1988). Except for body height and width, all characters can be defined as the shortest distance between two hard structures (bones, spines, etc.). Otolith weight was measured to 0.1 mg. The position of the morphometric measurements is shown in Fig. 2. A detailed definition of the different characters is given in Ekau (1988). They are derived from the systems of McCune (1981) and Fricke (1982).

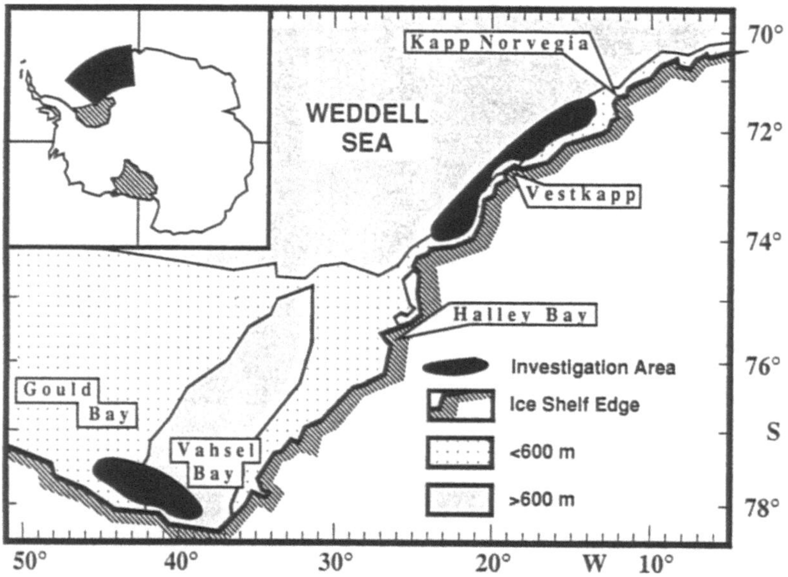

Fig. 1. Map of the Weddell Sea with areas of investigation during RV "Polarstern" cruises ANT III and V. Water depth less and deeper than 600-m is indicated. The 600-m isobath represents the shelf edge

The median line (ML) was taken to be the line between the tip of the snout and the middle of the caudal peduncle, which is comparable with the median line determined by Fricke (1982) for nototheniids. All measurements were related to standard length (SL) and otolith weight to fresh weight.

By using 27 morphometric characters, 11 different criteria could be quantified to rank the species on a scale: demersal-pelagic. These criteria were chosen after different authors:

1. Index of compression (BD/BW): maximum body depth divided by maximum body width (after Mahon 1984). This index should be negatively correlated with swimming speed and maneuverability.
2. Relative depth (BD): maximum body depth related to standard length (after Mahon 1984). This character is negatively correlated with swimming speed.
3. Mouth orientation (BDa/bML): this attribute is quantified here by the relation of the body depth above and the body depth beneath the median line. Mouth orientation indicates where the fish is feeding in the water column (Mahon 1984); Siewing (1980) indicates dorsal mouth for predators and pelagic feeders, terminal or ventral mouth for benthic feeders.
4. Position of the eye (IOW, OL): the interorbital width IOW, as the shortest distance between the eyes in relation to body width, should be narrow for benthic species and wide for pelagic and predatory species (Aleyev 1969; Gatz 1979; Mahon 1984). Orbital length OL divided by SL is related to depth and predatory character (Andriashev 1945; cited from Voskoboinikova 1980).

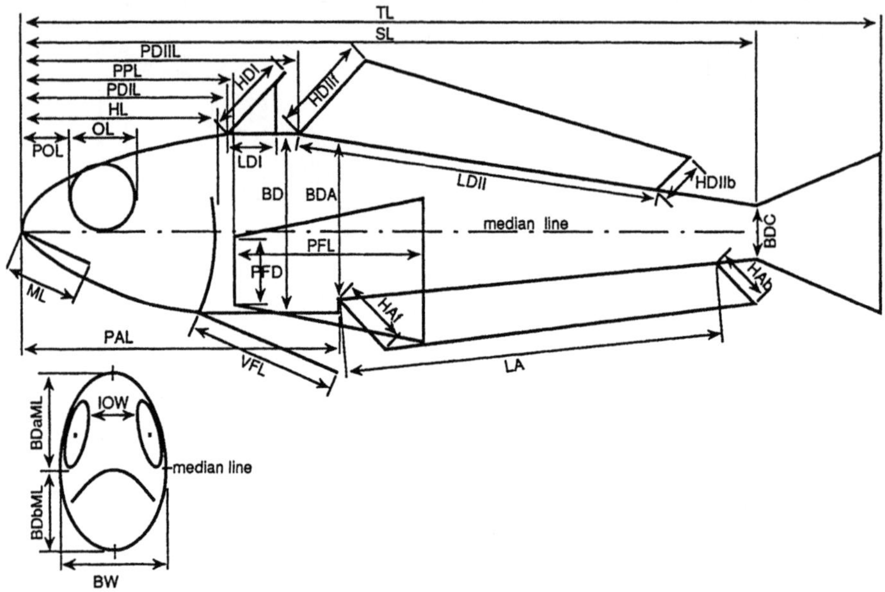

Fig. 2. Position of the morphometric measurements. *BD* body depth; *BDA* body depth at anus; *BDaML* body depth above median line; *BDbML* body depth beneath median line; *BDC* body depth at caudal peduncle; *BW* body width; *HAb* height of anal fin, back; *HAf* height of anal fin, front; *HDI* height of dorsal fin I; *HDIIb* height of dorsal fin II, back; *HDIIf* height of dorsal fin II, front; *HL* head length; *IOW* interorbital width; *LA* basal length of anal fin; *LDI* basal length of D I; *LDII* basal length of D II; *ML* maxillary length; *OL* orbital length; *PAL* preanal length; *PDIIL* predorsal length (DII); *PDIL* predorsal length (DI); *PFD* depth of pectoral fin; *PFL* length of pectoral fin; *POL* preorbital length; *PPL* prepectoral length; *SL* standard length; *TL* total length; *VFL* length of ventral fin

5. Body depth at caudal peduncle (BDC): measured as minimum depth of the caudal peduncle in relation to SL. A strong peduncle may affect fast–start performance (Webb 1977).
6. Otolith weight (OTW) : Eastman (1976) and Eastman and DeVries (1982,1985) observed reduced calcification in pelagic antarctic fish. Otolith weight as percentage of fresh weight times 10^{-3} was used as criterion for the degree of calcification.
7. Area of pectoral fin (APF) : relative pectoral fin area, estimated by maximum length of fin times maximum width, is great amongst slow swimmers, which use them for maneuvering and braking (Alexander 1967), and small amongst fast swimmers, which use them only as hydroplane. The area of fins is given in square units and are based on the measurements related to the standard length of the fish. The square unit therefore is called $(\%SL)^2$.
8. Area of dorsal and anal fin (ADII, AA) : areas were calculated as the product of length of the base line and the mean height of the fins. In slow-swimming species, the unpaired fins are used as oar and stabilizer and for swimming, therefore they have large areas. In fast-swimming species they are only used as stabilizer and area can be small (Gosline 1971; Moyle and Cech 1982). The anal fin should be smaller

in benthic species (relation ADII /AA > 1), because it is used as a support when they rest on the bottom.
9. Ventral fins (VFL) : in demersal species, ventral fins, together with the anal fin, are used as a tripod to rest on the bottom or substratum. Pelagic species have smaller and thinner fins (Gosline 1971). The length of the fin was used as criterion.
10. Length of gill rakers (LGR) : pelagic filter feeders should have long gill rakers while species feeding on benthic animals should have short gill rakers.
11. Color: this criterion is a qualitative one and can be related to depth and the relation to the substratum. Pelagic species normally are lighter (silvery) and more uniform in coloration than demersal ones (Aleyev 1969).

The analysis of the data was carried out in two different ways: (1) the species were ranked by their means along a scale demersal-pelagic according to the definition for the different criteria given above: rank 1 for a pelagic life style, rank 9 for a demersal one. If the mean values could not be separated by t-test ($\alpha = 0.05$), the species was given the same rank within this criterion. Finally, a mean rank was calculated for each species from its ranks in each criterion. The distribution of the criteria in the different species were also analyzed by a cluster analysis (the program was written by D. Piepenburg, Institute for Polar Ecology, Kiel). (2) The raw data, standardized by a z-transformation, were analyzed by factor analysis (using the program Systat on an Apple Macintosh computer). The data base for the factor analysis was a pairwise calculated correlation matrix of the ten morphometric characters: OL, BD/BW, BD, BDa/bML, IOW, BDC, OTW, ADII, ADII/AA, VFL (for abbreviations see above). The number of factors was chosen after their power to describe more than 70% of the variance in each variable (Flury and Riedwyl 1983). This method resulted in factors with eigen-values greater than 0.90. The Varimax Method was used to rotate the factors.

3 Results

3.1 Ranking After Mean Values of Morphological Criteria

The mean values of the morphological characters, used to rank the species with respect to demersal or pelagic life style, are given in Table 1.

The compression factor BD/BW ranges from 1.11 in *Pagothenia hansoni* to 1.67 in *Trematomus eulepidotus*. Good maneuverability is a prerequisite for demersal species, a low value against 1 is classified as demersal. Using the t-test, it is possible to divide the species into five groups: (1) *P. hansoni*, with the lowest value; (2) *P. bernacchii*, *T. centronotus* and *T. nicolai*, also with low values; (3) with a larger distance to the others, *P. borchgrevinki*, *T. loennbergi* and *T. scotti*; then (4) *T. lepidorhinus* and with the highest value, (5) *T. eulepidotus*.

As with the compression factor, body depth (BD) is related to water current or swimming speed. Fast-swimming or pelagic, respectively, species have a reduced body depth. The values range from 21.67 in *T. scotti* to 26.92% SL in *P. bernacchii*. Thus, two main groups, one containing *P. borchgrevinki*, *P. hansoni*, *T. eulepidotus*, *T. lepidorhinus*, *T. loennbergi* and *P. bernacchii*, *T. centronotus*, *T. nicolai*, and one with *T. scotti*, could be separated.

Table 1. Mean values x and standard deviations for the morphological criteria OL, BD/BW, BD, BDa/bML, IOW, BDC, OTW, ADII, ADII/AA, VFL, LGR and APF[a]

		P. bernacchii	*P. borchgrevinki*	*P. hansoni*	*T. centronotus*	*T. eulepidotus*	*T. lepidorhinus*	*T. loennbergi*	*T. nicolai*	*T. scotti*
BD/BW	x	1.21	1.45	1.11	1.21	1.67	1.54	1.41	1.24	1.43
	s	0.11	0.11	0.13	0.12	0.14	0.18	0.10	0.12	0.12
	n	14	3	16	39	134	201	75	18	121
	R	7	4	9	7	1	2	4	7	4
BD	x	26.92	23.45	23.78	25.06	23.95	22.73	23.16	25.98	21.67
	s	2.54	0.72	1.47	1.85	3.57	2.12	1.99	2.86	1.73
	n	14	3	17	47	164	176	75	19	122
	R	8	4	4	8	4	4	4	8	1
BDa/bML	x	1.15	0.55	1.12	1.21	0.72	1.23	1.30	0.76	1.42
	s	0.15	0.07	0.18	0.20	0.17	0.32	0.25	0.13	0.37
	n	7	3	11	45	92	127	59	14	100
	R	6	2	6	6	2	6	6	2	9
OL	x	8.11	7.47	7.25	9.63	9.58	8.94	8.70	9.39	10.11
	s	0.55	0.37	0.58	0.77	0.67	1.16	0.71	0.61	0.73
	n	13	3	17	50	42	216	84	19	120
	R	3	1.5	1.5	7	7	4.5	4.5	7	9
IOW	x	22.17	46.48	23.80	13.61	37.10	31.57	30.85	15.97	16.99
	s	3.18	6.93	3.62	2.21	5.49	6.00	3.53	2.11	3.63
	n	14	3	17	51	140	210	84	19	121
	R	5.5	1	5.5	9	2	3.5	3.5	7.5	7.5
BDC	x	6.88	8.40	7.13	6.83	7.25	7.09	7.06	7.72	7.79
	s	0.40	0.21	0.43	0.39	0.46	0.55	0.57	0.38	0.54
	n	13	3	18	50	140	216	83	19	122
	R	3.5	8	3.5	3.5	3.5	3.5	3.5	8	8
OTW	x	15.90	8.91	16.20	39.00	14.40	19.60	17.80	16.30	145.4
	s	7.30	0.40	9.20	3.30	3.60	7.00	8.40	6.80	34.10
	n	11	3	16	50	136	183	81	19	119
	R	4.5	1	4.5	8	4.5	4.5	4.5	4.5	9
ADII	x	560.6	522.3	570.8	614.8	484.8	489.5	498.1	700	594.9
	s	60.90	27.30	39.40	61.90	36.90	29.60	29.40	44.40	38.00
	n	11	2	15	49	106	131	76	16	36
	R	5.5	4	5.5	7.5	2	2	2	9	7.5
ADII/AA	x	1.70	1.27	1.62	1.48	1.33	1.15	1.10	1.80	1.38
	s	0.25	0.11	0.09	0.12	0.10	0.09	0.08	0.15	0.16
	n	11	2	15	49	104	129	74	16	101
	R	8	3	8	6	4	2	1	8	5
VFL	x	23.56	20.03	22.32	27.72	21.72	21.00	21.18	22.69	24.95
	s	1.79	0.85	2.32	1.74	1.48	1.99	1.60	1.99	2.19
	n	14	3	18	51	140	213	83	19	122
	R	5.5	2	5.5	9	5.55	2	2	5.5	8

Table 1. (continued)

		P. bernacchii	*P. borchgrevinki*	*P. hansoni*	*T. centronotus*	*T. eulepidotus*	*T. lepidorhinus*	*T. loennbergi*	*T. nicolai*	*T. scotti*
LGR	x	–	–	1.31	1.30	1.72	1.80	1.34	1.90	1.11
	s	–	–	0.15	0.25	0.16	0.33	0.32	0.55	0.40
	n	–	–	3	9	20	21	29	3	32
	R	–	–	5.5	5.5	2	2	5.5	2	5.5
APF	x	278.3	302.2	325.7	278	266.3	238.8	293.7	307.9	182.8
	s		21.70	31.85	21.08	51.52	21.62	25.51	21.17	21.64
	n	1	2	6	37	74	40	30	16	83
	R	4	7.5	7.5	4	4	2	7.5	7.5	1
Coloration		6.5	1	6.5	9	3	4	2	6.5	6.5
mean rank		5.58	3.25	5.58	6.88	3.42	3.23	3.85	6.35	6.23

[a] R is the rank of the species for the different criteria, n is the number of specimen investigated; *P* is for *Pagothenia*, *T.* for *Trematomus*.

In the mouth orientation (BDa/bML), three species, *P. borchgrevinki, T. eulepidotus*, and *T. nicolai*, had mean values < 1.0 (the range was 0.55–0.76), which indicates an orientation to the pelagic with respect to feeding and hydrodynamics. The other species had values slightly above 1 (1.12–1.30) and only *T. scotti* was distinctly separate. It had the highest value (1.42) and is oriented mostly to the bottom.

Orbital diameter OL ranged from 7.25 in *P. hansoni* to 10.01% SL in *T. scotti.* OL does not vary that much within the species, so that five groups of species could be distinguished by t-test.

The interorbital width IOW is more representative for classification into demersal/pelagic than OL. The mean values ranged from 13.61 in *T. centronotus* to 46.48% BW in *P. borchgrevinki.* Six groups could be separated by t-test.

The body depth at the caudal peduncle BDC was very similar in the nine species and ranged between 6.83 in *T. centronotus* and 8.40% SL in *P. borchgrevinki.* Only two groups could be separated: *P. borchgrevinki, T. nicolai* and *T. scotti* with strong caudal peduncles; the other species had lower values.

The largest otoliths (OTW as indicator for reducing calcification) were found in *T. scotti* with $145.4 \times 10^{-3}\%$ *P. borchgrevinki* has the lightest otoliths with $8.91 \times 10^{-3}\%$. The other species are very similar with values between 14.40 and $19.60 \times 10^{-3}\%$.

Using the area of the dorsal fin, the species could be divided into five groups. *T. eulepidotus, T. lepidorhinus*, and *T. loennbergi* had the smallest fins, with mean areas of 485–498 $(\% \text{ SL})^2$. They were followed by *P. borchgrevinki* (522; a single species in this group), *P. bernacchii, P. hansoni* (561–571), *T. centronotus* and *T.*

scotti (595–615), respectively. The largest fin area was found in *T. nicolai* with a value of 700, which was classified as highly demersal. A similar ranking of the species was attained by using the relationship between dorsal and anal fin area. The relationship is close to 1 in *T. loennbergi* and *T. lepidorhinus* and close to 1.3 in *P. borchgrevinki* and *T. eulepidotus*. These are followed by *T. scotti* (1.38) and *T. centronotus* (1.48). Highest values were found in *P. bernacchii, P. hansoni* and *T. nicolai* (1.62–1.70).

The shortest (and lightest) ventral fins were found in *P. borchgrevinki, T. lepidorhinus* and *T. loennbergi* (20.0–21.2% SL). *P. bernacchii, P. hansoni, T. eulepidotus* and *T. nicolai* were clustered in a second group (21.7–23.6 % SL). The longest ventrals were found in *T. scotti* (25.0 % SL) and *T. centronotus* (27.7 % SL).

Using the area of pelvic fins, four groups could be distinguished. The smallest pelvic fins were found in *T. scotti* [183 $(\% \text{ SL})^2$], followed by *T. lepidorhinus* (239). *P. bernacchii, T. centronotus* and *T. eulepidotus* could be grouped together (266–278) as well as *P. borchgrevinki, P. hansoni, T. loennbergi* and *T. nicolai* (294–326).

The length of gill rakers range between 1.1 and 1.9% SL. The differences are small and only the two following groups of species could be separated: (1) *T. eulepidotus, T. lepidorhinus,* and *T. nicolai*, with longer gill rakers; (2) *P. hansoni, T. centronotus, T. loennbergi*, and *T. scotti* with shorter ones.

The ranking within the criterion "coloration" was done subjectively after observations on fresh material. *P. borchgrevinki* has a very light and silvery color with weak brownish or bluish spots. The dorsal side is a bit darker (rank 1). *T. loennbergi* is also very uniformly colored, but darker than *P. borchgrevinki* (rank 2). *T. eulepidotus* and *T. lepidorhinus* are similar to *T. loennbergi*, but have more patterns on the sides (ranks 3 and 4). *P. bernacchii, P. hansoni, T. nicolai* and *T. scotti* are grouped together. They have brownish colors with different patterns (rank 6.5). The most conspicuous coloration was found in *T. centronotus* with yellow fins and white dots and stripes (rank 9).

The mean ranks giving a classification along the line from demersal to pelagic mode of life are shown in Table 1 and Fig. 3.

The lowest ranks were calculated for *P. borchgrevinki* and *T. lepidorhinus*, with 3.25 and 3.31, respectively. Very close to these values is *T. eulepidotus* with 3.50 and also *T. loennbergi* with 3.88. These four species have the widest adaptations to a pelagic life. The distance to the next species is great. Cluster analysis (Fig. 2), however, shows that there are large differences in particular ranks of the variables. The three *Trematomus* species are closely related, resulting in similar ranking for most of the variables. *P. borchgrevinki* is excluded because of ranks very different to those of the other species.

P. hansoni and *P. bernacchii* have mean ranks of 5.65 and 5.67, respectively. They are almost identical and are thus also grouped together in the cluster analysis.

There is a greater distance in the mean rank to the last three *Trematomus* species: *T. scotti* (6.31), *T. nicolai* (6.42) and *T. centronotus* (6.96). They are close together in their mean rank, but *T. scotti* is separated from the others by the cluster analysis.

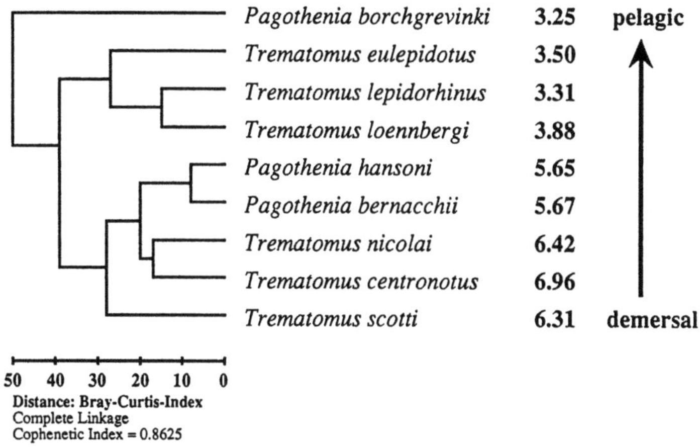

Fig. 3. Mean ranks and cluster dendrogram of the nine nototheniid species

3.2 Factor Analysis

3.2.1 Interpretation of the Principal Axes

The factor loadings indicate the degree of association between attribute and factor. Variables with high loadings in the same factor are highly correlated. The grouping of the ten characters used to describe pelagic or demersal life style by factor analysis produced four principal axes, which represent 79.3% of the total variance (Table 2).

Factor 1 describes the variables Orbital Length (OL), Area of Dorsal Fin II (ADII), Otolith Weight (OTW) and Length of Ventral Fin (VFL). This factor distinguishes between species or specimens according to their hydrodynamics and equilibrium sense. On one side would be the deep-living, slowly moving specimens resting mostly on the substratum. On the other hand would be the fast-swimming pelagic specimens, which need small fins for hydrodynamics, small otoliths for equilibrium, and small eyes.

Factor 2 describes variables which indicate the orientation of the fishes in the water column. At one extreme are nondepressed specimens with dorsal eyes and unpaired fins, which allow close affiliation to the bottom. At the other would be the compressed fish with lateral eyes and symmetric unpaired fins as active pelagic swimmers.

Factor 3 includes three variables, representing influence of body form and feeding orientation. On the one hand, we have deep-bodied fish close to the substratum, where feeding is directed upwards. On the other hand we find slender and active fish with feeding directed downwards.

Factor 4 was found to be dominated only by one variable, the strength of the caudal peduncle. This character separates between active swimmers and slow swimmers or fast-start swimmers, respectively.

Table 2. Correlation matrix (pairwise), factor loadings (Varimax rotation) and factor scores for the ten morphological characters of 405 specimens of 9 nototheniid species from the Weddell Sea.[a]

Correlation matrix

	OL	BD	BDC	VFL	BD/BW	BDa/bML	AdII/AA	OTW	IOW	ADII
OL	1.000									
BD	−0.277	1.000								
BDC	0.201	0.078	1.000							
VFL	0.516	−0.167	0.182	1.000						
BD/BW	−0.013	0.194	−0.045	−0.338	1.000					
BDa/bML	0.047	0.357	0.060	0.255	−0.229	1.000				
ADII/AA	0.120	0.321	0.177	0.306	−0.379	−0.135	1.000			
OTW	0.552	−0.533	0.352	0.555	−0.217	0.402	0.135	1.000		
IOW	−0.208	0.042	−0.227	−0.539	0.662	−0.351	−0.483	−0.551	1.000	
ADII	0.503	0.535	0.238	0.554	−0.099	0.324	0.020	0.851	−0.349	1.000
	1	2	3	4	5	6	7	8	9	10
Eigenvalues	3.838	1.879	1.309	0.906	0.634	0.478	0.396	0.316	0.168	0.077

	Rotated factor loadings				Rotated factor scores			
	Factor 1	Factor 2	Factor 3	Factor 4	Factor 1	Factor 2	Factor 3	Factor 4
OL	0.863	0.007	0.123	0.030	0.460	0.118	0.271	0.131
ADII	0.796	0.074	0.387	0.156	0.280	0.086	0.096	0.042
OTW	0.762	0.241	0.412	0.271	0.216	0.010	0.130	0.145
VFL	0.675	0.475	0.017	0.049	0.268	0.147	0.138	0.120
BD/BW	0.008	0.850	0.211	0.128	0.139	0.443	0.126	0.199
IOW	0.251	0.848	0.162	0.198	0.053	0.372	0.065	0.070
ADII/AA	0.123	0.686	0.507	0.160	0.058	0.316	0.358	0.012
BDa/bML	0.036	0.258	0.796	0.144	0.217	0.104	0.537	0.179
BD	0.455	0.095	0.690	0.280	0.137	0.081	0.337	0.279
BDC	0.170	0.053	0.003	0.932	0.081	0.099	0.045	0.878
Eigenvalues	2.729	2.279	1.772	1.151				
% Total variance expl.	27.291	22.791	17.721	11.509				

[a]ADII = Area of dorsal fin II; ADII/AA = relation of areas of dorsal fin II and anal fin;BD = body depth;BDC = body depth at caudal peduncle;BDa/bML = relation of body depth above median line to BD below ML;BD/BW = relation of body depth to body width;IOW = interorbital width;OL = orbital length;OTW = otolith weight;VFL = ventral fin length.

3.2.2 Ordination of the Specimens

With the aid of the factor scores (Table 2) the fish were ordinated along the four factors. Figures 4 and 5 present the results.

The scattergram of the factor scores (factors 1 and 4) of the specimens groups the species into different quadrants (Fig. 4). The first represents deep-living fish with the ability of fast starts (i.e., hunters). *Trematomus scotti* is the major species here, but *T. eulepidotus* also tends to fall into this combination of characters. Quadrant 2 shows those species with a more pelagic mode of life or good swimmers, but also with the ability of a fast start. *Pagothenia borchgrevinki* is found here, as well as *T. nicolai*. The third quadrant includes pelagic and enduring swimming fish, in

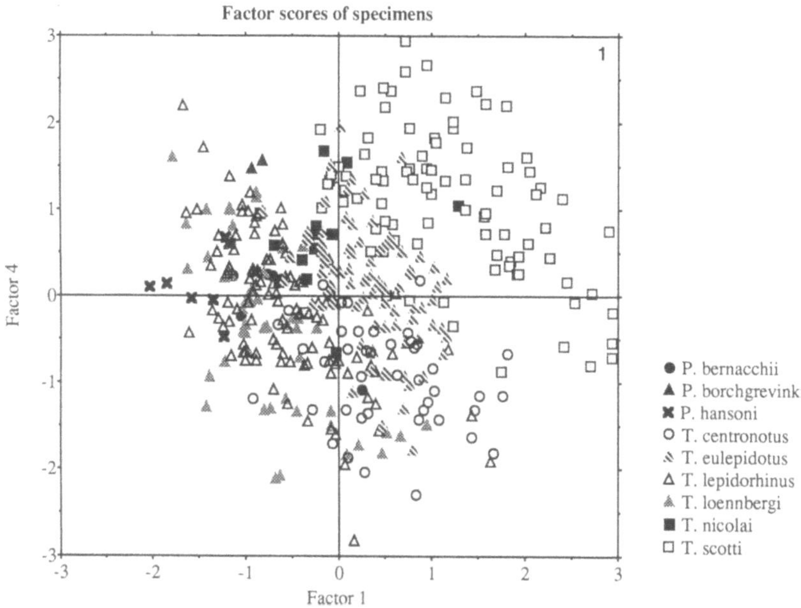

Fig. 4. Ordination of the specimens of the nine species according their scores in factors 1 and 4. Quadrants are numbered counter-clockwise: the quadrant above right is number 1

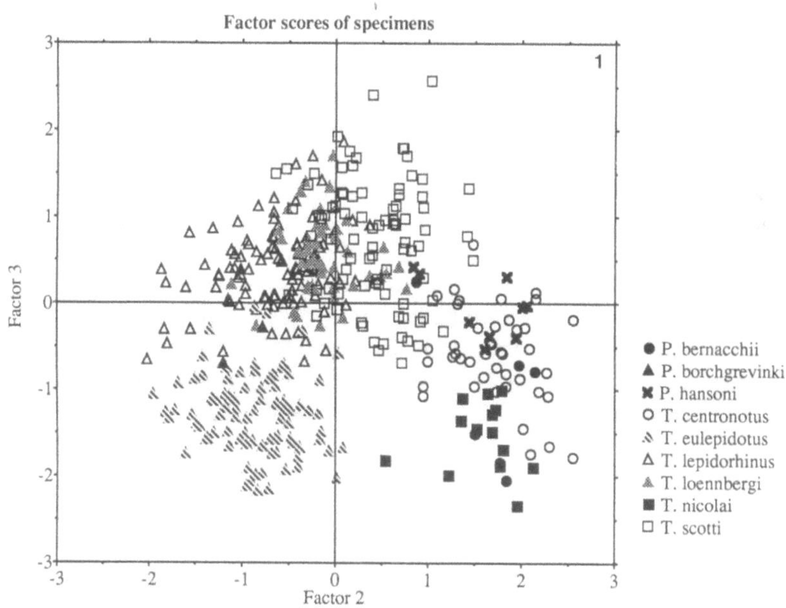

Fig. 5. Ordination of the specimens of the nine species according their scores in factors 2 and 3. Quadrants are numbered counter–clockwise: the quadrant above right is number 1

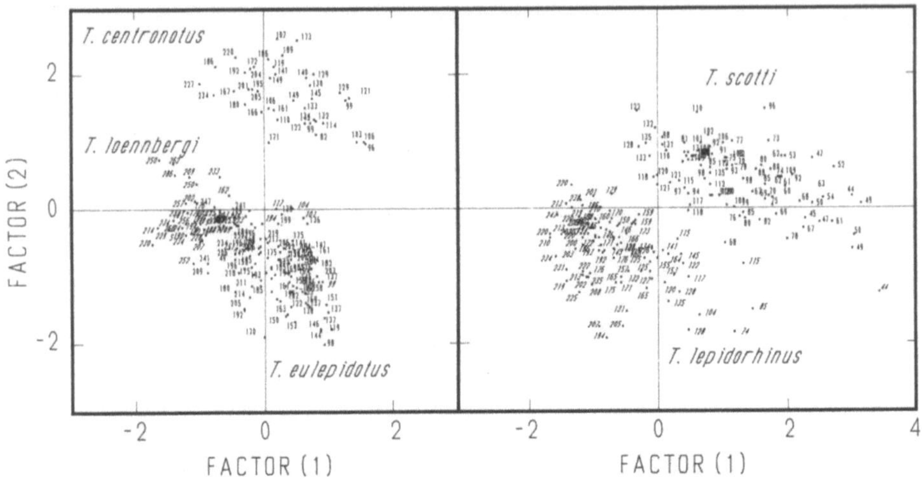

Fig. 6. Ordination of the specimens of five *Trematomus* species according their scores in factors 1 and 2. The *small points* give the ordinates of the factor scores, the *numbers* give the standard length of the specimens (*T. lepidorhinus* and *T. loennbergi* in *italics*)

this case *T. lepidorhinus* and *T. loennbergi*, but also some *T. centronotus*. The fourth quadrant includes deep-living demersal fish, which are active swimmers.

In the quadrants, outlined by factors 2 and 3 (Fig. 5), the following attributes can be assigned to the specimens: (1) quadrant 1 represents fish which sit on the bottom and also feed from the bottom. We find mainly *T. scotti* in this field. (2) The opposite is the quadrant 3 with actively swimming specimens feeding in the water column or at the surface, respectively. Here *T. eulepidotus* is mainly found. However, we find also some of *T. lepidorhinus* and *T. loennbergi* and the two specimens of *P. borchgrevinki*. (3) Quadrant 2 represents actively swimming fish, which feed from the bottom. Most of the *T. lepidorhinus* and *T. loennbergi* are found here, also some of *T. scotti*. (4) Quadrant 4 shows fish mainly sitting on the bottom, but feeding in the water column. *T. nicolai, T. centronotus, P. bernacchii*, and *P. hansoni* fall into this category. However, the orientation in feeding in *T. centronotus, P. hansoni*, and *P. bernacchii* is also to the bottom.

In factor 1, there was a strong length dependent shift from positive to negative values. Figure 6 demonstrates this tendency for the species *T. centronotus, T. eulepidotus, T. lepidorhinus, T. loennbergi*, and *T. scotti*. Factors 2, 3 and 4 are slightly length dependant to more positive values.

4 Discussion

Many authors have discussed the systematics and evolution of antarctic fish (Regan 1914; Norman 1938; Dewitt 1971; Andersen 1984; Ojeda 1986; Andriashev 1987; Miller 1987). It is accepted by these authors that the fish fauna has evolved from the bottom-living perciform species, but it is still not clearly understood why notothenioids adapted better to the cooling and isolating environment than other groups(Eastman and Grande 1989), resulting in their overall dominance and their high morphological variability. The intention of this paper is to analyze the diversification of the species in relation to their habitat and food with special emphasis to their mode of life.

The morphometric characters used for this work were chosen mainly after Mahon (1984). The interspecific variability was high enough to separate the species. Shape of caudal fin (Kramer 1960) or gut length (Ojeda 1986), for example, were too similar in all species to enable a separation. These would characterize nototheniids as slow-swimming carnivores. The ranges of values for the different characters in the different species corresponded well with data from the literature. This was demonstrated by Ekau (1988).

Different authors have already classified some of the nototheniid species as being pelagic, benthopelagic, cryopelagic, or demersal. Andriashev (1987) classifies *P. borchgrevinki* as being cryopelagic and *T. eulepidotus* pelagic, while *T. lepidorhinus* is called benthopelagic (DeWitt 1970). In Fig. 3, a group of four species have very similar mean ranks of between 3.25 and 3.88, which may be all classified as being more or less benthopelagic: *P. borchgrevinki, T. eulepidotus, T. lepidorhinus*, and *T. loennbergi*. The cluster analysis shows that the three are also very close together in the ranks of the different characteristics., but that *P. borchgrevinki* can be distinguished clearly.

All three *Trematomus* species should be classified as benthopelagic. However, *T. eulepidotus* is slightly different from the others with regard to its mouth orientation, orbital length, and ventral fin length. This indicates a different orientation in feeding ecology. While *T. eulepidotus* feeds mainly on euphausiids (Takahashi 1983; Schwarzbach 1988), which requires large eyes for hunting, *T. lepidorhinus* feeds mainly on benthic polychaetes and amphipods. By factor analysis, this species is classified as a good and preserving pelagic swimmer, which feeds off the bottom. On the other hand, *T. eulepidotus* is characterized as an active fish with fast-start ability and orientation to the pelagic. This could be essential for hunting euphausiids.

T. loennbergi is closely related to *T. lepidorhinus*. The mean rank is similar, the ranks in the different characters are also very similar, which is shown by the low interspecific distance in the cluster analysis, and in the factor analysis, the scattergrams of the two species cannot be separated in any factor (Figs. 4 and 5). This may be interpreted as being the result of occupying very similar ecological niches with respect to food, activity and habitat. While food overlap in summer between *T. eulepidotus* and *T. lepidorhinus* or *T. loennbergi* is only 25 or 13%, respectively, the overlap between *T. lepidorhinus* and *T. loennbergi* is 60% (Schwarzbach 1988). In fact, these species are found in different regions of the Weddell Sea: *T. eulepidotus* and *T. lepidorhinus* along the eastern shelf, *T. loennbergi* along the Filchner trench in the south.

An ontogenetic trend in morphology was found in all three *Trematomus* species. *T. lepidorhinus* and *T. loennbergi* have evolved hydrodynamic and equilibrium characteristics. As adults they may require this for improved capability to hunt pelagic euphausiids. *T. eulepidotus* becomes less pelagic with respect to orientation in the water column.

P. borchgrevinki has some demersal, but mostly pelagic characteristics. This results in a low mean rank, but it separates the species from the *Trematomus* species. The orientation of the characters seem to be indicative of an inverse benthic habitus. A "cryopelagic" mode of life means that the species is closely associated to the sea ice. Thus mouth orientation, color, otolith weight, and eye position are strongly pelagic, while characters of body shape are similar to benthic species.

P. bernacchii and *P. hansoni* are quite similar in their mean ranks and in the cluster analysis. In 8 of 13 characters they have the same rank. This results in the lowest interspecific distance in the cluster analysis. Their loadings in the factor analysis characterize them as predatory (big eyes), slowly moving species, often resting on the bottom. *P. bernacchii* may be a more active swimmer and feeding seems to be directed upwards. Both species prey on benthic animals (Hureau 1970; Eastman 1985; Schwarzbach 1988), but Heywood and Light (1975) and Moreno (1980) also found pelagic euphausiids in stomachs of *P. bernacchii*. *P. bernacchii* also appears more active of higher catch returns from traps (Ekau 1988), being attracted by the amphipods and isopods consuming the bait.

The species with the highest mean ranks between 6.3 and 7 are *T. scotti*, *T. nicolai*, and *T. centronotus*. These species are the most demersal of the nine species investigated. All species are closely affiliated to the substratum. Differences are found in the orientation and kind of feeding or hunting. *T. scotti* is a slow-swimming fish feeding from the bottom, but with fast-start capability, suggesting that it preys on fast-swimming organisms. *T. nicolai* is a good swimmer, directed upwards in feeding and hunting, and has also fast-start capability. Together with a very large maxillary length (Ekau 1988), these characters indicate a fish predator in the benthopelagic (Aleyev 1977). *T. centronotus*, as the species with the highest mean rank, can be classified as a slow but persevering swimmer with feeding directed forward. *T. scotti* and *T. centronotus* are found to be typical benthos feeders. Their food comprises mainly amphipods and polychaetes (Schwarzbach 1988). With increasing length they tend to orientate even closer to the botom; on the other hand, with respect to hydrodynamic and equilibrium adaptations, they are adapted to rise into the water column in order to feed.

Except for factor 1, the values of factor loadings have a slight positive tendency in the different species. This means a development from pelagic to more demersal characters, which may be explained by the demersal origin of the notothenioid fish. All notothenioid species, as far as is known, have benthic eggs, but many have pelagic larvae. In some species the pelagic phase may last several years (Hubold 1985; Hubold and Ekau 1987). *Notothenia rossii* turns to a demersal life within the first year and at a length of about 60 mm (Burchett 1983). *T. eulepidotus* of up to 60 mm length were found in pelagic trawls. They represent fish in the second summer. Specimens in the bottom trawls were usually larger than 70 mm. However, *T. lepidorhinus* > 40 mm was scarcely found in pelagic trawls, but frequently in bottom trawls (Hubold and Ekau 1987; Ekau 1988).

The advantage to adapt to a pelagic life seems apparent: short food chains, i.e., feeding on plankton, results in less waste of energy. Andriashev (1987) discussed food competition between the demersal species as one reason to orientate to the pelagic. Therefore, larvae should have better survival conditions when feeding on planktonic organisms. Kellerman (1986) and Slosarczyk (1983, 1986) found *T.*

eulepidotus, *T. scotti*, *P. bernacchii*, and *P. hansoni* larvae in plankton catches. The dominance of *P. antarcticum* in the pelagic also indicates the success of this strategy.

Why then do we find only one real pelagic species in high antarctic coastal waters? In the permanent pack-ice zone, the period of production in the pelagic is very short (about 3–4 months). Thus pelagic fish species have to migrate seasonally to seek other food resources during winter. *P. antarcticum*, for example, seems to disappear from the Vestkapp region during winter and returns only in December (Ekau 1990; Hain, pers. comm. 1990). Species remaining in the Weddell Sea or even confined to their special habitat would have to switch to benthic food during winter. Pelagic food would be only an additional source during a short season. Another reason could be the long-term seasonality in the ice coverage of the Antarctic Ocean. During several past glacial peroids, ice coverage in summer corresponded to the recent winter situation (Grobe 1986). Pelagic life was nearly impossible above the high antarctic shelves (Hughes 1981; Miller 1987) and species again had to orientate towards a demersal life. Although DeWitt (1970) postulates that the pelagic system in the High Antarctic is very young and incomplete in its evolution, it might be that the coastal ecosystem is quite well adapted to the short and long-term periodicities of ice coverage and productivity, which is expressed in a weak relation to pelagic food resources, relying more on the low but continuous benthic resources. This would also lead to slow growth rates and a relatively high specialization in food organisms and ecological niches, as is demonstrated in the morphology of the species.

Acknowledgment. My thanks go to G. Hubold for his critical discussion and to G. Dieckmann (AWI) for revision of the English version. This work was part of the fish program in the Alfred-Wegener-Institute and has the publication no. 326.

References

Alexander RM (1967) Functional design in fishes. Hutchinson, London; 160 pp

Aleyev YG (1969) Function and gross morphology in fish. Israel Program for Scientific Translations, Jerusalem

Aleyev YG (1977) Nekton. Junk, The Hague; 435 pp

Andersen NC (1984) Genera and subfamilies of the family Nototheniidae from the Antarctic and Subantarctic. Steenstrupia 10(1):1–34

Andriashev AP (1987) A general review of the antarctic bottom fish fauna. In: Kullander SO, Fernholm B (ed) Proc V Congr Europ Ichthyol, Stockholm 1985; pp 357–372

Barel CDN (1983) Towards a constructional morphology of cichild fishes (Teleostei, Perciformes). Neth J Zool 33(4):357–424

Barel CDN (1984) Form-relations in the context of constructional morphology: the eye and suspensorium of lacustrine Cichlidae (Pisces, Teleostei). Neth J Zool 34(4):439–502

Burchett MS (1983) The life cycle of *Notothenia rossii* from South Georgia. Bull Br Antarctic Surv 61:71–73

Davis WP, Birdsong RS (1973) Coral reef fishes which forage in the water column. Helgo Wiss Meeresunters 24:292–306

DeWitt HH (1970) The character of the midwater fish fauna of the Ross Sea, Antarctica. In: Holdgate(ed) Antarctic Ecology, vol 11. Academic Press, London, pp 305–314

DeWitt HH (1971) Coastal and deep-water benthic fishes of the Antarctic. Antarctic Map Folio Ser, Folio 15:1–10

Eastman JT (1976) The caudal skeleton of antarctic fishes (family Nototheniidae). Anat Rec 184(3): 582

Eastman JT (1985) The evolution of neutrally buoyant notothenioid fishes: their specializations and potential interactions in the antarctic marine food web. In: Siegfried WR, Condy PR, Laws RM(eds) Antarctic nutrient cycles and food webs. Springer Berlin Heidelberg New York Tokyo, pp 430–436

Eastman JT, DeVries AL (1982) Buoyancy studies of notothenioid fishes in McMurdo Sound, Antarctica. Copeia 2:385–393

Eastman JT, DeVries AL (1985) Adaptations for cryopelagic life in the antarctic notothenioid fish *Pagothenia borchgrevinki.* Polar Biol 4(1):45–52

Eastman JT, Grande L (1989) Evolution of the antarctic fish fauna with emphasis on the recent notothenioids. In: Crame JA (ed) Origins and evolution of the antarctic biota. Geol Soc. Spec. Publ.47:241–252

Ekau W (1988) Ecomorphology of nototheniid fish from the Weddell Sea, Antarctica. Ber Polarforsch 51:140 pp

Ekau W (1990) Demersal fish fauna of the Weddell Sea, Antarctica. Ant Sci 2(2):129–137

Flury B, Riedwyl H (1983) Angewandte multivariate Statistik. Fischer, Stuttgart, 187 pp

Fricke R (1982) Modifizierung und Anwendung von McCunes Körperform-Meßsystem für rezente benthische Fische (Pisces). Braunschw Naturk Schr 1(3):533–559

Gatz AJ (1979) Ecological morphology of freshwater stream fishes. Tulane Stud Zool Bot 21:91–124

Gosline WA (1971) Functional morphology and classification of teleosteen fishes. University of Hawaii,Honolulu, 208 pp

Grobe H (1986) Spätpleistozäne Sedimentationsprozesse am antarktischen Kontinentalhang vor Kapp Norvegia, östliche Weddell-See. Ber Polarforsch 27, 120 pp

Hempel G (1985) On the biology of polar seas, particularly the Southern Ocean. In: Gray JS, Christiansen ME (eds) Marine biology of polar regions and the effects of stress on marine organisms. Wiley, London, pp 3–33

Heywood RB, Light JJ (1975) First direct evidence of life under antarctic shelf ice. Nature (Lond) 254:591–592

Hubold G (1985) On the early life history of the high antarctic fish *Pleuragramma antarcticum.* In: Siegfried WR, Condy PR, Laws RM (eds) Antarctic nutrient cycles and food webs. Springer Berlin Heidelberg New York Tokyo, pp 445–451

Hubold G, Ekau W (1987) Midwater fish fauna of the Weddell Sea, Antarctica. In: Kullander SO, Fernholm B (eds) Proc V Congr Europ Ichthyol, Stockholm 1985, pp 391–396

Hughes TJ (1981) Numerical reconstruction of palaeo ice sheets. In: Denton GH, Hughes TJ (eds) The last great ice sheets. Wiley and Sons, New York; pp 221–261

Hureau JC (1970) Biologie comparée dec quelques poissons antarctiques (Nototheniidae). Bull Inst Oceanogr Monaco 68(2391):1–244

Kellermann A (1986) Zur Biologie der Jugendstadien der Notothenioidei (Pisces) an der antarktischen Halbinsel. Ber Polarforsch 31:148 pp

Kramer E (1960) Zur Form und Funktion des Lokomotionsapparates der Fische. Z Wiss Zool 163:1–36

Mahon R (1984) Divergent structure in fish taxocenes of north temperate streams. Can J Fish Aquat Sci 41(2):330–350

McCune AM (1981) Quantitative description of body form in fishes: implications for species level taxonomy and ecological inference. Copeia 1981(4):897–900

Miller RG (1987) Origins and pathways possible for fishes of the Antarctic Ocean. in: Kullander SO, Fernholm B (eds) Proc V Congr Europ Ichthyol, Stockholm 1985; pp 373–380

Moreno CA (1980) Observation on food and reproduction in T*rematomus bernacchii* from the Palmer Archipelago, Antarctica. Copeia 1:171–173

Moyle PB, Cech JJ (1982) Fishes: an introduction to ichthyology. Prentice Hall, Englewood Cliffs 593 pp

Norman JR (1938) Coast fishes Part III: The antarctic zone. Discovery Rep 18:1–105

Ojeda FP (1986) Morphological characterization of the alimentary tract of antarctic fishes and its relation to feeding habits. Polar Biol 5:125–128

Regan CT (1914) Fishes, British Antarctic ("Terra Nova") Expedition 1910. Nat Hist Rep Zool I:1–54

Schwarzbach W (1988) The demersal fish fauna of the eastern and southern Weddell Sea: geographical distribution, feeding of fishes and their trophic position in the food web. Ber Polarforsch 54:94 pp

Siewing R (ed) (1980) Lehrbuch der Zoologie, Bd 1 Allgemeine Zoologie, 3. Aufl. Fischer Stuttgart, 910 pp

Slosarczyk W (1983) Preliminary estimation of abundance of juvenile Nototheniidae and Channichthyidae within krill swarms east of South Georgia. Acta Ichthyol Piscatoria XIII(1):3–11

Slosarczyk W (1986) Attempts at a quantitative estimate by trawl sampling of distribution of post-larval and juvenile notothenioids (Pisces, Perciformes) in relation to environmental conditions in the Antarctic Peninsula region during SIBEX 1983–84. Men Natl Inst Polar Res Spec Issue 40:299–315

Takahashi M (1983) Trohic ecology of demersal fish community north of the South Shetland Islands, with notes on the ecological role of the krill. In:Nemoto T, Matsuda T (eds) Proc of the Biomass Colloqium in 1982. Mem Natl Inst Polar Res Spec Issue 27:183–192

Takahashi M, Nemoto T (1984) The food of the antarctic fish in the Wertern Ross Sea in Summer 1979. Polar Biol 3:237–239

Targett TE (1981) Trophic ecology and structure of coastal antarctic fish communities. Mar Ecol Progr Ser 4:243–263

Tiedtke JE, Kock K–H (1989) Structure and composition of the demersal fish fauna around Elephant Island. Arch Fischwiss 39 Beih 1:143–169

Voskoboinikova OS (1980) Visceral skeleton changes in the process of turning to pelagic mode of life in fishes of the family Nototheniidae. (russ). Akad Nauk SSSR, Zool Inst Trudy, 114:67–76

Voskoboinikova OS (1982) Reduction of the pterygoid bones of the visceral skeleton during evolution of the suborder Notothenioidei (Perciformes). J Ichthyol 22(5):105–111

Watson DJ, Balon EK (1984) Ecomorphological analysis of fish taxocenes in rainforest streams of northern Borneo. J Fish Biol 25:371–384

Webb PW (1977) Effects of median-fin amputation on fast-start performance of rainbow trout (*Salmo gairdneri*). J Exp Biol 68:123–135

Williams R (1985) Trophic relationships between pelagic fish and euphausiids in antarctic waters. In: Siegfried WR, Condy PR, Laws RM (eds) Antarctic nutrient cycles and food webs. Springer, Berlin Heidelberg New York Tokyo, pp 452–459

The Biological and Demographic Peculiarities of the Icefish *Champsocephalus gunnari* Lönnberg, 1905 from the Kerguelen Plateau

G. Duhamel

1 Introduction

The main fish species found in the catches of the Southern Ocean fishery since the beginning of exploitation in the 1970s is the channichthyid *Champsocephalus gunnari* (41.89% of the 2 678 100 tonnes reported between 1970 and 1989). Two main fishing grounds are known, the South Georgian and Kerguelen shelves, situated in the Atlantic and the Indian sectors respectively. The species is also recorded from other areas: South Shetland, S. Orkney, S. Sandwich, Bouvet, Heard Islands, and some submarine elevations near South Georgia and the Kerguelen/Heard shelves. However, it is notable that the species is absent between Bouvet and Kerguelen/Heard Islands and beyond. The distance, more than 2500 miles between the two latter islands, favors a concept regarding the separation of Indian and Atlantic populations. Differences reported in meristic character values (Kock 1980; Sosinski 1981) seem significant enough to consider the current impossibility in the meeting of the populations.

The Kerguelen and Heard shelves and the nearby seamounts are situated at the level of an important hydrological front, i.e., the polar front which crosses the Kerguelen plateau and meanders along the bottom topography. It is the northernmost area where the presence of the icefish *C. gunnari* has been recorded in the Southern Ocean; thus it seems of special interest to study the biological and demographic structure of the species in this area. A long-term scientific program conducted around the Kerguelen Islands is trying to clarify the most conspicuous peculiarities.

2 Material and Methods

A regular sampling of icefish in the catches of trawlers or scientific vessels operating off the Kerguelen Islands has been undertaken since 1979/80. Simultaneously, an ichthyological program investigated the inshore waters of the area. Quantitative (length frequency distributions, catch and fishing effort reports) and qualitative (gonads and stomach content analysis, and otolith sampling) data have been collected on a monthly basis to understand the biology and the dynamics of the species.

Museum National d'Histoire Naturelle, Laboratoire d'Ichtyologie générale et appliquée, 43 rue Cuvier, 75231 Paris Cedex 05, France

The sampling of fish has been made randomly and measurements reported in total length to the closest centimeter. The Everson scale (Everson 1977) has been used to determine the size of the first maturity, the maturation cycle, and the spawning period. The diet index Q (Hureau 1970) allows the evaluation of the relative parts of the prey items in the diet. A data base registering the results of some 35 000 trawls, conducted during the period 1979/80–1989/90, permits the spatiotemporal distribution of the exploited part of the population to be studied, the establishment of an index of abundance in the mean number of fish caught per fishing hour of the standardized bottom trawl, and the correlation of data with biological results.

Recent ichthyoplankton surveys with a Bongo sampler (net: 500 μM mesh) give the first data on the hatching time and place, the size of larvae, and values for the abundance of specimens collected per filtered volume.

3 Results

The spatial and temporal coverage of the Kerguelen shelf and surrounding banks (Skif in the SW and Kerguelen-Heard in the SE) is considered to be correctly investigated over the period 1979/80–1989/90 to avoid misinterpretation from dependable analysis. Each kind of data collection permits specific results and will be analyzed separately.

3.1 Length Frequency Distributions (LFD)

The monthly analysis of LFD shows the presence of unimodal distributions concerning fish of a limited range in length (5–10 cm in general), a limited maximal size (never more than 40 cm), and apparent groups of the same size followed over several years (Fig. 1). It is clear that the unimodal distributions correspond to groups of fish born during the same period and growing at the same rate from year to year. These age groups (cohorts) were followed easily during the 4–5-year period because the mean growth was quick and no new age group (recruitment) modified this aspect of the distribution for a long period of time.

In relation to the vast area studied, a geographical analysis of the LFD was undertaken. No significative difference in the mean length could be found to separate the shelf samples, except for rare distributions observed in the SW sector. No catches and related distributions were obtained on the Kerguelen-Heard banks during the period studied. Regular samples are recorded for the Skif bank (140 miles SW from the Kerguelen Islands) and this LFD never mixed with the shelf LFD (Fig. 1) during the same period of time. Thus, groups of fish can be considered as separated.

3.2 Growth

The use of otoliths (whole or cut) from *C. gunnari* in the Kerguelen area does not give satisfactory and reliable results; another method is necessary to evaluate growth. The mean length of the cohorts measured at regular intervals is commonly

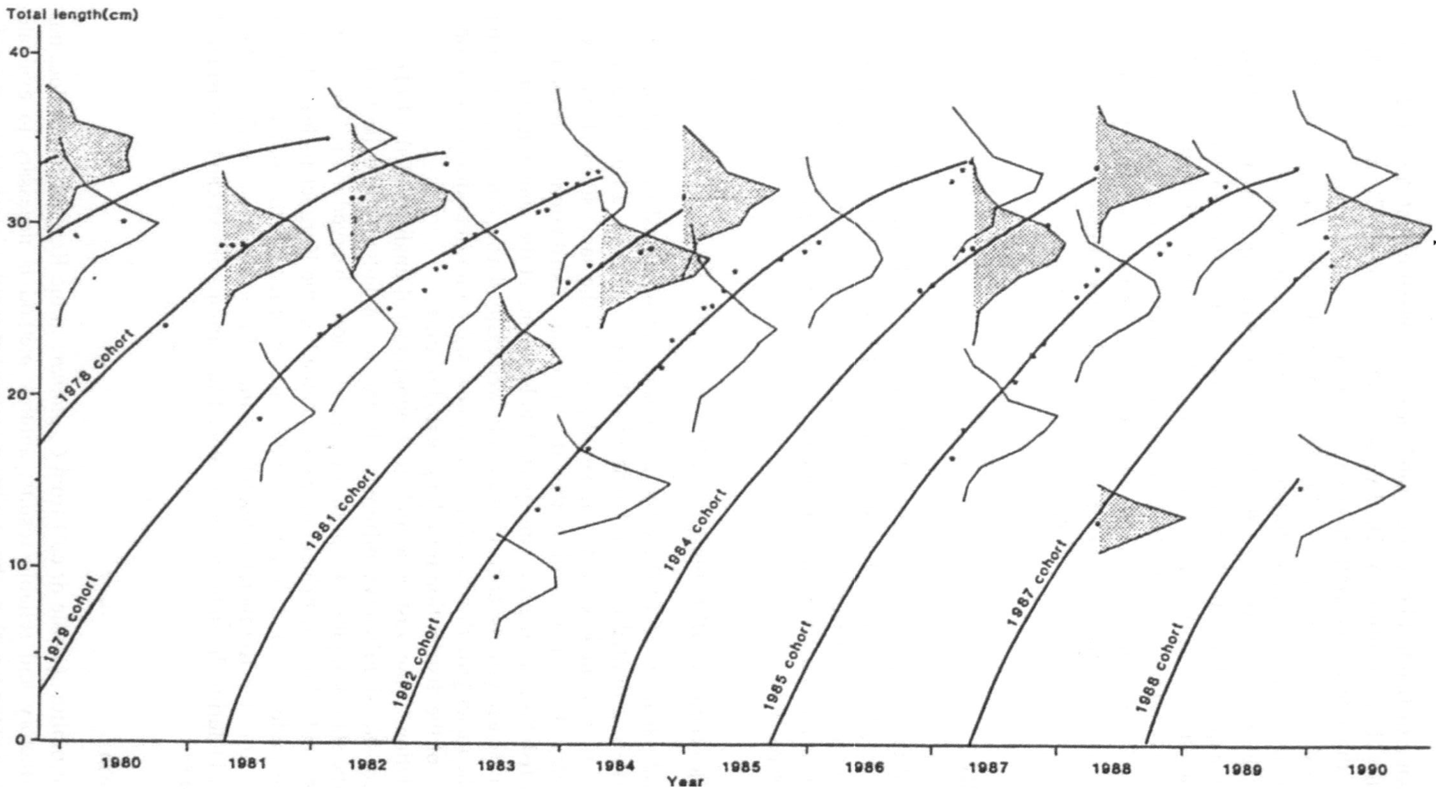

Fig. 1. Length frequency distributions (LFD) of icefish *Champsocephalus gunnari* observed from 1979/80 to 1989/90 in the Kerguelen Islands area. *Stars* represent monthly mean length; growth curves are plotted (*dotted* LFD are relative to the Skif bank, the other to the Shelf)

used for tropical fish stocks as an "integrated method" (Pauly 1980) or "modal progression analysis" (Sparre et al. 1989) and seems to be the most valuable method to evaluate growth in Kerguelen icefish. The time series is rather long, but the apparent lack of annual abundant recruitment permits the use of the method over a period of 4–5 years. The mean length observed during a similar period from 1 year to another and from cohort to cohort (Table 1) is stable enough in the LFD of the Kerguelen shelf and the Skif bank. The parameter L_∞ of the von Bertalanffy equation can be obtained by the Ford-Walford plot and gives a value of 39 cm (Duhamel

Table 1. Mean length of cohorts at age (0, 1, 2, 3, 4) from the length frequency distributions observed on the Kerguelen shelf and Skif bank from 1979/80 to 1989/90[a]

	Kerguelen shelf icefish group Summer mean total length (cm)				
	1979/80	1980/81	1981/82	1982/83	1983/84
1979 Cohort			24.2	29.0	32.4
			(02)	(02)	(02)
	1982/83	1983/84	1984/85	1985/86	1986/87
1982 Cohort	6.0	17.9	25.4	29.0	33.3
	(02)	(03)	(02)	(01)	(02)
	1985/86	1986/87	1987/88	1988/89	1989/90
1985 Cohort		16.5	26.0	30.7	33.0
		(02)	(02)	(01)	(02)
	1988/89	1989/90			
1988 Cohort		14.9			
		(11)			
	Skif bank icefish group Autumn mean total length (cm)				
	1977/78	1978/79	1979/80	1980/81	1981/82
1978 Cohort			24.1	28.9	32.0
			(10)	(03)	(03)
	1980/81	1981/82	1982/83	1983/84	1984/85
1981 Cohort			22.5	27.7	31.8
			(07)	(04)	(01)
	1983/84	1984/85	1985/86	1986/87	1987/88
1984 Cohort	4.6			28.7	33.3
	(10)			(03)	(04)
	1986/87	1987/88	1988/89	1989/90	
1987 Cohort		12.0		29.4	
		(04)		(02)	

[a]( ) Month of sampling.

1987). A mean growth curve can be plotted from the results to obtain the age and to estimate the birthday of each cohort by intersection with the time axis (see Fig. 1). If the growth rate of icefish seems similar between Kerguelen-shelf and Skif-bank cohorts, the translation of curves at similar months points out a difference of about 3 months between the growth rates to be superposed. The estimation of birth dates is also different.

From these results, a 3-year cycle of abundant recruitment appears for the two areas. The following cohorts have been detected during the period studied: (1) Kerguelen shelf, 1988, 1985, 1982, 1979, (1976); (2) Skif bank 1987, 1984, 1981, 1978, (1975).

3.3 Reproduction

The gonad analysis gives useful information on the reproductive cycle of icefish in the Kerguelen area. The first spawning is noted when the fish reach a mean size of about 25 cm, 34–35 months after hatching, which corresponds to the beginning of age 3 (July 1 is considered as the "birthday" for antarctic fish). The spawning time and place are very different for the Kerguelen shelf and Skif bank sectors. The spawning (stade IV of the Everson scale) has been observed during the winter and in shallow waters (100–140 m) on the shelf (NE sector), but occurs on the slope (280–340 m) of the Skif bank during spring. Near-bottom spawning concentrations and bottom egg-laying are characteristic of icefish in the area. An annual cycle of maturation seems to prevail after the fish have reached first maturity.

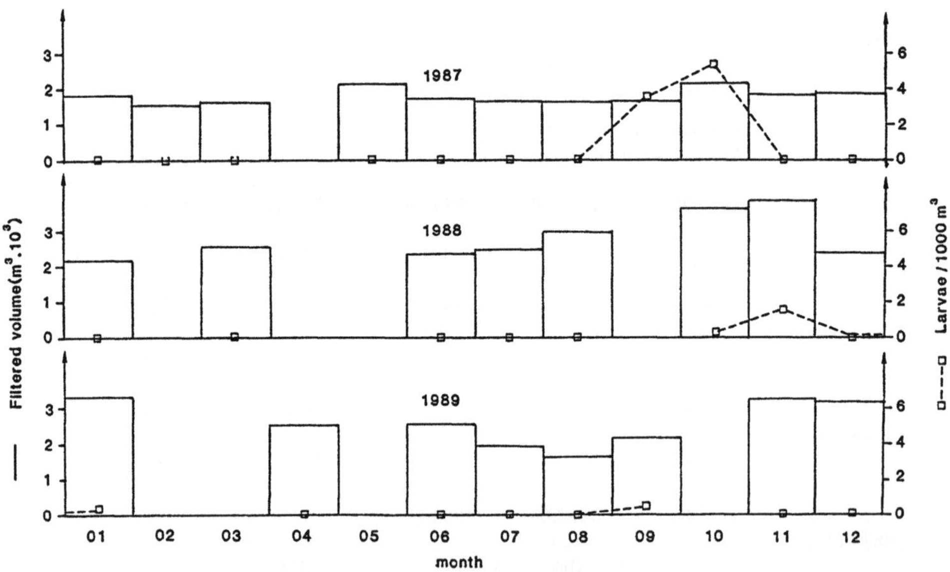

Fig. 2. Monthly abundance (number of specimens/1000 m^3) of *Champsocephalus gunnari* larvae from ichthyoplankton surveys in Morbihan Bay (east of the Kerguelen Islands) from 1987 to 1989. □, Number of larvae/1000 m3; boxes filtered volumes

3.3.1 Ichthyoplankton Phase

Results have been obtained for the Kerguelen shelf only, because surveys on the Skif bank are difficult to obtain regularly. In a time series of monthly sampling conducted in (Morbihan Bay in the eastern part of the islands from 1987 to 1989 (Fig. 2), larvae were detected at the end of winter/beginning of spring, which is in accordance with the premise of a winter spawning (probably July/August). In relation to the size of larvae (12.0–26.2 mm), hatching is evaluated to take place from the beginning of September to October, which corroborates the birthday date using the growth curve. The presence of larvae is noted annually and can be tentatively considered to belong the reproductions of the observed cohorts on the shelf (1987 hatching would result from spawning of the age 5, 1982 cohort; 1988 hatching age 3, 1985 cohort; 1989 hatching, age 4, 1985 cohort). The abundance is low but probably not representative in relation to a shallow sill at the entrance of the bay which disturbs fish migrations from the shelf and larvae flow with seawater masses. The surveys were extended to other, more open areas of the shelf during 1989 and were reduced to the most favorable period to resolve the problem of abundance estimations. The southern area (Audierne Bay) has not produced larvae,

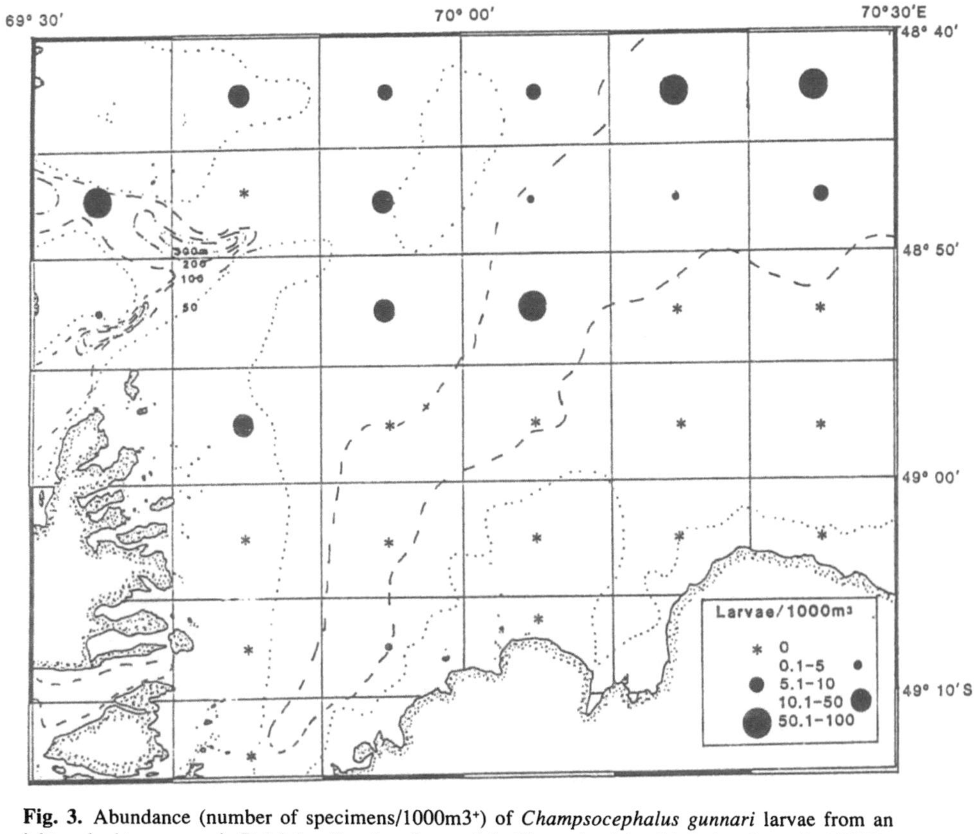

Fig. 3. Abundance (number of specimens/1000m3+) of *Champsocephalus gunnari* larvae from an ichtyoplankton survey in Baleiniers Bay (northeast of the Kerguelen Islands) during the period 29/10–01/11/1989

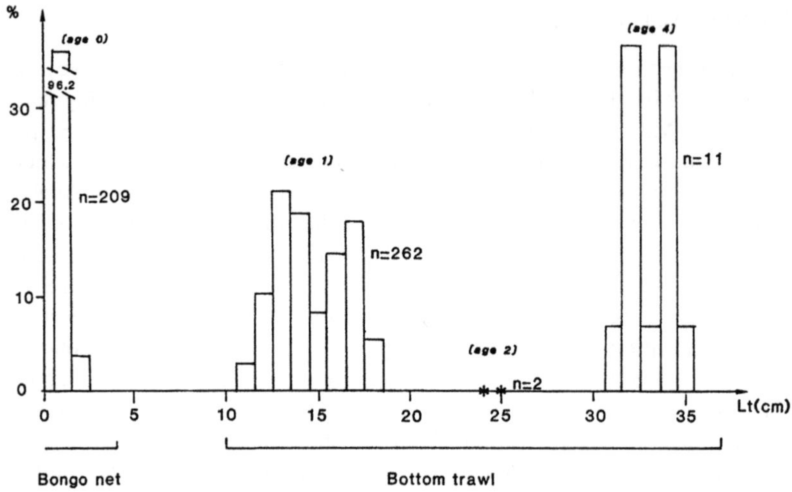

Fig. 4. Length frequency distribution of *Champsocephalus gunnari* from bongo net and bottom trawl samplings in Baleiniers Bay (northeast of the Kerguelen Islands) from 26/10 to 1/11/89. Lt Length; * samples 24 and 25 cm long

but the northeastern area (Baleiniers Bay) situated close to the spawning ground, found during a 1983 survey (Duhamel 1984), gives abundances reaching a maximum value of 97.7 larvae/1000m^3 (Fig. 3). The larvae (age 0+) originating from the 1989 spawning of the 1985 (age 4) cohort were recorded during the same period in Morbihan Bay. The distribution of larvae does not extend to shallow waters but probably projects beyond the studied area on the shelf. A very similar pattern of distribution with the spawning-ground boundary is noted, supporting the evidence of a recent hatching of eggs laid on the bottom. Simultaneously, a bottom trawl survey of this area allowed the estimation of the LFD of icefish present (Fig. 4). Three other age groups were recorded: age 1 of the 1988 cohort (juvenile); age 2 of the 1987 cohort (juvenile maturing); age 4 of the 1985 cohort (post-spawning specimens). These results support the premise that an annual recruitment occurs, because fish of various cohorts (1985, 1987, 1988, 1989) are present , but the results do not explain the 3-year cycle of the abundant cohorts observed on the shelf.

3.3.2 Spatiotemporal Distribution and Abundance

The distribution of the species on the Kerguelen shelf can be measured from portions of the cohorts analyzed using trawl catches (age 2 to 4), excluding the spawning period that involves migrations (autumn to winter) in the shallow part of the shelf (Fig. 5). A north movement, from the SE to the NE part of the shelf at an intermediate depth range (150–250 m), with growth, is obvious. A reduction of the south extension is noted from one cohort to the following during the period studied.

The data on the Skif bank are incomplete and concern mainly the prespawning and spawning period. The presence of the species on the bank seems to be irregular outside these periods, however, a clear pattern cannot be easily recognized.

The mean annual value of the abundance index of the cohorts which have been analyzed can be estimated in the two areas (Fig. 6). The abundance level is much lower for the Skif bank compared to that of the shelf at a similar age of the analyzed cohorts. A sharp decline in the abundance is noted with the growth of each cohort, reaching a negligible value at the age 4 for the two stocks. Finally, a negative tendency in the mean abundance at age 3 in the 1979 to the 1985 cohorts is detected for the shelf data.

3.4 Diet and Predation

The stomach content analysis of specimens older than 1 year places *C. gunnari* from the Kerguelen area within the ranks of specialized planktonivorous predators. Only euphausiids (mainly *Euphausia vallentini*) and hyperiid amphipods (mainly *Themisto gaudichaudi*) are common prey. Fish (Myctophidae and fingerlings of its own species) are also sometimes found. An analysis of the diet index (Q) in the 1979 cohort, during similar periods of the year, shows a relative change in the diet composition with age (Fig. 7). The relationships between predators and prey of icefish are summarized in Fig. 8, which indicates the importance of the species in the trophic chain of the Kerguelen Islands marine ecosystem.

a

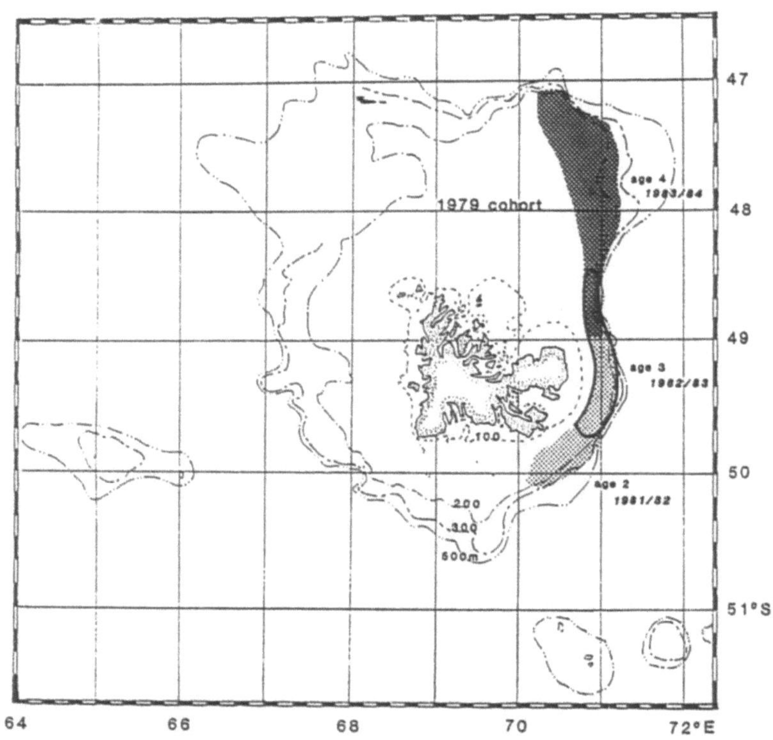

Fig. 5a-c. Annual (summer-period) distribution of three successive cohorts of *Champsocephalus gunnari* on the Kerguelen shelf (1981/82 to 1988/89)

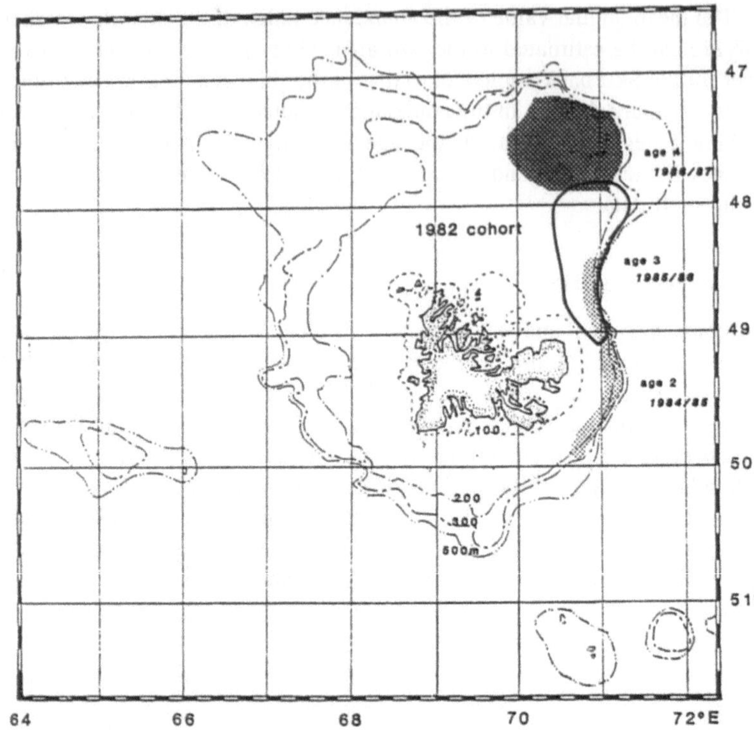
47
48
49
50
51°S
64
66
68
70
72°E
age 4
1986/87
1982 cohort
age 3
1985/86
age 2
1984/85
100
200
300
500m

Fig. 5b

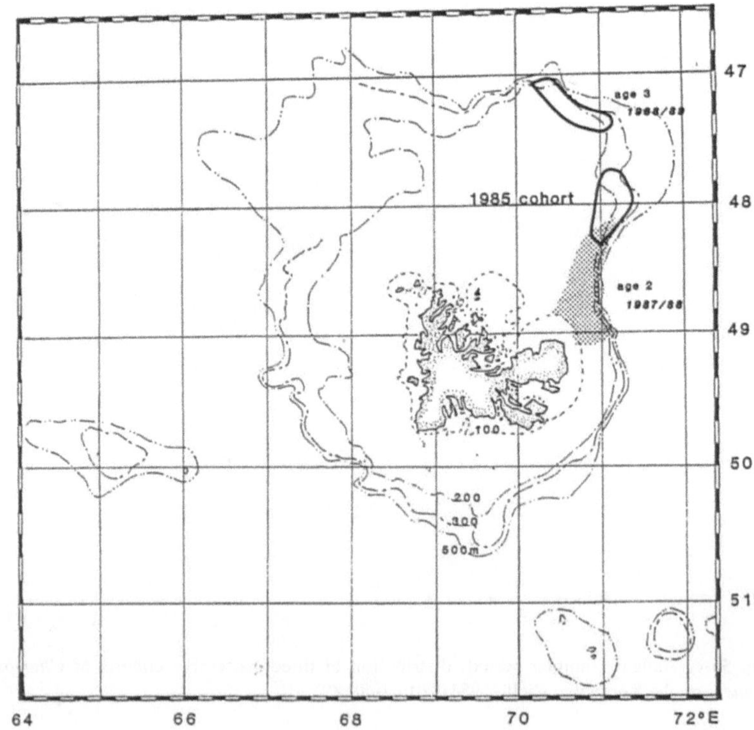
47
48
49
50
51°S
64
66
68
70
72°E
age 3
1988/89
1985 cohort
age 2
1987/88
100
200
300
500m

Fig. 5c

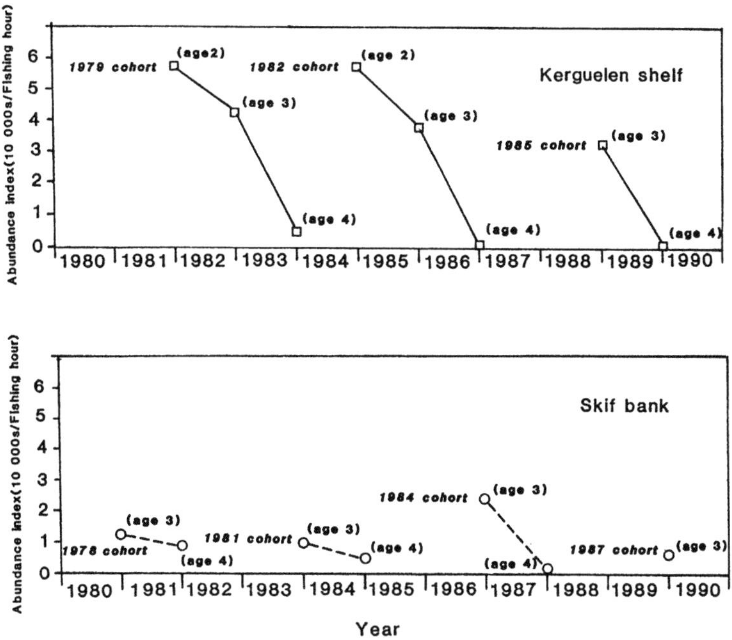

Fig. 6. Annual mean abundance (number of fish caught per fishing hour of bottom trawling) of successive cohorts of *Champsocephalus gunnari* on the Kerguelen shelf and the Skif bank from 1980 to 1990

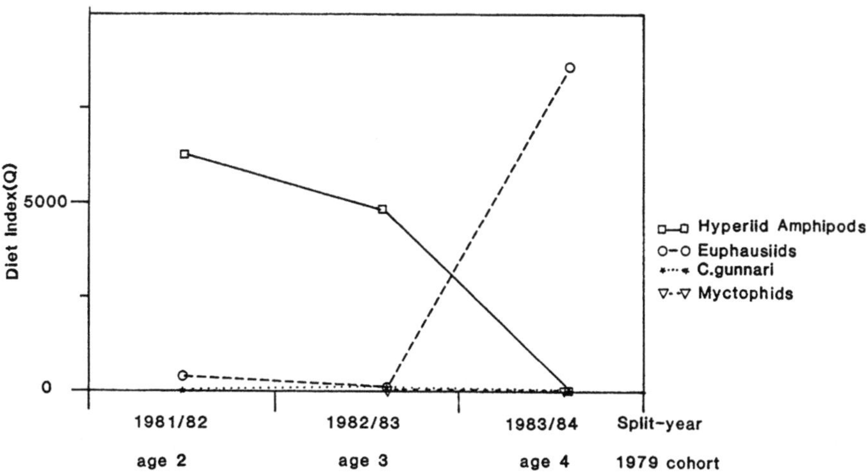

Fig. 7. Annual variation in the diet index (Q) for the main preys of the *Champsocephalus gunnari* Kerguelen-shelf 1979 cohort

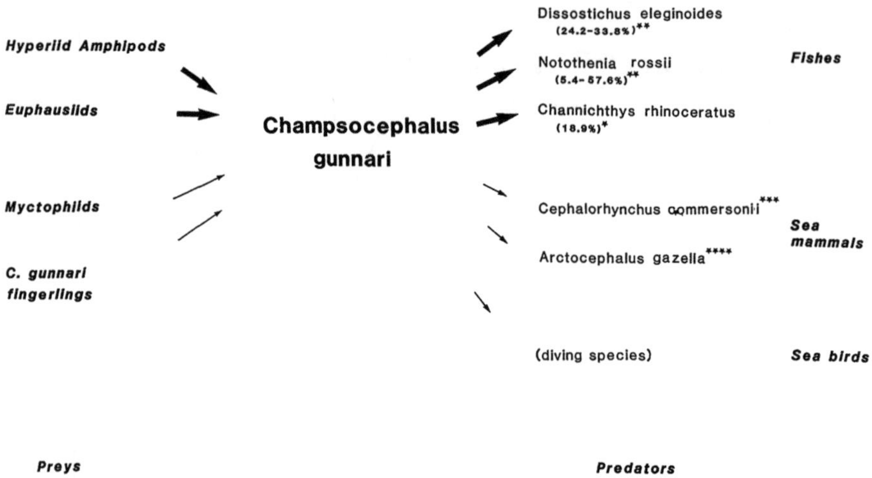

Fig. 8. Place of *Champsocephalus gunnari* in the trophic marine system of the Kerguelen Islands area. *()* percentage occurrence of *C. gunnari* in the diet of the species * from Duhamel (1981); ** from Duhamel (1987); *** from Robineau and Duhamel (1984); **** from Green *et al.* (1989) at Heard Island

4 Discussion

A biological and demographic study of icefish in the Indian sector of the Southern Ocean would be incomplete if the data available for the nearby Heard shelf and the associated banks were not included. In addition, comparisons with the Atlantic populations must be taken into account.

The differences in the observed values of meristic characters between the Atlantic and Indian sectors are certainly related to a clinal variation in the most northern population (Kerguelen Islands), where the species live under warmer hydrological conditions (summer sea-surface temperature: 6 °C) than those in the southern area of distribution (South Shetland: 0 °C) as noted by Sosinski (1981) and observed for other species (Duhamel 1987).

It is helpful that the LFD allow us to study growth on the basis of a long time series, because contradictory results are obtained from the growth patterns of otoliths (Olsen 1955; Kock 1981; Sosinski 1981; Kochkin 1982) and other anatomical structures (Gubsch 1980; Kochkin 1982; Trecul 1983). The unimodal structure of the LFD is not clearly observed for the main population of the Atlantic sector (South Georgia), which has a probable relationship to annual recruitments, but it is clearly visible in the LFD reported for South Orkney (Kock et al. 1985; Kock and Köster 1989). In this area, for example, a single cohort can be easily followed from 1981/82 to 1985/86.

The LFD data obtained for the Kerguelen are show for the first time that two groups, corresponding probably to two separate populations, are observed in the same geographical area with distinct spawning periods and depth range of egg-laying. The data published prior to the period studied (Slosarczyk and Wysokinski

1980; Sosinski 1981; Duhamel 1987) allows the corroboration of the 3-year cycle. The LFD are available for two periods (1974/75 and 1977/78) and are summarized as follows:

1. Skif bank
   a) April 1975: 1972 cohort;
   b) February 1978: 1975 cohort; observed also at the end of 1979 (Fig. 1),
2. Kerguelen shelf
   a) April 1974: 1973 cohort;
   b) March/April 1975: 1973 cohort;
   c) December 1977: 1973 cohort.

Some LFD are available (Slosarczyk and Wysokinski 1980; Sosinski 1981) for another area: the Kerguelen/Heard bank. Several cohorts (1969 and 1970, at least) were observed in April 1975 and the cohort 1973 during 1977/78. Very few specimens were collected afterwards. Two specimens of 28–29 cm (probably from the 1982 cohort) were recorded during January 1985 (Duhamel 1987), and about 20 specimens of 18–20 cm (1985 cohort) during a joint USSR/France scientific cruise (March 1987). The LFD, mean length, and cohorts are similar to those observed on the Kerguelen shelf during the same period. Sosinski (1981) also noted that the run of maturation of gonads is equal in the Kerguelen shelf and in the Kerguelen-Heard bank, this would imply that these two (sub)-populations are identical.

The historical data for icefish on the Heard shelf are anecdotic. Sosinski (1981) gives a shaded area of distribution on a map; cruise MD03 of R.V. "Marion-Dufresne" collected some specimens (20-23 cm in range, probably 1971 cohort) during April 1973 in shallow waters NE of the islands (Duhamel 1987). A recent investigation (Williams and Ensor 1988) during autumn/winter 1987 has led to the detection of concentrations of the species at the level of the slope in the northeastern part of the shelf and a nearby bank (30 miles to the last area). The observed LFD (Williams, pers. comm.) are different in the two sectors. Three age groups have been identified on the shelf: 0+ (Lt = 6.8 cm); 1 (Lt = 18.9 cm) and 4 (Lt = 34.9 cm). The similarity with the mean length (Lt) of the Kerguelen shelf LFD is surprising: age 1 (Lt = 18.3 cm) and age 4 (Lt = 32.8 cm) for a summer sampling (age 0+ is not sampled in commercial trawling). On the other hand, the observed mean values on the nearby bank are very different (Lt = 11.2, 22.8 and 36.4 cm) and not comparable to other data, the Skif bank included. The relationships between size groups of the Kerguelen and Heard areas need further examination to confirm separation, or not, of populations. A special survey of the potential spawning grounds would probably help to elucidate this problem.

The general distribution of abundant shelf-icefish groups seems to be related to the highest productive area in the eastern part of the Kerguelen islands. An "island mass effect" is favorable to the productivity and phytoplankton biomass in the sector (Plancke 1977). Chlorophyll a reaches values up to 2 mg m^{-3} in northeastern part of the shelf (Jacques et al. 1982) and primary production is high in this sector: 140–420 mg C $m^{-3}J^{-1}$ (Tréguer 1987). The zooplankton productivity is also higher than in the other areas, with a mean biomass of 4300 mg m^{-2} for the upper 100 m (Razouls and Razouls 1982). The exclusive zooplankton diet composition of the icefish explains its concentration in these areas. The absence of krill, *Euphausia superba*, in the area explains the difference in the diet composition compared to the Atlantic sector, where this euphausiid is the main prey of icefish (Permitin and Tarverdieva 1972; Komposki 1980; Kock 1981).

The variation in abundance observed for a series of cohorts is related to the combined effects of natural mortality (predators) and fishing mortality, beginning after use in the fishing industry (ages 2 or 3 following the cohorts). The current proportion of the species in the catches since exploitation by the fishing industry (1979/80) is important (55.64% of the 185 320 tones reported) and shows that *C. gunnari* is still a target species. The fishing effort contributes to the rapid vanishing of the cohorts in the two areas studied. The origin of the 3-year cycle of abundant cohorts would be related to a high mortality rate, natural or amplified by exploitation. The new cohorts actually originate from the first reproduction of the preceding one (i.e., the 1985 cohort originates from the first spawning of the 1982 cohort reaching age 3) and the low level in abundance of the spawners at age 4 does not allow production of a second abundant recruitment. However, the presence of annual spawning is proved by ichthyoplankton surveys. The hypothesis of a fishery mortality effect would be supported by the apparent decrease in the abundance of new cohorts and a reduction in the total area of distribution. The management of the species needs further study of the recruitment jointly organized off the Kerguelen and the Heard Islands. Finally, it will be impossible to reach the observed maximum size of the Atlantic population (64 cm) in South Georgia (Permitin 1973) because 45 cm is the maximum size registered in the Kerguelen/Heard area. This emphasizes once more the difference between the Atlantic and Indian icefish populations.

Acknowledgements. Fieldwork was supported by the Mission de Recherche of Terres Australes et Antarctiques Francaises. I thank the crew of scientific vessels and trawlers for their help on board. Special thanks are due to fishery observers and P. Koubbi, N. de Mahuet, C. Meurou for their assistance in collecting data and samples. I am grateful to J.C. Hureau who read the manuscript.

References

Duhamel G (1981) Caractéristiques biologiques des principales espèces de poissons du plateau continental des iles Kerguelen, Cybium 5(1):19–32

Duhamel G (1984) Campagne Austral/83 à bord du chalutier Austral. 21 mai au 27 juin 1983. Les rapports des campagnes à la mer. Mission de Recherche des TAAF 83/05:1–58

Duhamel G. (1987) Ichtyofaune des secteurs indien occidental et atlantique oriental de l'océan Austral. Biogéographie, cycles biologiques et dynamique des populations. Thèse de Doctorat d'Etat. Univ Paris VI. Microfiche SN 87 200 447 Inst d'Ethnol, pp 512

Everson I (1977) The living resources of the Southern Ocean. FAO GLO/50/77/1 Rome, pp 1–156

Green K, Burton HR, William R (1989) Diet of antarctic fur seals *Arctocephalus gazella* (Peters) during the breeding season at Heard Island. Ant Sci 1(4):317–324

Gubsch G (1980) Studies on age determination and growth of the icefish *Chaenocephalus aceratus* (Lönnberg), ICES GMG 47:1–12

Gubsch G (1982) Zur Verbreitung und zur Biologie der Eisfische (Chaenichthyidae) im atlantischen Sektor der Antarktis. Fisch Forsch 20(2):29–47

Hureau JC (1970) Biologie comparée de quelques poissons antarctiques (Nototheniidae). Bull Inst Océanogr Monaco 68 (1391):1–244

Jacques G, De Billy G, Panouse M (1982) Biomasse et production primaire dans les secteurs antarctique et subantarctique de l'océan indien (mars 1980). CNFRA 53:87–99

Kochkin PN (1982) Study on age and growth of pike glassfish *Champsocephalus gunnari* Lönnberg. J Ichthyol 22 (3):37–47

Kock KH (1980) Graphical analysis of length frequency of *Champsocephalus gunnari* Lönnberg (Channichthydae) from South Georgia. Cybium 8:33–42

Kock KH (1981) Fischereibiologische Untersuchungen an drei antarktischen Fischarten: *Champsocephalus gunnari* Lonnberg, 1905, *Chaenocephalus aceratus* (Lönnberg 1906) und *Pseudochaenichthys geogianus* Norman, 1937 (Notothenioidei, Channichthyidae). Mitt Inst Seefisch Hamburg 32:1–226

Kock KH, Köster FW (1989) The state of exploited fish stocks in the Atlantic sector of the Southern Ocean in 1988. Mitt Inst Seefisch, Hamburg 46:1–73

Kock KH, Duhamel G, Hureau JC (1985) Biology and status of exploited antarctic stocks. Biomass Sci Ser 6:1–143

Komposki A (1980) On feeding of *Champsocephalus gunnari* Lönnberg, 1905 (Pisces, Chaenichthyidae) off South Georgia and Kerguelen Islands. Acta Ichthyol Piscat 10 (1):25–43

Olsen S (1955) A contribution to the systematics and biology of Channichthyid fishes from South Georgia. Nytt Mag Zool 3:79–93

Pauly D (1980) A selection of simple methods for the assessment of tropical fish stocks. FAO Fisheries Circular 729:1–54

Permitin YY (1973) Fecundity and reproductive biology of icefish (Channichthyidae), fish of the family Muraenolepidae and dragonfish (Bathydraconidae) of the Scotia sea (Antarctica). J Ichthyol 13 (2):204–215

Permitin YY, Tarverdieva MI (1972) Feeding of some species of Antarctic fishes in the South Georgia Island area. J Ichthyol 12 (1):104–114

Plancke J (1977) Phytoplankton biomass and productivity in the subtropical convergence area and shelves of the western Indian subantarctic islands In: Llano GA (ed) Adaptations within antarctic ecosystems. Proc 3rd SCAR Symp on Artarctic Biology. Smithonian Inst, Washington, pp 51–73

Razouls C, Razouls S (1982) Eléments du bilan énergétique du mésozooplancton antarctique. CNFRA 53:131–141

Robineau D, Duhamel G (1984) Régime alimentaire du dauphin de Commerson *Cephalorhynchus commersonii* (lacépède, 1804) aux îles Kerguelen pendant l'été Austral. Bull Mus Nat Hist Nat Paris. 4ème sér 6. sect A (2):551–559

Slosarczyk W, Wysokinski A (1980) Ichthyological and fisheries studies of the shelf fishing grounds in the region of Kerguelen Islands (Antarctic). Pol Polar Res 1 (4):173–190

Sosinski J (1981) Comparative biology of antarctic icefish *Champsocephalus gunnari* Lönnberg, 1905 from the Antarctic (in Polish). Studia Materialy Ser B 48:1–91

Sparre P, Ursin E, Venema SC (1989) Introduction to tropical fish stock assessment Part 1 - Manual. FAO Tech Pap 306/1–337

Trecul F (1983) Histologie osseuse et marques de croissance cyclique chez deux téléosteens antarctiques, *Notothenia squamifrons* et *Champsocephalus gunnari* (Notothoenioidei, Perciformes). Mém DEA Univ Paris VII: pp 67

Tréguer P (1987) Comparative physical and chemical characteristics of coastal waters from Kerguelen and Heard Islands and their oceanic surroundings during summer, relation to phytoplankton growth. CNFRA 57:17–26

Williams R, Ensor PH (1988) Extensive research needed to assess fish resources around Heard Island. Aust Fish 47 (2):2–4

Is the Growth of Polar Fish Limited by Temperature?

A. Clarke and A.W. North

1 Introduction

It is a widespread observation that polar marine invertebrates and fish grow slowly. Although a few species have been reported as growing rapidly (for example the sponge *Mycale acerata* at McMurdo Sound: Dayton et al. 1974), it is generally true that most polar marine invertebrates and fish grow more slowly than related species from warmer water (reviewed in Clarke 1983, 1988).

There have been three basic explanations for this slow growth, namely:

1. A direct limitation by temperature.
2. Adaptation to cold water necessitates an elevation of metabolic rate with a concomitant reduction in the energy available for growth.
3. Seasonal resource limitation.

The first explanation is intuitively appealing (everybody knows things slow down in the cold). There is also an extensive literature documenting the effects of low temperature on the growth of farmed fish. Here the effect is essentially one of straightforward thermodynamics, although it may well be compounded by the effect of temperature on factors such as appetite, behaviour and coordination. However, it is not valid to extend the example of the effect of temperature change on an individual organism or species to the more general consideration of the adaptation of organisms to temperature (although this is frequently done). At the very least it ignores the possibility that over the course of evolutionary time the fish may have been able to achieve some degree of compensation for the rate-limiting effects of low temperature.

The second argument is based on the concept of *metabolic cold adaptation* (for discussions of this influential idea see Holeton 1974; Clarke 1980, 1983, 1987, 1991; Wells 1978; Cossins and Bowler 1987). The central idea is that adaptation to polar temperatures involves an elevation of metabolic rate. As a result, relatively less of the energy intake is available for growth, which is consequently reduced (Dunbar 1968). Since there would seem to be nothing in principle to prevent a polar animal from increasing its energy intake to increase growth, were food available (up to maximal levels where other constraints limit year-round food intake), this argument is in essence one of resource limitation.

The third explanation invokes no effect of temperature at all. It is assumed that the machinery of growth has evolved compensation for temperature over the course

British Antarctic Survey, High Cross, Madingley Road, Cambridge, CB3 OET, UK of evolutionary time (though this compensation may not be perfect), and that annual growth is limited by a severely seasonal pattern of resource availability. Since growth is limited to short periods when feeding is possible (or energetically worthwhile), overall annual growth is slow. A key point here is that when growth is actually underway, it may proceed rapidly. This argument is reviewed extensively in Clarke (1988) and Clarke and Peck (in press).

1.1 Distinguishing the Competing Hypotheses

How can we decide which (if any) of these three explanations is the correct one? Clearly the first two hypotheses are based on the assumption that in some way, either directly or indirectly, low temperatures lead inevitably to a reduced growth rate. They would therefore be disproved (or at least severely tested) by any observations of polar fish growing rapidly.

The second explanation has, in addition, received a setback in that more recent physiological work has largely failed to substantiate the early observations of a significantly elevated metabolic rate in polar marine invertebrates and fish.

The third hypothesis is more difficult to test. It assumes that polar fish have evolved a capacity to grow rapidly. However, it does not predict that rapid annual growth rates will necessarily be observed, for the seasonality of the polar marine ecosystem may be preventing this. Since this point is crucial to any test of the hypothesis, it requires a more detailed examination.

1.2 Temperature Compensation and Growth

Growth is a complex process, subject to a number of controls and limitations at a variety of levels. Purely for the sake of illustration, let us equate growth with protein synthesis. This is a simplification, but a reasonable one for the development of the argument. The central thesis is illustrated diagrammatically in Fig. 1.

The upper trace represents the synthetic activity of a polyribosome in a fish living at 10 °C. The height of the bar represents the rate at which the ribosome is making polypeptide, and the gaps indicate periods when there is no synthetic activity. During the period Δt the ribosome synthesizes 16 units of protein. Let us now assume that a related polar species of fish living at 0 °C produces only 8 units of protein during the same period. This corresponds to a Q_{10} of 2, which is quite reasonable biologically and might suggest a direct rate limitation by temperature.

There are two possible extreme explanations for the slower rate observed in the polar fish. The fish is that there is something in the process of protein synthesis which is limited by temperature, and even when working maximally the polar fish can still produce only 8 units of protein. This is illustrated in the middle trace where the height of the bar is reduced and there are no gaps. The ribosome is working continuously but the rate of polypeptide elongation (represented by the height of the bar) is reduced by the low temperature.

An alternative explanation, however, is that the ribosomal machinery can compensate fully for temperature and when it is operating it produces protein just as rapidly as the temperate species (and hence the height of the bar is the same), but other factors dictate that only 8 units of protein are synthesized. These might

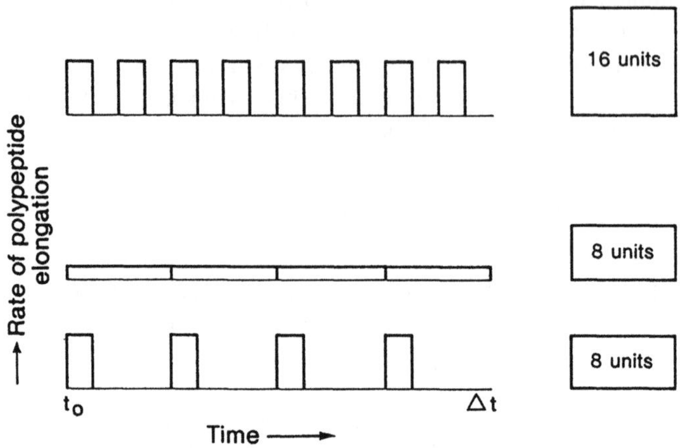

Fig. 1. A diagrammatic representation of polypeptide elongation at 10 °C (*upper trace*), at 0 °C with no temperature compensation (*middle trace*), and at 0 °C with complete temperature compensation but with resource limitation (*lower trace*). The height of the *bars* indicates the rate of polypeptide elongation and the *boxes* show the relative amounts of protein synthesized in time Δt. After Clarke, 1991

include resource limitation (amino acids or energy), or a reduced requirement for protein turnover at low temperature. This situation is illustrated by the bottom trace.

It is possible of course that the real explanation lies somewhere between these two extremes. The important point is that it is impossible to distinguish between these two explanations simply by a measure of overall growth rate. *One cannot look at a polar fish, see that it grows slowly and conclude simply that the growth rate is slowed by low temperature*. These arguments are described in the wider context of cold adaptation by Clarke (1991).

The argument outlined above for protein synthesis can be extended to whole-organism growth, as shown in Fig. 2. Here a hypothetical temperate fish is shown with some seasonality of growth and a summer growth rate of 2 g per month (Fig. 2a). The annual growth rate of the polar fish illustrated in Fig. 2b is exactly half that of the temperate fish. The upper trace shows a fish with both a reduced overall annual growth rate and little seasonality. Clearly in this polar fish, the annual growth rate is lower than either the annual growth and peak summer growth of the temperate species, for in the year mass increased by only 8 g, compared with 16 g for the temperate fish. However, from the argument outlined above it is not possible to conclude that this represents direct temperature limitation.

The second polar fish in Fig. 2b (lower trace) shows a distinct seasonality of growth with essentially no growth in winter. Overall, however, annual growth is similar to the first polar fish (that is, an annual growth increment of 8 g). More importantly, however, summer growth is at a rate fully comparable with peak summer growth in the temperate fish (2 g/month). Annual growth rate however, is kept low by a much reduced summer growth period and nonexistent winter growth.

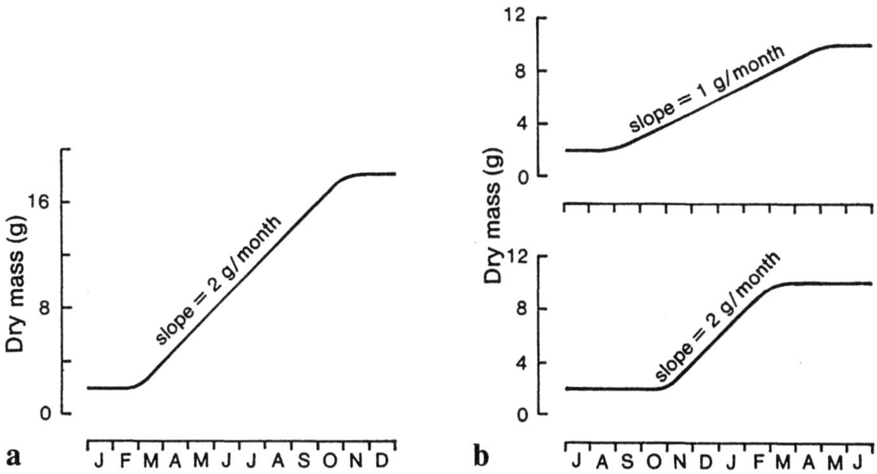

Fig. 2. A diagrammatic representation of growth in temperate and polar fish. **a** Annual growth in a representative temperate-water fish showing some seasonality. **b** Annual growth in representative polar fish. The *upper trace* shows a polar fish with slow overall growth and little seasonality. The *lower trace* shows a polar fish with pronounced seasonality but with a summer growth rate identical to that of the temperate species in **a**. Note that both polar fish have the same annual growth increment (8 g) which is only half that of the temperate fish

The lower trace in Fig. 2b exemplifies the possibility of a fish whose growth machinery is fully compensated for temperature when compared with a temperate relative, but whose annual growth rate is limited by the seasonality imposed by the polar marine ecosystem.

Figures 1 and 2 are purely theoretical constructs and they exemplify the hypothetical organisms beloved of armchair ecologists. They are of value only in focussing our attention on the important questions to be tackled by fieldwork in the real world. In the case of temperature limitation of growth, they suggest three key questions:

1. Are the annual growth rates of all polar fish slow?
2. Do polar fish show seasonality of growth, and is this more or less severe than in temperate and tropical species?
3. What are the maximum growth rates achieved by polar fish, and how do these compare with fish from warmer waters?

We will attempt to answer these in turn, although it is only the third question which tackles the heart of the problem of temperature limitation of growth in polar fish.

2 Annual Growth Rates of Adult Antarctic Fish

Several studies have shown that adult tropical fish grow faster than temperate fish, which in turn tend to grow faster than polar fish (see Pauly 1980). These comparisons are complicated by effects of size, and also various corrections for temperature that are often applied. However, when a more detailed examination of the data is undertaken, the basic trend remains: adult fish growth rates tend to increase in the order polar-temperate-tropical (North, in prep.). At any one temperature there are, of course, wide variations in growth rate associated with size, ecological niche, activity pattern and so on. Nevertheless there appears to be an underlying positive relationship between growth rate and temperature.

3 Seasonality of Growth of Adult Antarctic Fish

Although there have been few detailed studies of the growth of antarctic fish in different seasons, the available evidence suggests that growth is seasonal. For example, the otoliths and scales of many species of antarctic fish have annuli comprised of a fast-growing and slow-growing zones (Olsen 1954, 1955; Wohlschlag 1961; Everson 1970, 1984; Hureau 1970), and it has been demonstrated that in some species, the fast-growing zone in otoliths is formed during summer and autumn (North 1988). Seasonal patterns in relative liver mass, and total mass in relation to length in antarctic fish suggest that the energy available for growth varies during the year. For juvenile *Notothenia rossii* at South Georgia, the mass of liver as a proportion of total body mass, and the condition index both varied with season (Burchett 1983). Similar patterns were also found in adult antarctic fish (Everson 1970; Hureau 1970). Although in adult fish the annual cycle of gonad development, lowered feeding rates associated with the spawning period (Sakakibara et al. 1989), and egg guarding in some species (Daniels 1978; Hourigan and Radtke 1989) are likely to reduce somatic growth rate for part of the year, and thus mask the effects of temperature or food availability, there is slight evidence that some species may lose weight during winter and early spring (Kawaguchi et al. 1989).

Unfortunately, there are no data with which to test the hypothesis that maximum summer growth rates are similar to those of warmer-water species. We are therefore unable to choose between the competing hypotheses on the basis of data from adult fish. This difficulty is due in part to the slow growth of adult fish, for this means that serial samples (for example, monthly size/frequency distributions) will show little or no discernible change. Comparisons in adult fish are also confounded by the influence of reproduction in growth rate. We therefore need to look at life stages where changes might be seen in sequential samples: larval stages.

4 Growth Rates of Larval Fish

An overview of the growth and ecology of larval antarctic fish is presented elsewhere (North, this vol.). In this chapter we summarize just those aspects that are of relevance to the question of possible growth limitation by temperature.

Growth rates are obtained by plotting the mean or median size of samples of larvae against time. The relationship is frequently linear, at least over limited periods of time, and the growth rate can be expressed as the slope of the regression line. A representative data set for the mackerel icefish, *Champsocephalus gunnari*, is shown in Fig. 3.

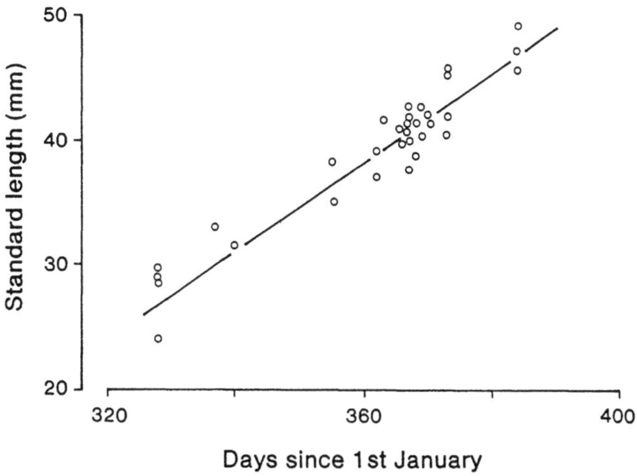

Fig. 3. Growth of larvae (1981 cohort) of the mackerel icefish *Champsocephalus gunnari* at South Georgia in the 1981/82 austral summer. Each data point represents the mean standard length (mm) of a sample of larvae (n=7–219). On some days several samples were taken from different locations, but all were within Cumberland Bay. The slope of the regression line shown represents a mean overall growth rate of 0.354 mm/day

Repeated studies have shown that, as would be expected, the growth rate of larvae varies from year to year. Larvae will grow more rapidly in warmer years with good feeding than in cooler years with poor food availability. Such variation in growth rate does not affect the hypothesis being tested, which is whether the growth machinery has evolved compensation for the low temperature overall. Clearly local variations around the long-term mean temperature may still have accelerating or decelerating affects on growth rate (though these can be difficult to distinguish from concomitant effects of variation in food availability).

Data are now available for larval growth rate in fish from the tropics to the polar regions. The data used in this study include all studies of the larval growth marine teleost fish published in the last 10 years or so, total of 123 studies. Older data and studies reported in the grey literature (such as ICES reports) have not been included. These data (the complete data set) contain multiple data points for eight species, ranging from two estimates for sole, *Solea solea*, to nine estimates for herring, *Clupea harengus*. Although the different estimates represent real biological variation (the herring data, for example, include larval growth rates for Baltic, Atlantic and Pacific herring) there are statistical problems in analyzing a mixed data set containing replicate values for some species but not others. A new data set was therefore constructed by eliminating all such duplications. Since the hypotheses under examination involved a comparison of maximum growth rates, the fastest growth rate was selected and all other replicates eliminated. This new data set (the reduced data set) contained 78 studies. Both data sets were subjected to the same statistical analyses (and in only one case did the significance level of the result change).

When larval growth rate was plotted as a function of habitat temperature, a significant positive relationship was obtained(complete data set, $F = 32.72$; reduced

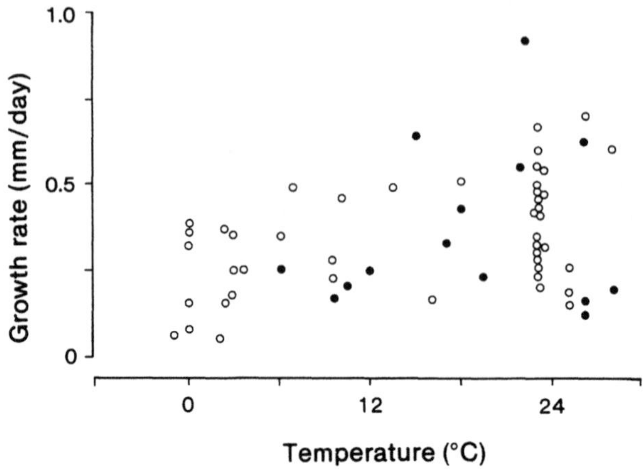

Fig. 4. Relationship between larval growth rate (mm/day) and environmental temperature for marine teleost fish. ○ Field data; ● laboratory data. This plot shows only the reduced data set (n = 78), although a plot of the full data set (n = 123) shows a closely similar pattern. Seventeen studies were not plotted because precise temperature data were unavailable; in five cases, two data points were coincident and only one has been plotted for clarity. The overall slope is 7.35×10^{-3}, equivalent to a Q_{10} of 1.24 over the range 0–30 °C

data set, F = 10.89; both P < 0.01). There was, however, a great deal of variation, with some polar species growing faster than some tropical species (Fig. 4). As with all data sets collated from the literature, part of this variation will be due to differences in technique or approach and some will reflect genuine biological variation. Some of the studies were field data, and part were the result of laboratory

Table 1. Growth rates of larval fish (mm/day)[a]

	n	Growth rate (mm/day)	
		Mean	SE
Reduced data set			
Polar	13	0.229	0.034
Temperate	31	0.407	0.038
Tropical	34	0.388	0.028
Complete data set			
Polar	26	0.194	0.021
Temperate	60	0.350	0.026
Tropical	36	0.374	0.028
	ANOVA:	F = 4.87, $P < 0.01$ (reduced)	
		F = 9.13, $P < 0.001$ (complete)	

[a]Mean and standard error (SE) have been calculated for polar (< 4 °C), temperate (4–23 °C) and tropical zones (> 23 °C). F = variance ratio for one-way ANOVA of growth rate on zone; P = probability of this result arising by chance. Results presented for both the complete and reduced data sets.

experimental work. The problems of mixing field and experimental data are exemplified by the data obtained for larval cod *(Gadus morhua)* by Radtke (1989). Larvae were held in culture and fed *ad libitum*; the resultant mean growth rates (1.2 – 1.3 mm/day) were much higher than any observed for cod in the wild, and the data have therefore been excluded from all statistical analyses. These data are important, however, in indicating the maximum growth rates that can be achieved under favourable circumstances, thereby suggesting that the growth rates of many fish larvae in the wild may be food limited.

For further analysis, the data were divided into three temperature zones: polar(ambient temperature < 4 °C), temperate (temperatures between 4 and 23 °C) and tropical (temperatures > 23 °C). Larval growth rate/frequency distributions for each zone are shown in Fig. 5. There was a substantial overlap in the larval growth rates from polar, temperate and tropical zones. Analysis of variance, however, indicated that the larval growth rate varied significantly with zone, although the difference between temperate and tropical data was small (Table 1). The maximum

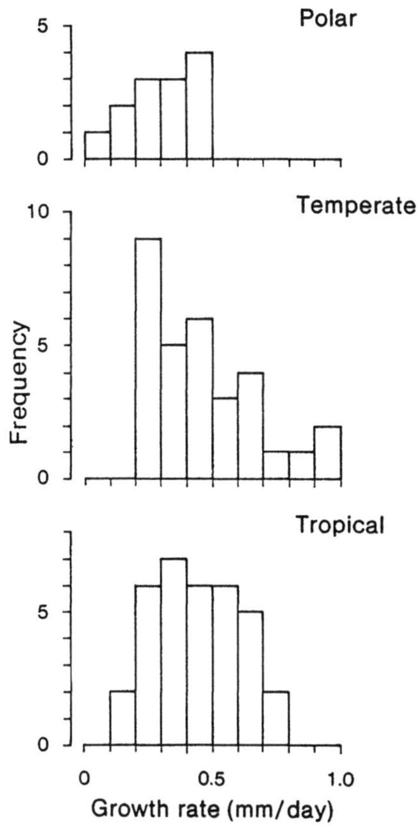

Fig. 5. Frequency histograms showing maximum rates of larval growth (mm/day) reported for 78 species of fish classified as polar, temperate or tropical species. The full data set shows a closely similar pattern

larval growth rate observed was also lower in polar species (0.37 mm/day) than in temperate (0.92 mm/day) or tropical (0.70 mm/day) species.

These data would thus suggest that larval growth is limited by temperature. The problem with such a simplistic analysis is the confounding effect of larval size. In general, larval growth increases with initial size. Since the mean size of larvae varied significantly with zone, polar larvae being on average four times larger than tropical larvae (Table 2), it was clearly necessary to correct for the effects of size. A traditional way to do this is to calculate relative growth rate (as growth rate/initial size). However this measure itself varied with size(complete data set, F = 69.3; reduced data set, F = 42.8; both P < 0.001), and so this approach did not help.

We have attempted to correct for the effect of larval size in two different ways. The first was to examine the relationship between growth rate and size separately for larvae from each of the three zones. These regressions were then used to estimate the growth rate for a larva of standard size (5-mm, chosen as a size represented in all three zones: Table 2). The individual regressions for each zone are given in Table 3. Although the predicted growth rate of a 5-mm larva increased with temperature from 0.17 mm/day in a polar species to 0.53 mm/day in a tropical species, the regression data indicated that this comparison must be interpreted with care. Not only did the slope of the growth rate/size regression increase by a factor of 8 (x25 in the full data set) in the direction polar-temperate-tropical, but so did the significance of the regression (indeed there was no significant relationship between size and growth rate in either polar or temperate fish for the complete data set). Thus, although mean larval size and the predicted growth rate for a 5-mm larva changed with zone, so did the nature of the growth rate/size relationship. A direct comparison such as that in Table 3 is thus suspect, for it is not comparing like with like.

Table 2. Sizes of larval fish[a]

	n	Initial length (mm)			
		Mean	SE	Min	Max
Reduced data set					
Polar	13	11.2	1.91	4.0	25.0
Temperate	31	5.4	0.61	1.5	16.0
Tropical	34	3.2	0.22	1.1	6.2
Complete data set					
Polar	26	13.0	1.34	4.0	25.0
Temperate	61	5.8	0.42	1.5	16.0
Tropical	36	3.1	0.22	1.1	6.2

ANOVA: F = 22.84, $P < 0.001$ (reduced)
F = 49.07, $P < 0.001$ (complete)

[a] Mean and standard error (SE), maximum and minimum have been presented for polar (< 4 °C), temperate (4–23 °C) and tropical (> 23 °C) fish. All data are for initial length (mm) at the start of the growth period. F = variance ratio for one-way ANOVA of size on zone; *P* = probability of this result arising by chance. Results presented for both the complete and reduced data sets.

Table 3. Predicted growth rates (mm/day, mean, ±95% confidence intervals where relationship between growth rate and size is significant) for larval fish of length 5mm, classified into polar (> 4 °C), temperate (4–23 °C) and tropical (> 23 °C) zones [a]

	Regression					Predicted growth	
	n	slope[b]	SE[b]	F	*P*	Mean	95% CL
Reduced data set							
Polar	13	10.26	4.35	5.55	< 0.05	0.17	±0.09
Temperate	31	29.71	10.22	8.45	< 0.01	0.40	±0.07
Tropical	34	81.49	17.23	22.38	< 0.001	0.50	±0.08
Complete data set							
Polar	26	3.34	3.14	1.13	NS	(0.17)	–
Temperate	60	15.04	8.36	3.24	NS	(0.34)	–
Tropical	36	85.98	15.63	30.27	< 0.001	0.54	±0.07

[a] For each zone a regression line was calculated for growth rate on initial length (mm). Regression data presented are slope, standard error of the slope (SE), variance ratio for the regression (F) and the probability of this result arising by chance (*P*). NS = not significant ($P > 0.05$). Results presented for both the complete and reduced data set.
[b] $\times 10^3$ for clarity of presentation.

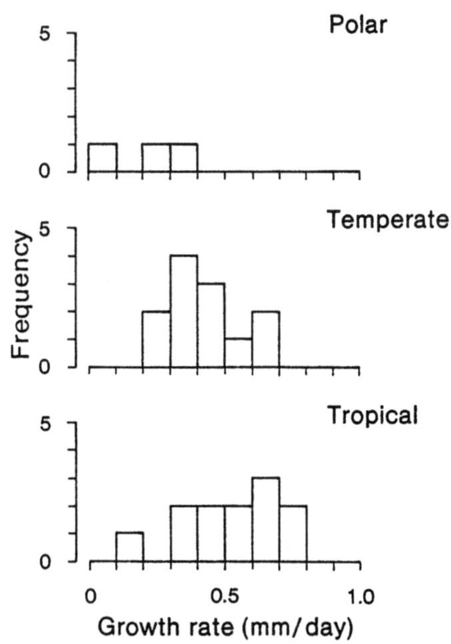

Fig. 6. Frequency histograms showing maximum rates of larval growth (mm/day) for larvae in the range 4–6.2 mm initial length (reduced data set). Classification and representation as for Fig. 5. The full data set shows a closely similar pattern

The second approach was to limit the comparison to those larvae within the range of sizes represented in all three zones. This was 4.0–6.2 mm initial size (Table 2), and limiting analysis to these larvae reduced the data sets to 44 (full) and 28 (reduced) examples. Although there was still a considerable overlap in the spread of growth rates (Fig. 6), there was nevertheless a statistically significant variation of growth rate with zone (Table 4). The maximum growth rate observed decreased with latitude (polar, 0.25; temperate, 0.85; tropical 0.70 mm/day), and the ratio of the mean growth rate for tropical and polar larvae was > 3, in both data sets (Table 4). This is larger than the ratios calculated from the larval fish of all sizes (reduced data set, 1.7; complete data set, 1.9; see Table 1), as would be expected since we were attempting to control for the effects of size where polar larvae are on average larger than tropical larvae ; growth rate is in general a positive function of larval size. The difficulty with this type of analysis is that we may not be comparing truly representative fish, because we have selected larvae from the lower end of the size range of polar species, the middle of the temperate size range, and from the upper end of tropical size range. Although the growth-rate distributions of the reduced data set appeared to be reasonably representative of the overall distributions (see Fig. 5), there was still the more subtle problem that the fish selected may not be taxonomically representative.

These data thus suggested that there is a positive effect of temperature on growth rate in larval fish. Although controlling for the confounding effect of larval size is not easy, in all the analyses undertaken, both mean growth rate and the maximum growth rate observed increased in the direction polar-temperate-tropical. However, we are left with the problem that even the maximum observed field growth rates may not be representative of the actual capacity for rapid growth. If growth rates in the wild are always food limited, then we will not be able to answer

Table 4. Growth rates of larval fish (mm/day)[a]

	n	Growth rate (mm/day)	
		Mean	SE
Reduced data set			
Polar	3	0.152	0.058
Temperate	13	0.394	0.051
Tropical	12	0.464	0.051
Complete data set			
Polar	5	0.127	0.036
Temperate	27	0.303	0.039
Tropical	12	0.464	0.051

ANOVA: $F = 3.86$, $P < 0.05$ (reduced)
$F = 6.43$, $P < 0.01$ (complete)

[a]Mean and standard error (SE) have been calculated for larval fish ranging from 4.0–6.2 mm in initial size, and classified into zones as for previous tables. F = variance ratio for one-way ANOVA of growth rate on zone; *P* = probability of this result arising by chance. Results presented for both complete and reduced data sets.

the question of whether temperature imposes an upper limit on growth rate from field data alone. The only way to resolve this question will therefore be through laboratory experiments. The growth rates obtained by Radtke (1989) for cod larvae fed *ad libitum* in the laboratory were higher than any of the growth rates observed in the field. This suggests strongly that field growth rates of larval fish are indeed limited either by food availability, or the ability of the larva to obtain it. A valid test of temperature limitation of growth rate can only take place in the laboratory.

5 Why is Food Seasonally Limiting?

A seasonal availability of food for larval fish is easy to understand. Larval fish are essentially visual predators and hence would have a reduced ability to locate prey in the long winter nights at high latitudes. Also they prey extensively on the younger stages of zooplankton, especially copepods (North and Ward, in press), as do the postlarval and juvenile stages of many species (Hubold and Ekau 1990). Since secondary production of these herbivorous prey species is constrained to the summer months of enhanced primary production(reviewed in Clarke 1988) there will clearly be a corresponding seasonality in food availability for larval fish.

Zooplankton and benthic invertebrates are frequently long-lived in polar waters. This will therefore tend to smooth out any seasonality in total biomass. Foxton(1956, 1964) showed that the total zooplankton biomass in surface waters (both 0–50 and 50–100-m depth horizons) of the subantarctic and antarctic regions increased from a minimum value in winter(roughly June to September) to a peak in summer. In the 0–50-m horizon, biomass increased more than sixfold in subantarctic waters and almost as much in antarctic waters. Peak biomass was reached in February in the subantarctic, but not until April in truly antarctic waters, reflecting the progressively later summer season at higher latitudes. Although part of this increase must be due to new production (copepods, euphausiids, salp budding) the total zooplankton biomass integrated over the top of 1000 m of the water column remained roughly constant throughout the year. The bulk of the seasonal variation in surface water zooplankton biomass is thus due to a seasonal vertical migration (Foxton 1956; Hopkins 1971; Voronina 1972). However, inshore (where fish larvae are most abundant; North, this vol.) the biomass of zooplankton may be four to five times lower in winter than in summer (Ward 1989).

Predatory species which follow the seasonal migration of copepods(for example, chaetognaths) have an essentially year-round supply of food, and consequently can feed and grow throughout the year. Growth is, however, faster in summer (David 1955, 1965).

These observations would suggest that (contrary to the conclusions drawn from studies of larval fish), adult fish taking zooplankton should have a year-round supply of food and hence show a reduced seasonality of growth. It is possible that some food organisms may migrate into habitats where they are less accessible to predation. These might include those species which live close to the underside of ice floes. Since few species of antarctic fish are neutrally buoyant (Eastman and DeVries 1982; Eastman 1985) such organisms may be safe from predation (although some species utilize caverns in the underside of sea ice to rest).

However, for species which live demersally and take a substantial proportion of benthic prey,there would seem to be food available year-round. It is possible that seasonal variations in the activity of some epibenthic and infaunal species may make them less conspicuous or less available in winter. Overall, however, it is not clear why demersal fish should show a seasonality in food intake. Nevertheless studies of seasonal variation in condition index and stomach contents indicate that such a seasonality does occur (Burchett 1983; Everson 1970; Hureau 1970; Sakakibara et al. 1989). This intriguing problem needs further work.

6 Cues and Proximate Causes

If the explanation for the reduced growth rate in winter is indeed simply resource limitation, then the obvious experiment is to provide excess food in winter and see if growth rate responds. Interestingly, there is frequently no response, usually because the feeding rate is reduced rather than food being ingested at summer rates and unused. Such observations highlight the difficulties of much experimental work on seasonal aspects of the biology of marine organisms. Although it is easy to control for the obvious physical factors such as temperature, oxygen, photoperiod or food, it is very difficult to allow for internal factors such as hormonal state. It is quite possible that the lack of response to excess food in polar organisms held under simulated winter conditions may be related to higher level controls having switched off appetite and/or the synthetic machinery. Until we have a better understanding of such factors we cannot interpret such experiments (Clarke 1991).

Experimental work has shown that the feeding activity of *Harpagifer antarcticus* is tied closely to temperature and photoperiod (Targett et al.1988). This does not mean, however, that the feeding process exhibits a peculiar sensitivity to temperature. Rather it is that the fish are using temperature and photoperiod as seasonal clues ; given the narrow seasonal range of sea-water temperature at inshore sites in Antarctica (Littlepage 1965; Clarke et al. 1988) it is perhaps not surprising that the fish have evolved a precise sensitivity to temperature. Nevertheless this appears to be simply the use of temperature as a proximate cue, not an innate sensitivity of the feeding mechanism itself to temperature. Care must be taken in distinguishing proximate cues from ultimate causes. Ceasing to feed simply because the temperature has fallen below a certain level would have no adaptive value in itself. However, it may be that the various effects of temperature on digestive enzyme activity, cost of feeding and so on, mean that the cost to benefit ratio of foraging in winter is not favourable. The adaptive value then comes in using seasonal changes in photoperiod and temperature as a cue for the onset of winter.

7 Seasonality or Temperature?

Although it is tempting to explain many features of the biology of polar organisms in terms of the low temperature, this apparently simple explanation is often misleading. The evidence from larval Antarctic fish (and from polar marine invertebrates: Clarke and Peck, in press) suggests that it is seasonal food availability and not the low temperature that normally limits growth rate. The evidence from adult fish is more equivocal.

Nevertheless, we suggest that it would be unwise to use temperature as a variate in the way it is often used in predictive ecological models. Several recent models have attempted to describe global trends in fish population growth by simple equations including a temperature term. Examples include those Caddy and Sharp (1986) and Palomares and Pauly (1989). The latter model, solved for a carnivorous fish of 500-g mass, predicts a decrease in food consumption from 2.20 (% body mass/day) at 21 °C to 1.48 at 11 °C and only 0.34 at 1 °C for a carangiform swimmer of caudal-fin aspect ratio 2. (Since most antarctic fish are notothenioids which use predominantly labriform swimming, this equation is potentially misleading for the Southern Ocean.) The corresponding Q_{10} values are 1.49 and 4.35 (compared with Q_{10} of 1.24 for the over-all relationship in Fig. 4). Although these equations are intended solely as broad descriptors of the marine ecosystem for fisheries management and not as explanatory models, they nevertheless imply a role for temperature which may not exist in reality.

The problem is that these equations and models(like the relationships explored in this work) are based solely on correlations with temperature. Temperature and the seasonality of primary production covary broadly with latitude across the globe, and it may be that the general decrease in mass-specific food consumption from tropics to poles reflects this increasing seasonality permeating to higher levels in the marine food web (Clarke 1988).

Although the results of the analysis undertaken in this study imply that temperature acts as a constraint on the growth rate of larval fish, we cannot regard this as a firm conclusion. Laboratory studies of cod larvae fed *ad libitum* have suggested that larval fish may have capacities for rapid growth that are rarely (if ever) realized in the wild. If this is indeed so, then the results of this study may reflect only a latitudinal variation in food availability for fish larvae. The only way that we will be able to resolve the question of whether temperature sets an upper constraint on the growth rate of larval (or adult) fish will therefore be carefully controlled laboratory experiments on representative species from polar, temperate and tropical seas. Field data will always be equivocal.

References

Burchett MS (1983) Age and growth of the antarctic fish *Notothenia rossii* from South Georgia. Bull Br Antarct Surv 60:45–61

Caddy JF, Sharp GD (1986) An ecological framework for marine fishery investigations. FAO Fisheries Tech Pap 283, 152pp

Clarke A (1980) A reappraisal of the concept of metabolic cold adaptation in polar marine invertebrates Biol J Linn Son 14:77–92

Clarke A (1983) Life in cold water : the physiological ecology of polar marine ectotherms. Oceanogr Mar Biol Ann Rev 21:341–453

Clarke A (1987) The adaptation of aquatic animals to low temperatures. In:Grout BWW, Morris GJ (eds) The effects of low temperatures on biological systems. Arnold, London, pp 315–348

Clarke A (1988) Seasonality in the antarctic marine environment. Comp Biochem Physiol 90B:461–473

Clarke A (1991) What is cold adaptation and how should we measure it? Am Zool 31:

Clarke A, Peck LS (in press) The physiology and energetics of polar zooplankton. Polar Res

Clarke A, Holmes LJ, White MG (1988) The annual cycle of temperature, chlorophyll and major nutrients at Signy Island, South Orkney Islands, 1969-1982. Bull Br Antarct Surv 80:65–86

Cossins AR, Bowler K (1987) Temperature biology of animals. Chapman and Hall, London

David PM (1955) The distribution of *Sagitta gazellae* Ritter-Zahoney. Discovery Rep 27: 235–278

David OM (1965) The Chaetognatha of the Southern Ocean. In: Van Mieham J, Oye P (eds) Biogeography and ecology in Antarctica. Junk, The Hague, pp 296–323

Dayton PK, Robilliard GA, Paine RT, Dayton LB (1974) Biological accommodation in the benthic community at McMurdo Sound, Antarctica. Ecol Monogr 44:105–128

Dowels, RA (1978) Nesting behaviour of *Harpagifer bispninis* in Arthur Harbour, Antaretic Peninsula. J. Fish Biol 12:465–474

Dunbar MJ (1968) Ecological development in polar regions: a study in evolution. Prentice-Hall, Englewood Cliffs, New Jersey 119 pp

Eastman JT (1985) The evolution of neutrally buoyant notothenioid fishes: their specializations and potential interactions in the antarctic marine food web. In: Siegfried RW, Condy PR, Laws RM (eds) Antarctic nutrient cycles and food webs. Proc 4th SCAR Symp on Antarctic Ecology. Springer, Berlin Hiedelberg New York Tokyo, pp 430–436

Eastman JT, DeVries AL (1982) Buoyancy studies of notothenioid fishes in McMurdo Sound, Antarctica. Copeia 1982:385–393

Everson I (1970) The population dynamics and energy budget of *Notothenia neglecta* Nybelin at Signy Island, South Orkney Islands. Bull Br Antarct Surv 23:25–50

Everson I (1984) Fish Biology In: Laws RM (ed) Antarctic ecology II Academic Press, London pp 491–532

Foxton P (1956) The distribution of the standing crop of zooplankton in the Southern ocean. Discovery Rep 28:191–236

Holeton GF (1974) Metabolic cold adaptation of polar fish: fact or artefact? Phys Zool 47:137–152

Holm-Hansen O, Mitchell BG, Hewes CD, Karl DM (1989) Phytoplankton blooms in the vicinity of Palmer Station, Antarctica. Polar Biol 10:49–57

Hopkins TL (1971) Zooplankton standing crop in the Pacific sector of the Antarctic. In : Llano GA, Wallen IE (eds) Biology of the antarctic seas IV. Antarctic Res Ser Am Geophys Union, Washington DC 17:347–362

Hourigan TF, Radtke RL (1989) Reproduction of the antarctic fish *Nototheniops nudifrons*. Mar Biol 100:277–283

Hubold G, Ekau W (1990) Feeding patterns of post-larval and juvenile notothenioids in the southern Wedell Sea (Antarctica). Polar Biol 10:255–260

Hureau JC (1970) Biologie comparee de quelques poissons antarctiques (Nototheniidae). Bull Inst Oceanogr Monaco 68 (1391), 244 pp

Kawaguchi K, Ishikawa S, Matsuda O, Naito Y (1989) Tagging experiment of nototheniid fish, *Trematomus bernacchii* Boulenger under the coastal fast ice in Lutzow-Holm Bay, Antarctica. Proc NIPR Symp Polar Biol 2:111–116

Kellerman A (1986) On the biology of early life stages of notothenioid fishes (Pisces) off the Antarctic Peninsula. Ber Polarforsch 31:1–149

Littlepage JL (1965) Oceanographic investigations in McMurdo Sound, Antarctic. In: Llano GA (ed) Biology of antarctic seas II. Antarctic Res Ser, Am Geophy Union, Washington DC 5:1–37

North AW (1988) Age of antarctic fish : validation of the timing of annuli formation in otoliths and scales. Cybium 12:107–114

North AW, White MG (1987) Reproductive strategies of antarctic fish. In: Kullander SO, Fernholm B (eds) Proc V Congr Europ Ichthyol, Stockholm 1985. Swed Mus Vertebrate Zool, Stockholm , pp 381–390

North AW, Ward P (in press) The feeding ecology of larval fish in an antarctic fjord, with emphasis on *Champsocephalus gunnari*. In: Kerry K, Hempel G (eds) Ecological change and conservation of antarctic ecosystems. Springer, Berlin Heidelberg New York Tokyo

Olsen S (1954) South Georgia cod (*Notothenia rossii marmorata* Fischer). Norsk Hvalfangsttid 7 (43):373–382

Olsen S (1955) A contribution to the systematics and biology of chaenichthyid fishes from South Georgia. Nytt Mag Zool 3:79–93

Palomares ML, Pauly D (1989) A multiple regression model for predicting the food consumption of marine fish populations. Aust J Mar Freshwater Res 40:259–273

Pauly D (1980) On the interrelationships between natural mortality, growth parameters, and mean environmental temperature in 175 fish stocks. J Cons Int Explor Mer 39(2):175–192

Radtke RL (1989) Larval fish age, growth, and body shrinkage: information available from otoliths. Can J Fish Aquat Sci 46:1884–1894

Sakakibara S, Kondo Y, Tobayama T, Hoshiai T (1989) Growth of the nototheniid fish, *Trematomus bernacchii* and *Pagothenia borchgrevinki* reared in aquarium. Proc NIPR Symp Polar Biol 2:105–110

Targett TE, Young KE, Konecki JT, Grecay PA (1988) Research on wintertime feeding in antarctic fishes. Ant J US 23(5):211–214

Voronina NM (1972) Vertical structure of the pelagic community in the antarctic. (in russ). Okeanologiya 12:492–296

Ward P (1989) The distribution of zooplankton in an antarctic fjord at South Georgia during summer and winter. Ant Science 1:141–150

Wells RMG (1978) Respiratory adaptation and energy metabolism in antarctic nototheniid fishes. NZ J Zool 5:813–815

Wohlschlag DE (1961) Growth of an antarctic fish at freezing temperatures. Copeia 1 (1961):11–18

Review of the Early Life History of Antarctic Notothenioid Fish

A.W. North

1 Introduction

The Antarctic cooled rapidly about 38 million years ago, when significant amounts of sea ice began to form. It became isolated in the last 30–14 million years when the Drake Passage opened to form a circum-Antarctic Ocean, and has generally cooled since then (Kennett 1977). Some key features of the Southern Ocean include low mean temperatures, which decrease and have a smaller annual range further south (+3.5±5 °C at 50 °S to –1.8 ± 0.2 °C at 78 °S; Littlepage 1965; Duhamel 1987), and the increased seasonality of daylight and primary productivity with increasing latitude. Continental shelves are deep, with a mean depth of 250 m at the South Shetland Islands and 450 m at the southern Antarctic Peninsula (DeWitt 1971). Circumpolar water masses circulate around the continent, which is ringed by glaciers, and in many places by permanent floating or grounded ice shelves. In winter, 1/15 of the world's sea surface is covered by floating antarctic sea ice (Gordon 1967).

The antarctic fish fauna has undergone a dramatic reduction in diversity since the early Tertiary period (Eastman and Grande 1989); lacking any evidence, the reasons for this are open to speculation. It is suggested that important factors during the evolution of the ecosystem were the progressive cooling, reduction in shelf area and increasing shelf depth that have resulted in a low diversity of invertebrate species (Eastman and Grande 1989). In the Antarctic there are around 204 demersal fish species and 75 mesopelagic species, including 96 species of notothenioids. Although only about 34% of the present species belong to the Perciform suborder Notothenioidei, by number of individuals and biomass they dominate the demersal fish communities of the continental shelves of the Southern Ocean (Andriashev 1965, 1987). There are at least 110 species of Notothenioidei which are restricted to the colder (< 10°C) waters of the southern hemisphere, including southern South America and New Zealand. They grow slowly and occupy ecological niches which involve a low energy flux. The adults of many species are inactive sedentary forms which swim at low speeds using pelvic fins and have a limited fast-swimming endurance (MacDonald et al. 1987; Johnston 1989). They have evolved glycoprotein antifreezes to survive at temperatures below the normal freezing point (ca. –0.8 °C) of fish blood (DeVries and Wohlschlag 1969). Fish of one family, the

British Antarctic Survey, Natural Environment Research Council, High Cross, Madingley Road, Cambridge CB3 OET, UK

Channichthyidae (icefish), have no functional haemoglobin (Ruud 1954) and thereby reduced blood viscosity (Twelves 1972), and a large heart and blood volume (Hemmingsen and Douglas 1970; Holeton 1970). Six families of Notothenioidei are found south of the polar frontal zone: Notothenidae, Artedidraconidae, Bathydraconidae, Channichthyidae, and Harpagiferidae (Fisher and Hureau 1985). The other Family, Bovichthyidae, has two antarctic species (the larvae of which are undescribed). An introduction to the biology of antarctic fish may be obtained from various reviews (Andriashev 1965, 1987; DeWitt 1971; DeVries and Eastman 1981; Everson 1984; Fischer and Hureau 1985; Kock 1985; Eastman and DeVries 1986; MacDonald et al. 1987; North and White 1987; di Prisco et al. 1988; Johnston 1989). Kellermann (1989a) has provided a catalogue of the early life stages which summarizes their development and ecology.

The larval stages (from newly hatched yolk-sac larvae to larvae metamorphosing to the early juvenile) of few antarctic fish were described until comparatively recently (Table 1). Larvae of 11 species of antarctic notothenioid were illustrated up to 1978, 8 of which were correctly identified. In the last 11 years, the larvae of a further 46 species have been described, and the ecology of some has been investi-

Table 1. The number of species of the larval stages of Notothenioidei first described and the number illustrated by each author(s) with year

First description	Total illustrated	Source
1	1	Pappenheim (1912)
3	5	Regan (1916)
1	1	Nybelin (1951)
2	5	Everson (1968)
	1	Hureau (1970)
1+1[a]	1+1[a]	Ciechomski and Weiss (1976)
	1	Daniels (1978)
6	6	Efremenko (1979b)
4	6	Efremenko (1979a)
1[a]	1[a]	Robertson and Mito (1979)
1	1	Moreno (1980)
1	1	White et al. (1982)
7	24	North and White (1982)
8	27	Efremenko (1983)
	1	Slosarczyk (1983)
1	10	Efremenko (1984)
1	1	Camus and Duhamel (1985)
1	2	Kellermann (1986a)
	3	Duhamel (1987)
1	1	Gon (1987)
	1	Burren (1988)
1	4	Gon (1988)
	1	Ekau (1989)
	1	Konecki and Targett (1989)
14	49	Kellermann (1989b); North and Kellermann (1989)
56 Total species (54 antarctic and 2 temperate species)		

[a]Temperate notothenioidei.

gated. There are, for example, some reports on the growth rates of larval notothenioids (White et al. 1982; Keller 1983; Kellermann 1986a; Slosarczyk 1987; Radtke et al. 1989; North 1990). This review focuses on development, morphology, pigmentation, community structure, diet and growth. In addition, the early development of the Notothenioidei is briefly compared with fish from warmer regions.

2 Egg Development and Hatching Periods

Some species of Notothenioidei lay demersal eggs in a nest under stones or inside the oscular cavity of sponges (Arnaud 1974; Daniels 1978; Moreno 1980; Hourigan and Radtke 1989; Konecki and Targett 1989). In *Pagothenia bernacchii* and *Harpagifer antarcticus* at the Antarctic Peninsula, the female guards the nest of eggs (Daniels 1978; Moreno 1980), whereas in *Nototheniops nudifrons* it is the male (Hourigan and Radtke 1989). In *H. antarcticus* at the South Orkney Islands, however, there may be an interchange of guarding by the male and female, or both may guard the nest (Burren 1988). Some species, for example, *Notothenia neglecta*, have pelagic eggs (White et al. 1982; North 1988). It is now possible to identify the eggs of seven notothenioid species from their surface morphology (Riehl and Kock 1989). Notothenioid eggs are yolky and large, between 2–6 mm in diameter depending on species (Marshall 1953; Everson 1984). They develop over a long period of about 2–6 months from spawning to hatching (Daniels 1978; White et al. 1982; Burchett et al. 1983; Camus and Duhamel 1985; Kellermam 1986a; North and White 1987; Kock 1989).

It used to be thought that the larvae of antarctic fish hatched in spring or summer (Efremenko 1983), to synchronize with the peak abundance of larval copepods, their principal food items (Marshall 1953; Everson 1984). However, as studies have been extended to cover the full seasonal cycle, it has been shown that some species hatch in winter (Burchett et al. 1983; Ekau 1989; Kellermann 1989b; North and White 1987). It is possible that further studies may find some species hatching in autumn. There is a succession of species hatching throughout winter to summer, with each species hatching during a period that may be as short as 2 months (in a particular year and region) or more than 3 months. At South Georgia, the order of hatching was familial. In early winter a bathydraconid species hatched first, followed by three channichthyid species, and then a progression of nototheniid species from early spring through summer (Burchett et al. 1983; North and Ward 1989; North and White 1987). The first species to hatch preceded the spring increase in copepod numbers by 5 months. There is also a temporal succession of the species hatching from winter to summer at the Antarctic Peninsula (Kellermann 1989a; Kock 1989) and during spring and summer in the southern Weddell Sea (Hubold, in press). In a few species, including *Nototheniops nudifrons* and *Pagothenia hansoni*, there appears to be a long period of spawning and hatching (Hourigan and Radtke 1989; North 1990).

3 Larval Development

Notothenioid larvae are large (5–20 mm SL) at hatching (Marshall 1953; Burchett et al. 1983; North and White 1987; North 1990; Kellermann 1989a). At hatching, the larvae are not advanced in terms of fin and skeletal development (North and White 1987; Kellermann 1989a). The fins of newly hatched larvae consist of a primordial fin fold and pectoral fins with no fin rays developed; the pelvic fins are not apparent except in the channichthyids (Regan 1916; North and White 1987; Kellermann 1989a).

However, the mouth is well developed at hatching and the early yolk-sac larvae can feed (Regan 1916; Kellermann 1986a; North and Ward 1989), which is an advanced state of development (Marshall 1953). Teeth are present in yolk-sac larvae of the channichthyids, two species of the subfamily Gymnodraconinae (Bathydraconidae) and in a few nototheniid species (e.g., *Dissostichus eleginoides*); however, teeth are not apparent in the majority of notothenioid species. Early development of teeth in the channichthyids is thus precocious compared to related Perciform fish (North 1990).

The duration of yolk-sac absoption is about 2–5 weeks in small species such as *H. antarcticus* (Daniels 1978; Burren 1988), and up to 3 months in larger species, such as the channichthyids and *Parachaenichthys georgianus* (North, BAS unpubl. data).

In many notothenioids, development of fin rays proceeds first in the caudal and pectoral fins, followed by the dorsal and anal fins, and then the pelvic fins. The development of the pelvic fins in the channichthyids is precocious (Regan 1916); although these fins also develop early in some species of Trachinidae and Gobiidae (Russell 1976). Development of the pectoral fins in the Notothenioidei (and some Blenniidae) is somewhat precocious compared to related Perciform families (North 1990).

Most species of Notothenioidei do not develop to the early juvenile stage (all fin rays formed) until 6–12 months after hatching (Kellermanm 1989a, b; North 1990). Metamorphosis for the larger notothenioid species is generally at a greater size than smaller species (e.g., 80 mm for *Pagothenia hansoni*, 40 mm for *Notothenia angustifrons* and 25 mm for *Harpagifer georgianus*; North 1990). The larval length span is the length range of the larvae of each species from hatching to metamorphosis. In antarctic Notothenioidei, the larval length span is large and generally increases with adult size; for example, it is about 40 mm (5.5–7.5 to 45–50 mm) for N. angustifrons and about 68 mm (12–13 to 80 mm) for *P. hansoni* (North 1990).

4 Morphology

Larval Notothenioidei have a general body form and a pattern of fin and teeth development which is characteristic for each family or subfamily (Regan 1916; North and White 1982; Kellermann 1989a). The diversity in larval morphology of the antarctic notothenioid families is illustrated in Fig. 1. Larval body shape in the

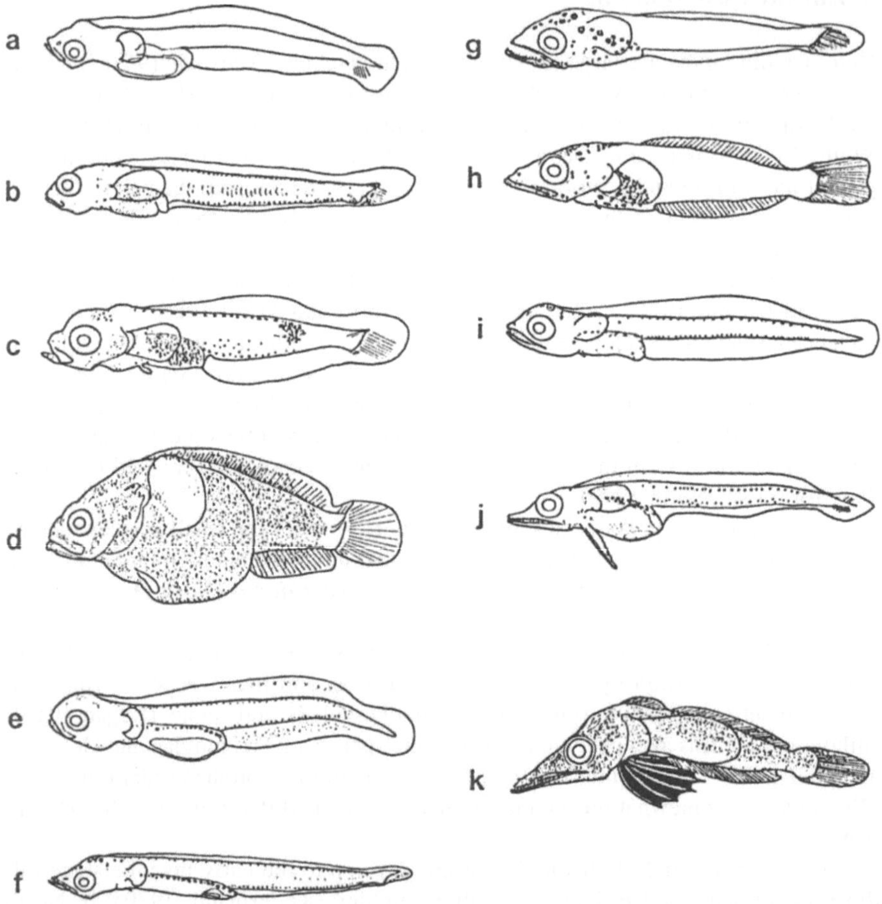

Fig. 1. Larvae of the common families of antarctic Notothenioidei (after North 1990). **a** Nototheniidae, *Pagothenia hansoni*, yolk-sac larva 12.1 mm total length (TL). **b** Nototheniidae, *Trematomus newnesi*, post-flexion larva 13.5 mm TL. **c** Harpagiferidae, *Harpagifer georgianus*, post-flexion larva 13.5 mm TL. **d** Artedidraconidae, *Pogonophryne marmorata*, post-flexion larva 27 mm TL. **e** Bathydraconidae, *Parachaenichthys georgianus*, yolk-sac larva 15 mm TL. **f** Bathydraconidae, *Parachaenichthys georgianus*, postflexion larva 22.5 mm TL. **g** Bathydraconidae, *Psilodraco breviceps*, post-flexion larva 18.3 mm TL. **h** Bathydraconidae, *Psilodraco breviceps*, post-flexion larva 35 mm TL. **i** Channichthydae, *Champsocephalus gunnari*, preflexion larva 12.5 mm TL. **j** Channichthydae, *Chaenocephalus aceratus*, larva 21 mm TL. **l** Channichthydae, *Pseudochaenichthys georgianus*, juvenile 71.5 mm TL

older larvae corresponds to that of the juveniles and adults of each family. Larval nototheniids have a fusiform shape. Channichthyids have a large abdomen, head and jaws with well-developed teeth, and a large pelvic fin (Regan 1916). Bathydraconids are long and slender, except the Gymnodraconinae which resemble the channichthyids, although they do not have large pelvic fins. Artedidraconids are short and deep with a large abdomen. Harpagiferids have a fairly large head and abdomen and are somewhat short. Larvae of the temperate nototheniid *Eleginops maclovinus* (Ciechomski and Weiss 1976, Fig. 1b) are superficially similar to the harpagiferids. Early juveniles of *Bovichthys variegatus*, Bovichthyidae (Robertson and Mito 1979, Fig. 7) are superfically similar to fingerling *Notothenia rossii* and *N. neglecta* (North and White 1982, Figs. 11, 12) although B. *variegatus* has an opercular spine.

5 Pigmentation

The larvae of the most closely related species of the Nototheniidae from the Kerguelen Island region have similar pigment patterns (Koubbi, Duhamel, Camus, pers. comm.). In the nototheniids from further south, there is some tendency for larvae of closely related species to have similar general pigmentation; however, there are also distinct differences between some species within a subgenus. Some examples are given below.

The subgenus Lepidonothen includes *Notothenia kempi* and *N. squamifrons* whose larvae both have melanophores above the abdomen and a precaudal vertical band of melanophores (see Efremenko 1983, Figs. 12, 14). However, *N. squamifrons* has a ventral pigment row which is lacking in *N. kempi*. The Gobionotothen includes *Notothenia angustifrons* and *N. gibberifrons* whose larvae are similar in possessing a dorsal and ventral pigment row and dorsal abdominal melanophores (Efremenko 1983, Figs. 10, 11). They differ in that preflexion larvae of *N. gibberifrons* have a mediolateral pigment row which does not develop until the early juvenile stage in *N. angustifrons*.

Besides the subgenera, there are also parallels in pigment pattern within genera. For example, the genus *Trematomus* also contains several species with similar larval pigmentation. *T. newnesi, T. eulepidotus and T. centronotus* are similar to *N. gibberifrons*, with dorsal, ventral and mediolateral pigment rows; *T. nicolai* is similar, although the mediolateral pigment is restricted to the caudal third of the postanal section. Species with different pigmentaton in the genus are *T. lepidorhinus*, which only has a few posterior dorsal melanophores, and *T. scotti*, which has no dorsal pigment row; the former has lateral melanophores, whereas the latter does not (Kellermann 1989a, Figs. 36, 37, 40, 42)

There are also trends in the pigment patterns of other families. The 11 genera of Channichthyidae form 7 phylogenetic groups (Iwami 1985). Similarities in pigmentation can be found among the larvae from some of these groups, notably between the group containing *Pseudochaenichthys georgianus*, *Pagetopsis macropterus* and *Pagetopsis maculatus* (Kellermann 1989a, Figs. 19, 20, 21). Pigmentation patterns of the larvae within some other groups are, however, different. For example, the *Cryodraco, Chionobathyscus, Chaenocephalus* group (Kellermann 1989a, Figs. 12, 15, 17).

Pigment patterns may be involved in camouflage or deception by masking the fishes' outline, concealing conspicuous prey in the stomach, or creating a distracting flickering effect during swimming, and in species recognition to facilitate the formation of schools (Moser 1981). It is tempting to suggest that species with an almost identical pigment pattern are the most closely related, and that similar pigmentation of species in different families implies a similar function. Before valid conclusions can be made, however, the taxonomic significance and function of these pigment patterns must be clarified.

6 Distribution and Communities

The methods used to investigate larval notothenioids may underestimate their abundance and provide biased data on their size (Kellermann 1986a). There is considerable net avoidance by larval Notothenioidei during daylight and a large variance between samples taken at the same station (North 1988, 1990; North and Murray, BAS unpubl. data). To estimate abundance at a station, it is therefore recommended that samples should be taken at night and that a series of replicate samples are analyzed.

Many studies have described the occurrence of larval Notothenioidei in the Southern Ocean. The surveys of the international BIOMASS programme have provided much data on the distribution of the early stages of antarctic fish in various localities including the Antarctic Peninsula, Prydz Bay region (70° E at the Antarctic Continent), and in the Indian Ocean Sector (BIOMASS 1987, 1988). Similar surveys have also been made at South Georgia (North 1988). Most of these studies sampled at various times of day, using one or two net hauls at each of the stations which were often > 40 km apart. Estimates of larval fish abundance from such surveys are at best, only semiquantitative.

The Southern Ocean may be divided into many subsystems according to geographical region, the prevailing water masses, bathymetry and depth in the water column. However, the information on larval fish communities is limited and it is only possible to describe the most common species in some regions.

At Kerguelen Island the larvae of *Notothenia cyanobrancha, Notothenia squamifrons, Notothenia acuta* and *Harpagifer kerguelensis* were most common (Duhamel 1987). The larvae of *Champsocephalus gunnari* may also be abundant (Duhamel, pers. comm.)

Near the coast at South Georgia the most abundant larvae in September (early spring) were those of *C. gunnari, Pseudochaenichthys georgianus, Chaenocephalus aceratus, Nototheniops nudifrons* and *Parachaenichthys georgianus* (North 1988, 1990). In January (summer) the most abundant larvae were those of *Notothenia gibberifrons, Nototheniops larseni, Pagothenia hansoni* and *C. gunnari*. Larvae of *Notothenia angustifrons* were common in the immediate sublittoral zone in December.

Kellermann and Kock (1988) have reported that in summer at the Antarctic Peninsula there is considerable interannual variation in the median larval abundance of the common species. The most abundant larvae were those of *N. larseni, Notothenia kempi, N. gibberifrons, Trematomus newnesi* and *Chionodraco antarcticus*; and the larvae of *Pleuragramma antarcticum* were particularly abundant (Kellermann 1986b). Further surveys with wider seasonal coverage have shown that the larvae of *N. nudifrons* and *Trematomus scotti* may also be common (Kellermann 1989b).

There is still relatively little known about the Prydz Bay area. Preliminary indications suggest that the larvae of *Notothenia acuta, Notothenia squamifrons, Notothenia mizops, Notothenia kempi* and *P. antarcticum* are the most abundant, particularly the latter species (Williams et al. 1986; Biomass 1988).

The pelagic fish fauna of the Weddell Sea is dominated by P. *antarcticum*, and this is also true of the larval fish community (Hubold 1984, 1990; Hubold and Ekau 1987). The remaining species in this community include 18 notothenioids (Hubold 1990; Hubold and Ekau 1990) whose older stages form the demersal fish community (Kock et al. 1984; Ekau 1988).

It appears that the larvae of *P. antarcticum* are often an important component of the pelagic micronekton in waters south of 60 °S latitude.

The early stages of non-notothenioid fishes are also found in the pelagic antarctic marine ecosystem. At South Georgia, the Antarctic Peninsula, the Weddell Sea and Prydz Bay the larvae of mesopelagic fish, especially *Electrona* spp., *Gymnoscopelus* spp. (Myctophidae) and *Notolepis* spp. (Paralepididae), were also common, particularly over water depths > 500 m (Efremenko 1983; BIOMASS 1988; Gon 1988; Kellermann and Kock 1988; North 1988; Kellermann 1989b; Hubold and Ekau 1990; Hubold 1990).

7 Habitat Partioning

Larval notothenioidei are pelagic mainly in the upper 250 m of the water column. In summer, larval *P. antarcticum* occured in the upper 100 m of the water column in the Weddell Sea (Hubold 1984, 1985) and in the upper 135–200 m at the Antarctic Peninsula (Kellermann 1986b, Kock 1986). At South Georgia, fish larvae occured throughout the water column of a 265-m-deep fjord in early spring but were restricted to the upper 140 m during summer (North 1988; North and Ward 1990). In summer the larvae of several species at South Georgia undertook diurnal vertical migrations and that of *C. gunnari* was the most distinct (North 1988 1990).

Different life-history stages of the notothenioids live at different mean depths, with the adults usually occupying the deepest waters. As many species are demersal as adults, this also results in horizontal separation. There is an ontogenetic vertical and horizontal zonaton of life-history stages in *P. antarcticum*, which live at deeper depths and over deeper water with increasing size (Hubold 1985; Kellermann 1986b). An ontogenetic migration also occurs in *N. neglecta* and *N. rossii*; at South Georgia the larvae of these two species are found offshore, whereas at the fingerling phase they migrate inshore (White et al. 1982).

The larvàl stages of several species of Notothenioidei are most abundant close to the coast at South Georgia. Larval fish abundance was estimated quantitatively in summer along two transects off the northeast coast. Three common species were most abundant within 10 km of the coast (North 1990), including larval *Champsocephalus gunnari*, the adults of which are most common over the shelf of the island to depths > 500 m (Kock 1981). Preliminary investigations in late winter suggest that most notothenioid larvae (including those of *C. gunnari*) were within 5 km of the coast of the island (White and North, BAS unpubl. data). Large shoals of larvae were observed in sublittoral waters (< 40 m depth) at South Georgia

(North 1990). Efremenko (1983) suggests that cyclonic currents retain the larvae in neritic waters, and if these gyres are absent there are effects on recruitment to the juvenile and adult populations.

As described earlier, there is a temporal succession in the abundance of the early larvae of different species during the year. This increases the complexity of the ecosystem and will reduce intra- and interspecific competition for food by spreading predation on mesozooplankton more evenly throughout the year. The combination of larval temporal succession with some vertical and horizontal ontogenetic separation results in a partitioning of the fish component of the ecosystem by time and space. This ecological niche partitioning may increase the carrying capacity of the ecosystem (Kellermann 1989a,b; North and Ward 1989; North and White 1987).

8 Food

The larvae of most notothenioids feed on pelagic mesozooplankton (mainly crustacea) in the neritic zone. Larval fish, including notothenioids, swallow their prey whole. Consequently, their mouth width limits the maximum width of prey they can consume (Hunter 1981; Balbontin et al. 1986; North and Ward 1989). Investigations by Kellermann (1986a, 1987), Balbontin et al. (1986), North and Ward (1989) and North (1990) have shown that the early larvae of many notothenioids feed on prey 0.05–0.75-mm wide, with most prey consisting of copepods and euphausiids around 0.3–0.5-mm wide. Exceptions are the larvae of most members of the Channichthyidae and *Psilodraco breviceps* (Bathydraconidae) which have large mouths; they are piscivorous, feeding on the early larvae of other species (> 0.8 mm wide) in addition to the early stages of euphausiids, and copepods (Regan 1916; Kellermann 1986a, 1987, 1989c; Balbontin et al. 1986; Hubold and Ekau 1990; North 1990; North and Ward 1990).

The range of food types in the diet of larval notothenioids also includes amphipods, molluscs and fish. Larval *P. antarcticum* fed on pelagic molluscs (*Limacina spp.*) in the Weddell Sea (Hubold 1984, 1985). At the Antarctic Peninsula their diet was mainly copepods (*Oncea* spp.), calanoid copepods eggs, tintinnids and the eggs of *Euphausia superba* (Kellermann 1987). Larvae of *Nototheniops larseni* and *Notothenia kempi* fed on zooplankton, mainly calanoid eggs, cyclopoid copepods and tintinnids, except those of *C. rastrospinosus* which fed on euphausiids and fish larvae (Kellermann 1986a, 1987, 1989c). Balbontin et al. (1986) reported a similar range of crustacean prey for the larvae of *Notothenia kempi, Trematomus newnesi, Nototheniops larseni* and *P. antarcticum* at the Antarctic Peninsula. At South Georgia, the neritic copepod *Drepanopus forcipatus* and other copepod species were the main food of fish larvae (North and Ward 1990). In winter, early larvae fed on the later stages of copepods, including ovigerous female *D. forcipatus* (North and Ward 1989). The channichthyids, *Pseudochaenichthys georgianus* and *Chaenocephalus aceratus*, were wholly or partly piscivorous.

Kellermann (1987) reported an inverse relationship between food abundance and selective feeding by *P. antarcticum* larvae on the copepod *Oithona* spp. In winter, the early larvae of *C. gunnari* consumed a higher proportion of ovigerous female *Drepanopus forcipatus* than were present in the plankton (North and Ward 1989, 1990).

In winter, the seasonality of food availability, day length and light intensity combine to reduce feeding opportunities for polar invertebrates (Clarke 1980, 1988) and fish larvae (North and Ward 1990). Notothenioid larvae feed mainly during daylight (Kellermann 1986a; North and Ward 1990). Therefore, during the antarctic winter, feeding by the larvae will be limited, by day length which shortens dramatically with increasing latitude, and by low light levels due to sea ice and overlying snow. Compared to summer, zooplankton biomass in the epipelagic zone (< 100 m depth) of the Southern Ocean is about halved in winter (Foxton 1956; Andrews 1966), and inshore at South Georgia it is 4–5 times lower (Ward 1989). Furthermore, the energy (lipid) content of zooplankton is lower in late winter/early spring (Clarke 1988; Hagen 1988). Therefore, in winter the likely food energy intake of antarctic fish larvae will be lower than in summer (North 1990).

9 Predators

Almost nothing is known about the predators of Antarctic fish larvae except, as previously mentioned, the larvae of several species of notothenioidei, mainly Channichthyidae, feed on fish larvae. The importance of piscivory within the larval fish communities may be indicated by the ratio of the abundance of piscivorous species to that of other species. In early spring at a South Georgia fjord, the ratio for the abundance of larval *Chaenocephalus aceratus* and *Pseudochaenichthys georgianus* to other species is about one-tenth (data from North 1990). In mid-November to mid-December at the Antarctic Peninsula, the ratio of the number of *Chionodraco rastrospinosus* to those of other species was about 1/14 (data from Kellermann and Kock 1988). These ratios suggest that there are likely to be interactions, between the first-year production of each cohort of the piscivorous species and their prey species.

10 Parasites

The parasites of larval antarctic fish have rarely been studied, although some harbour digenetic trematodes from an early age. Early larvae of *Notothenia* sp., *Pagothenia hansoni*, *Parachaenichthys georgianus* and *Pseudochaenichthys georgianus*, taken during winter from the nearshore environment at South Georgia (North and Ward 1989), contained gut parasites, including *Elytrophalloides oatesi* (Zdzitowiecki 1988). The larvae of *Champsocephalus gunnari* from South Georgia were also observed to have gill, oral/buccal cavity and gut parasites but these have not yet been formally identified (North, BAS unpubl. observ.).

11 Growth

There are seven published studies of the growth of antarctic fish larvae, based on sequential samples of fish size, which cover a total of 14 species (White et al. 1982; Keller 1983; Hubold 1985; Kellermann 1986a; Slosarczyk 1987; Radtke et al. 1989; North 1990). These studies give growth rates up to 0.37mm SL day^{-1} for periods in spring to early autumn.

Larval *Pleuragramma antarcticum* (size range 9–21 mm SL) grew at 0.19–0.25 mm SL day^{-1} (Keller 1983; Hubold 1985). Growth rates of *Nototheniops larseni, P. antarcticum, Notothenia kempi* and *Chionodraco rastrospinosus* at the Antarctic Peninsula varied from 0.08 mm SL day^{-1} for *N. larseni* of 6–25 mm SL in spring to fall, to 0.37 mm SL day^{-1} for *C. rastrospinosus* of 25–50 mm SL in late summer (Kellermann 1986a). At the Antarctic Peninsula, the growth rate of larval *C. aceratus* of 15–40 mm SL was about 0.11–0.16 mm SL day^{-1} (Slosarczyk 1987). The growth rate of *Trematomus newnesi* (15–25 mm mean SL) was about 0.37 mm SL day^{-1} at the Antarctic Peninsula (Radtke et al. 1989). North (1990) studied ten species at South Georgia. The fastest growth rates were for *Notothenia gibberifrons* (8–33 mm SL) and *C. gunnari* (25–50 mm SL) during summer, with 0.37 and 0.35 mm SL day^{-1}, respectively. There was a slight decrease in the mean size of larval (4.5–10 mm SL) *Harpagifer georgianus* in samples taken over a seven-month period (May to December). This may be the explained by degrowth (or a long hatching period). An initial period of slow growth (0.07 mm SL day-1) in *Notothenia angustifrons* (4.8–12 mm SL), was followed by growth at 0.25 mm SL day^{-1} (size range 7–29 mm SL).

Larval growth rates determined from daily rings in otoliths are in a similar range (Radtke et al. 1989; Radtke and Kellermann, this Vol.). The maximum observed linear growth rates of late larval/early juvenile nototheniids and channichthyids, of up to 0.37 mm SL day^{-1} (Kellermann 1986a; Slosarczyk 1987; North 1990), are similar to those of the larvae.

12 Physiology

Little is known about the physiology of the pelagic early life history stages. Some observations have been reported on routine metabolism, muscle physiology and development of organs associated with the immune response.

The early stages of notothenioids have a better muscle arrangement for cruise swimming than the adults, in which slow muscles are only a minor component of the trunk muscles (Johnston and Camm 1987). The pelagic early stages of six species have a superficial layer of slow muscle, one to two fibres thick, on the entire lateral surface of the trunk, suggesting a moderate capacity for continuous sub-carangiform swimming (Dunn et al. 1989). Johnston (1989) has found that the maximum tail-beat frequency of early juvenile *Notothenia neglecta* is very modest, compared to adults.

During the metamorphosis of *Pagothenia hansoni*, there was no evidence for a shift in the relationship between the mass specific rate of oxygen uptake with size (Morris and North 1984).

The lymphoid organs of larval *H. antarcticus* from the South Orkney Islands have a basic structure and development comparable with that of warm water teleosts (O'Neill 1989).

13 Comparison With Fish From Other Regions

To make valid regional comparisons between fish, it is necessary to use those most closely related. The families with the closest taxonomic affinities to the Notothenioidei are the Blenniidae, Callionymidae, Gobiidae, Trachinidae, and Uranoscopidae (Greenwood et al. 1966; Eakin 1981; Andersen 1984). Larval stages of the Blennioidea and Trachinoidea seem most similar to larval Notothenioidei (Stevens et al. 1984) and this concurs with the placing of the notothenioid suborder between that of the Trachinoidei and Blennioidei by Greenwood et al. (1966).

As described earlier, antarctic fish larvae are not especially advanced in overall development at hatching (North and White 1987), although they are somewhat precocious because they can soon feed. Some temperate and many tropical species, including some members of the Blenniidae, hatch with poorly developed jaws (Russell 1976; Thresher 1984).

In marine fish, there is a general latitudinal trend from the equator to the poles of increasing egg and larval size (Marshall 1953), and a similar trend in the size at metamorphosis (North 1990). The difference in length between the larvae at hatching and the metamorphosing late larval/early juvenile stage (larval length span) also increases towards the poles. Depending on species, the larval length span typically ranges between 20–80 mm in antarctic Notothenioidei (North 1990), 9–45 mm in northern temperate fish (Russell 1976) and 7–21 mm in tropical fish (Thresher 1984; Thresher and Brothers 1989). Therefore, a typical notothenioid fish grows over a greater length range from hatching to metamorphosis than a typical tropical fish.

To compare growth rates with fish from other regions it is necessary to choose fish growing over a similar size range, because growth is likely to be length- and mass-specific (von Bertalanffy 1938). A simple model, assuming a linear increase in mean absolute length with time, has been used to estimate the growth of fish larvae (North 1990). It was observed that summer larval growth rates (0.12–0.37 mm length day $^{-1}$) of antarctic fish and of arctic cod, at temperatures < 4 °C are within the range reported for the larvae of temperate species at temperatures > 14 °C; and it was therefore concluded that temperature does not severely limit the growth rate of antarctic fish (North 1990).

Growth rate was further examined using both a complete data set from a total of 122 studies of marine teleost larvae (reference list available), and a reduced data set (78 studies) using only the fastest growth rates reported for each species (Clarke and North, this Vol.). This confirmed that polar fish larvae grow at rates within the range of those from temperate and tropical regions; however, when larvae of the same initial size were compared, on average, polar fish larvae do grow significantly more slowly than those from warmer regions.

It seems likely that the annual growth rate of larval fish is limited by the increased seasonality of food availability and day length towards the poles, and that there is an upper constraint to growth rate set by habitat temperature (Clarke 1980, 1988; Clarke and North, this Vol.).

The period of development from the larval to the juvenile stage in antarctic fish is usually 6–12 months, because of slow growth rates combined with a large larval length span; whereas periods of weeks or days are common for temperate and tropical marine fish, respectively (North 1990).

14 Future Studies

This review raises some questions which may stimulate further studies. Why have antarctic Notothenioidei evolved to have a life-history strategy which involves big eggs and larvae and a large size at metamorphosis to the early juvenile stage? What is the species diversity and seasonal biomass of fish larvae around the coasts of Antarctica, particularly in the Ross Sea, the shallow inshore waters (< 50 m depth), and the epibenthic environment? Why do some species hatch and begin feeding in winter when the availability of epipelagic mesozooplankton is at a minimum? What is the small-scale distribution of fish larvae and how does this relate to that of their potential food? How do the locomotory and respiratory systems change during development, especially in the haemoglobinless icefishes? What are the maximum potential growth rates of the larvae?

Acknowledgements. I thank A. Kellermann, M.G. White, A. Clarke and R.B. Heywood for their constructive comments on the manuscript.

References

Anderson NC (1984) Genera and subfamilies of the family Nototheniidae (Pisces, Perciformes) from the Antarctic and Subantarctic. Steenstrupia 10:1–34

Andrews KJH (1966) The distribution and life-history of *Calanoides acutus* (Giesbrecht). Discovery Rep 36:117–162

Andrieshev AP (1965) A general review of the antarctic fish fauna. In: van Mieghem J, van Oye P (eds) Biogeography and ecology in Antarctica. Junk, The Hague, pp 491–550

Andrieshev AP (1987) A general review of the antarctic fish bottom fauna. In: Kullander SO, Fernholm B (eds) Pro V Congr Europ Ichthyol, Stockholm 1985. Swed Mus Nat Hist, Stockholm, pp 357–372

Arnaud PM (1974) Contribution a la bionomie marine benthique des régions antarctiques et subantarctiques. Téthys 6:465–656

Balbontin F, Garreton M, Neuling J (1986) Composicion del alimento y tamano de las presas en larvas de peces del estrecho Bransfield (SIBEX-Fase II - Chile). Ser Cient INACH 35:125–144

BIOMASS (1987 Report of the Post-SIBEX fish data evaluation workshop: phase 1, validation. Cambridge, UK. 6–17 October 1986. BIOMASS Rep Ser 52 46 pp

BIOMASS (1988) Report of the Post-SIBEX fish data evaluation workshop: phase 2, evaluation. Cambridge, UK. 17–28 August 1987. BIOMASS Rep Ser 54 38 pp

Burchett MS, Sayers PJ, North AW, White MG (1983) Some biological aspects of the nearshore fish populations at South Georgia. Bull Br Antarct Surv 59:63–74

Burren PJ (1988) Reproductive biology of *Harpagifer* sp. at Signy Island, South Orkney Islands. M.Sc. Dissertion, Univ North Wales, 52 pp

Camus P, Dumahel G (1985) Ponte et développement embryonnaire de *Notothenia rossii rossii* (Richardson, 1844) Nototheniidae del Iles Kergulen. Cybium 9:283–293

Ciechomski JD, Weiss G (1976) Desarollo y distribucion de postlarvas del robalo *Eleginops maclovinus* (Valencienes, 1830) Dollo, 1904; de la merluza negra *Dissostichus eleginoides* Smitt, 1899 y de las Nototenias *Notothenia* spp. Pisces, Nototheniidae. Physis A 35(91):115–125

Clarke A (1980) A reappraisal of the concept of metabolic cold adaptation in polar marine invertebrates. Biol J Linn Soc 14:77–92

Clarke A (1988) Seasonality in the antarctic marine environment. Comp Biochem Physiol 90B(3):461–473

Daniels RA (1978) Nesting behaviour of *Harpagifer bispinis* in Arthur Harbour, Antarctic Peninsula. J Fish Biol 12:465–474

DeVries AL, Eastman JT (1981) Physiology and ecology of notothenioid fishes of the Ross Sea. J Roy Soc NZ 11(4):329–340

DeVries AL, Wohlschlag DE (1969) Freezing resistance in some antarctic fishes. Science NY 163:1073–1075

DeWitt HH (1971) Coastal and deep-water benthic fishes of the Antarctic. Antarctic map folio ser, Folio 15. Am Geogr Soc New York, 10 pp

di Prisco G, Maresca B, Tota B (1988) Marine biology of antarctic organisms. Proc Int Conf Mar Biol Antarctic. Comp Biochem Physiol 90B:459–460

Duhamel G (1987) Ichtyofaune des secteurs Indien Occidental et Atlantique Oriental de l'Ocean Austral: biogeographie, cycles biologiques et dynamique des populations. Thèse de Doctorat d'Etat dès Sciences Naturelles, Univ Pierre et Marie Curie, Paris, 687 pp

Dunn JF, Archer SD, Johnston IA (1989) Muscle fibre types and metabolism in post-larval and adult stages of notothenioid fish. Polar Biol 9:213–223

Eakin RR (1981) Osteology and relationships of the fishes of the antarctic family Harpagiferidae (Pisces, Notothenioidei). Antarct Res Ser 31:81–147

Eastman JT, DeVries AL (1986) Antarctic fishes. Sci Am 255:96–103

Eastman JT Grande L (1989) Evolution of the antarctic fish fauna with emphasis on the recent notothenioids. In: Crame JA (ed) Origins and evolution of the antarctic biota. Geol Soc Spec Pub 47:241–252

Efremenko VN (1979a) Lichiniki shesti vidov ryb sem. Nototheniidae iz morya Skotia. Vop Ikhtiol 19(6): 1068–1078 [The larvae of six species of the family Nototheniidae from the Scotia Sea. J. Ichthyol 19(6):95–104]

Efremenko VN (1979b) Opisanie lichinok shesti vidov ryb semesjstva Chaenichthyidae iz morya Skotia. Vop Ikhtiol 19(3):458–469. [Description of the larvae of six species of the family Chaenichthyidae from the Scotia Sea. J Ichthyol 19(3):65–75]

Efremenko VN (1983) Atlas of fish larvae of the Southern Ocean. Cybium 7(2):1–74

Efremenko VN (1984) Larvae of the family Nototheniidae from the Scotia Sea. J Ichthyol 24:34–43

Ekau W (1989) Ökomorphologie nototheniider fische aus dem Weddellmeer, Antarktis. Ber Polarforsch 51:140 pp

Ekau W (1989) Egg development of *Trematomus eulepidotus* Regan, 1914 (Nototheniidae, Pisces) from the Weddell Sea Antarctica. Cybium 13:213–219

Everson I (1968) Larval stages of certain antarctic fishes. Bull Br Antarct Surv 16:65–70

Everson I (1984) Fish biology. In: Laws RM (ed) Antarctic ecology II. Academic Press, London, pp 491–532

Fischer W, Hureau J-C (eds) (1985) FAO species identification sheets for fishery purposes. Southern Ocean (CCAMLR Convention Area Fishing areas 48, 58 and 88), II. Food and Agriculture Organization of the United Nations, Rome, pp 233–470

Foxton P (1956) The distribution of the standing crop of zooplankton in the Southern Ocean. Discovery Rep 28:191–236

Gon O (1987) A description of the postlarva of *Cygnodraco mawsoni* Waite, 1916 (Pisces, Bathydraconidae) from the Southern Ocean. S Afr J Zool 22(4):231–322

Gon O (1988) The fishes collected during the South African SIBEX I + II expeditions to the Indian Sector of the Southern Ocean (60–66° S, 48–64° E). S Afr J Antarct Res 18(2):55–70

Gordon AL (1967) Structure of the antarctic waters between 20° W and 170° W. Antarctic map folio series, Folio 6. Am Geogr Soc New York, 14 pp

Greenwood PH, Rosen DE, Weitzman SH, Myers GS (1966) Phylogenetic studies of teleostean fishes with a provisional classification of living forms. Bull Am Mus Nat Hist 131(4):339–456

Hagen W (1988) On the significance of lipids in antarctic zooplankton. Ber Polarforsch 49:1–129

Hemmingsen EA, Douglas EL (1970) Respiratory characteristics of the hemoglobin-free icefish *Chaenocephalus aceratus*. Comp Biochem Physiol 33:733–744

Holeton GF (1970) Oxygen uptake and circulation by a hemoglobinless antarctic fish (*Chaenocephalus aceratus* Lönnberg) compared with three red-blooded antarctic fish. Comp Biochem Physiol 34:457–471

Hourigan TF, Radtke RL (1989) Reproduction of the antarctic fish *Nototheniops nudifrons*. Mar Biol 100:277–283

Hubold G (1984) Spatial distribution of *Pleuragramma antarctium* (Pisces: Nototheniidae) near the Filchner- and Larsen ice shelves (Weddell Sea/Antarctica). Polar Biol 3:231–236

Hubold G (1985) The early life history of the high antarctic silverfish, *Pleuragramma antarcticum*. In: Siegfried WR, Condy PR, Laws RM (eds) Antarctic nutrient cycles and food webs. Springer, Berlin, Heidelberg New York Tokyo, pp 445–451

Hubold G (1990) Seasonal patterns of larval fish distribution in the zooplankton community of the southern Weddell Sea. In: Kerry K, Hempel G (eds) Ecological change and conservation of antarctic ecosystems. Proc SCAR Symp Antarctic Biol Springer, Berlin Heidelberg New York Tokyo, pp 149–158

Hubold G, Ekau W (1987) Midwater fish fauna of the Weddell Sea, Antarctica. In: Kullander SO, Fernholm B (eds) Proc V Congr Europ Ichthyol, Stockholm 1985. Swed Mus Nat Hist, Stockholm, pp 391–396

Hubold G, Ekau W (1990) Feeding patterns of post-larval and juvenile notothenioids in the southern Weddell Sea (Antarctica). Polar Biol 10:255–260

Hunter JR (1981) Feeding ecology and predation in marine fish larvae. In: Lasker R (ed) Marine fish larvae. Washington Sea Grant Program, Univ Washington, Seattle, pp 33–77

Hureau J-C (1970) Biologie compareé de quelque Poissons antarctiques (Nototheniidae). Bull Inst océanogr Monaco 68(1391) 244 pp

Iwami T (1985) Osteology and relationships of the family Channichthyidae. Mem Natl Inst Polar Res Ser E 36 69 pp

Johnston IA (1989) Antarctic fish muscles – structure, function and physiology. Ant Sci 1:97–108

Johnston IA, Camm J-P (1987) Muscle structure and differentiation in pelagic and demersal stages of the antarctic teleost *Notothenia neglecta*. Mar Biol 94:183–190

Keller R (1983) Contributions to the early life history of *Pleuragramma antarcticum* Boul. 1902 (Pisces, Nototheniidae) in the Weddell Sea. Meeresforsch 30:10–24

Kellermann A (1986a) On the biology of early life stages of Notothenioid Fishes (Pisces) off the Antarctic Peninsula. Ber Polarforsch 31:149 pp

Kellermann A (1986b) Geographical distribution and abundance of postlarval and juvenile *Pleuragramma antarcticum* (Pisces, Notothenioidei) off the Antarctic Peninsula. Polar Biol 6:111–119

Kellermann A (1987) Food and feeding ecology of postlarval and juvenile *Pleuragramma antarcticum* (Pisces; Notothenioidei) in the seasonal pack ice zone off the Antarctic Peninsula. Polar Biol 7:307–315

Kellermann A (1989a) Catalogue of early life stages of Antarctic notothenioid fishes. In: Kellermann A (ed) Identification key and catalogue of larval antarctic fishes. BIOMASS Sci Ser 10:45–136

Kellermann A (1989b) The larval fish community in the zone of seasonal pack–ice cover and its seasonal and interannual variability. Arch FischWiss 39:81–109

Kellermann A (1989c) Food and feeding of early stage *Chionodraco rastrospinosus* DeWitt and Hureau 1979 (Pisces; Notothenioidei) off the Antarctic Peninsula. Pesq Antart Bras 1:25–30

Kellerman A, Kock K–H (1988) Patterns of spatial and temporal distribution and their variation in early life stages of antarctic fish in the Antarctic Peninsula region. In: Sahrhage D (ed) Antarctic Ocean and resources variability. Springer, Berlin, Heidelberg New York, Tokyo pp 147–159

Kennett JP (1977) Cenozoic evolution of antarctic glaciation, the Circum-Antarctic Ocean, and their impact on global paleoceanography. J Geophys Res 82(27):3843–3860

Kock K-H (1981) Fischereibiologische Untersuchungen an drei antarktischen Fischarten: *Champsocephalus gunnari* Lönnberg, 1905, *Chaenocephalus aceratus* (Lönnberg, 1906) und *Pseudochaenichthys georgianus* Norman, 1937 (Notothenioidei, Channichthyidae). Mitt Inst Seefisch Hamburg 32 226 pp

Kock K-H (1985) Marine habitats-antarctic fish. In: Bonner WN, Walton DWH (eds) Key environments Antarctica. Pergamon Press, Oxford, pp 173–192

Kock K-H (1986) The state of exploited fish stocks in the Scotia Arc region during SIBEX (1983–1985). Arch FischWiss 37(1):129–186

Kock K-H (1989) Reproduction in fish around Elephant Island. Arch FischWiss 39:171–210

Kock K-H, Schneppenheim R, Siegel V (1984) A contribution to the fish fauna of the Weddell Sea. Arch FischWiss 34(2/3):103–120

Konecki JT, Targett TE (1989) Eggs and larvae of *Nototheniops larseni* from the spongocoel of a hexactinellid sponge near Hugo Island, Antarctic Peninsula. Polar Biol 10:197–198

Littlepage JL (1965) Oceanographic investigations in McMurdo Sound, Antarctic. Antarct Res Ser 5:1–37

MacDonald JA, Montgomery JC, Wells RMG (1987) Comparative physiology of antarctic fishes. In: Blaxter JHS, Southward AJ (eds) Advances in marine biology, vol 24. Academic Press, London, pp 321–388

Marshall NB (1953) Egg size in arctic, antarctic and deep-sea fish. Evolution 7:328–341

Moreno CA (1980) Observations on food and reproduction in *Trematomus bernacchii* (Pisces: Nototheniidae) from the Palmer Archipelago, Antarctica. Copeia 1:171–173

Morris DJ, North AW (1984) Oxygen consumption of five species of fish from South Georgia. J Exp Mar Biol Ecol 78:75–86

Moser HG (1981) Morphological and functional aspects of marine fish larvae. In: Lasker R (ed) Marine fish larvae. Washington Sea Grant Program, Univ Washington, Seattle, pp 89–131

North AW (1988) Distribution of fish larvae at South Georgia: horizontal, vertical and temporal distribution and early life history relevant to monitoring year-class strength and recruitment. Scientific Committee of the Convention for the Conservation of Antarctic Marine Living Resources Selected Scientific Papers 1987, (WG–FSA–87/16) pp 105–142

North AW (1990) Ecological studies of antarctic fish with emphasis on early development of inshore stages at South Georgia. PhD thesis, CNAA, UK, 319 pp

North AW, Kellermann A (1989) Key to the early stages of antarctic fish. In: Kellermann A (ed) Identification key and catalogue of larval antarctic fishes. BIOMASS Sci Ser 10, Cambridge, SCAR, pp 1–44

North AW, Ward P (1989) Initial feeding by antarctic fish larvae during winter at South Georgia. Cybium 13:357–364

North AW, Ward P (1990) The feeding ecology of larval fish in an antarctic fjord, with emphasis on *Champsocephalus gunnari*. In: Kerry K, Hempel G (eds) Ecological change and conservation of antarctic ecosystems. Proceedings of the 5th SCAR Symposium on Antarctic Biology. Springer, Berlin, Heidelberg New York, Tokyo, pp 299–307

North AW, White MG (1982) Key to fish larvae from the Scotia Sea, Antarctica. Cybium 6:13–32

North AW, White MG (1987) Reproductive strategies of antarctic fish. In: Kullander SO, Fernholm B (eds) Proc V Congr Europ Ichthyol, Stockholm 1985. Swed Mus Nat Hist, Stockholm, pp 381–391

Nybelin O (1951) Subantarctic and antarctic fishes. Scientific Results of the Brategg Expedition 1947–48, 2, 32 pp

O'Neill JG (1989) Ontogeny of the lymphoid organs in an antarctic teleost, *Harpagifer antarcticus* (Notothenioidei: Perciformes). Dev Comp Immunol 13:25–33

Pappenheim P (1912) Die Fische der Deutschen Südpolar–Expedition 1901–1903. Die Fische der Antarktis und Subantarktis. Dtsch Südpol Exped 1901–03 13 Zool 5:162–184

Radtke RL, Targett TE, Kellermann A, Bell JL, Hill KT (1989) Antarctic fish growth: profile of *Trematomus newnesi*. Mar Ecol Prog Ser 57:103–117

Regan CT (1916) British Antarctic 'Terra Nova' Expedition, 1910. Larval and post-larval fishes. Nat Hist Rep Br antarct Terra Nova Exped Zool 1(4):125–156

Riehl R, Kock K-H (1989) The surface structure of antarctic fish eggs and its use in identifying fish eggs from the Southern Ocean. Polar Biol 9:197–203

Robertson DA, Mito S (1979) Sea surface ichthyoplankton off southeastern New Zealand, summer 1977–78. NZ J Mar Fresh Res 13:415–425

Russell FS (1976) The eggs and planktonic stages of british marine fishes. Academic Press, London, 524 pp

Ruud JT (1954) Vertebrates without erythrocytes and blood pigment. Nature (Lond) 173:848–850

Slosarczyk W (1983) Juvenile *Trematomus bernacchii* and *Pagothenia brachysoma* (Pisces, Nototheniidae) within krill concentrations off Balleny Islands (Antarctic). Pol Polar Res 4(1–4):57–69

Slosarczyk W (1987) Contribution to the early life history of Channichthyidae from the Bransfield Strait and South Georgia (Antarctica). In: Kullander SO, Fernholm B (eds) Proc V Cong Europ Ichthyol, Stockholm 1985. Swed Mus Nat Hist, Stockholm, pp 427–433

Stevens EG, Watson W, Matarese AC (1984) Notothenioidea: development and relationships. In: Moser GH, Richards WJ, Cohen DM, Fahay MP, Kendall AW, Richardson SL (eds) Ontogeny and systematics of fishes. ASIH, Spec Publ, 1, Kansas, USA, pp 561–564

Thresher RE (1984) Reproduction in reef fishes. TFH Publ, Ascot, 393 pp

Thresher RE, Brothers EB (1989) Evidence of intra- and inter-oceanic regional differences in the early life history of reef associated fishes. Mar Ecol Prog Ser 57:187–205

Twelves EL (1972) Blood volume of two antarctic fishes. Br Antarct Surv Bull 31:85–92

von Bertalanffy L (1938) A quantitative theory of organic growth. Human Biol 10:181–213

Ward P (1989) The distribution of zooplankton in an Antarctic fjord at South Georgia during summer and winter. Ant Sci 1:141–150

White MG, North AW (1987) Postlarval Nototheniodei and midwater fish collected during the SIBEX cruise by British Antarctic Survey, 1985. In: Kullander SO, Fernholm B (eds) Proc V Congr Europ Ichthyol, Stockholm 1985. Swed Mus Nat Hist, Stockholm, pp 405–411

White MG, North AW, Twelves EL, Jones S (1982) Early development of *Notothenia neglecta* from the Scotia Sea, Antarctica. Cybium 6:43–51

Williams R, Kirkwood JM, O'Sullivan DB (1986) ADBEX I cruise zooplankton data. ANARE Res Notes 31:107 pp

Zdzitowiecki K (1988) Occurence of digenetic trematodes in fishes off South Shetlands (Antarctic). Acta Parasit Pol 33(3):155–167

Age Determination in Antarctic Fish

M. G. White

1 Introduction

The determination of age is an essential component in understanding the life history of fish, and is important in the assessment of the population dynamics of species to provide the data necessary to enable effective conservation and fisheries management.

In aquatic ecosystems, where there are distinct seasonal variations of such factors as rainfall, temperature, food availability and day length, it is implicit to most age determination methods that these extrinsic events regulate the growth and reproductive patterns of the fish, resulting in growth being episodic, and reproductive events occurring at specific intervals during the year. In fish from temperate waters, the well-defined seasonal periodicity of the environment result in the major growth increments having an annual cycle. In non-antarctic species of fish, structures such as otoliths, scales, fin rays, and various skeletal bones have all been shown to exhibit incremental structures that can be interpreted as being related to seasonal changes in growth patterns (cf. reviews by Bagenal 1974; Chilton and Beamish 1982; Prince and Pulos 1983; Pentilla and Dery 1988). In addition, on a smaller scale otoliths have been demonstrated to exhibit daily growth increments (Pannella 1971).

A variety of techniques have been developed to reveal increments and interpret growth patterns in otoliths. This include the "crack and char" method (Christensen 1964), staining (Albrechtsen 1968), thin section (Rauck 1975) and the use of multiple thin sections embedded in pigmented resin (Williams and Bedford 1974; Bedford 1983). Methods have been developed to examine scales (Tesch 1971) and use semi-automated techniques for interpretation of both scales and otoliths (Fawell 1974; Mason 1974). X-ray techniques and chemical analytical procedures (Dannevig 1956; Casselman 1983) have been used to examine the internal structure of calcareous skeletal structures and otoliths. The presence of daily increments in otoliths (Pannella 1971, 1974) provides a time-consuming but potentially precise method for validating gross structures such as seasonal growth increments and age determination, particularly where the seasonal increments are difficult to interpret. The examination of daily increments also offers the opportunity to quantify short-

British Antarctic Survey, Natural Environment Research Council, High Cross, Madingley Road, Cambridge, CB3 OET, UK term changes in growth patterns and is of value when investigating larval development. Chemical markers such as tetracycline (Kobayashi et al. 1964) or biological stains (Tsukamoto et al. 1989; Richter and McDermott 1990) have been used to time-mark seasonal growth patterns and acetazolamide can be used to introduce checks among daily increments (Mugiya 1977).

The techniques to determine age in non-antarctic freshwater and marine fish have been reviewed (Bagenal 1974; Chilton and Beamish 1982; Pentilla and Dery 1988; Prince and Pulos 1983). It is clear from these reviews that the structures and techniques used for any given species have to be determined by detailed appraisal of the methodology and the biology before a successful procedure can be adopted to age individuals with confidence (Chilton and Beamish 1982). Even the daily ring techniques have been demonstrated to exhibit significant errors when the precision of estimates is critically tested by the use of samples from known age individuals and exchange between readers (Moksness and Campana, in press).

The antarctic marine environment is characterized by low but stable temperatures and a markedly seasonal production cycle mediated by the seasonal changes in day length and the shading effects of sea ice. In this environment, biological events such as growth and reproduction have been shown to be seasonal (White 1977; Clarke 1988) and this is well represented among the life histories of antarctic fish (Everson 1984; Kock 1985). Therefore, because the polar marine environment is markedly seasonal, it might be anticipated that growth increments will be readily determined in the traditional structures such as otoliths, scales and bones.

The development of extensive commercial-scale fisheries in antarctic waters, and the need to manage these stocks, as well as the scientific interests arising from international research initiatives such the BIOMASS programme, have stimulated a series of investigations to develop reliable methods for determining the age of antarctic fish.

2 Morphology

The identification and quantification of residual structures from fish in the diet of predators is well established and in particular, the use of the sagittal otoliths is important because these otoliths can be identified to species level. Of the 200 or so species of antarctic fish (Andriashev 1987), the otoliths of nearly 100, including the majority of the most abundant species, have been described (Hureau 1962; Yukhov 1971; North et al. 1984; Hecht 1987). This has enabled the fish occurring in the diet of birds (Croxall et al. 1988), seals (North et al. 1983) and squid (Kear 1989) to be identified and their relative contribution to the diet of predators to be estimated.

The use of otoliths and other residual structures to determine the size of fish can be problematical. In the diet of predators this difficulty can be exacerbated by erosion of the otoliths during digestion in the intestine. A further difficulty is that the residence time in the gut of predators is usually unknown.

Experiments to quantify the effects of digestion using piscivorous birds suggest that caution is necessary when using otoliths to estimate the size of fish predated. Feeding experiments with the blue-eyed shag, *Phalacrocorax atriceps*, demonstrated that up to 70% of the mass of otoliths could be lost during a two-day residence time in the gut (Price, BAS unpubl. data).

The relationship between the size of otoliths and the size of individual fish has been dependent on the use of mass and/or a series of linear measurements. Statistical methods are then applied to derive a relationship with defined variables. The advent of image analysis systems has refined this procedure so that the size of a fish can be related to the dimensions of an uneroded otolith with precision (Radtke 1990). Nevertheless, the use of otolith/fish size relationships for use as a method for determining age remains questionable for Antarctic species of fish. This is especially so for species that exhibit marked sexual dimorphism and for estimates of age in species that are longevous.

3 Age/Length Distribution

Length frequency distributions can be used to identify cohorts within samples of a species. In aquatic ecosystems that are subject to markedly seasonal changes, these cohorts often represent year classes. The fish populations in the Southern Ocean are characterized by slow but seasonal growth rates and annual reproductive cycles. These characteristics, largely as a result of the highly seasonal polar marine environment, should result in distinct modes being revealed, representing year classes when length frequency distributions are plotted. Using this technique, age/length relationships have been calculated for several species (Kock 1980; Kock et al. 1985). The determination of age from length frequency distributions has limited use for the analysis of antarctic fish populations. This is because the early stages, particularly the "0"-and "1"-group fish are underrepresented in samples, as many species are pelagic at this stage in their life history (Freytag 1980a; Burchett 1983a). In addition, antarctic fish grow slowly, have delayed maturation and are long-lived; therefore, although length frequency distributions may be used to determine the age classes of juveniles, the length/age classes overlap considerably and the classification is usually impractical for adult fish. The use of size frequency techniques to measure the age of antarctic fish is largely confined to the evaluation of the population dynamics of fast-growing species such as *Champsocephalus gunnari* (Duhamel 1987; Duhamel, this Vol.) or used to investigate growth in more slow-growing species during the first 4–5 years such as juvenile *Notothenia rossii* (Burchett 1983b; Fig. la,b). However, the method lacks precision for age determination, because even when there are distinct size frequency groups it can be shown that there is still considerable overlap among the age classes (Fig. 1b,d).

4 Seasonal Structures in Otoliths, Scales, Fin Rays and Bones

It is clear that morphological and length frequency distribution techniques have limited utility for determining the age of individual fish, although these procedures may be useful as a means of deriving a first approximation of age distribution within a sample of fish. However, it is evident that the interpretation of growth

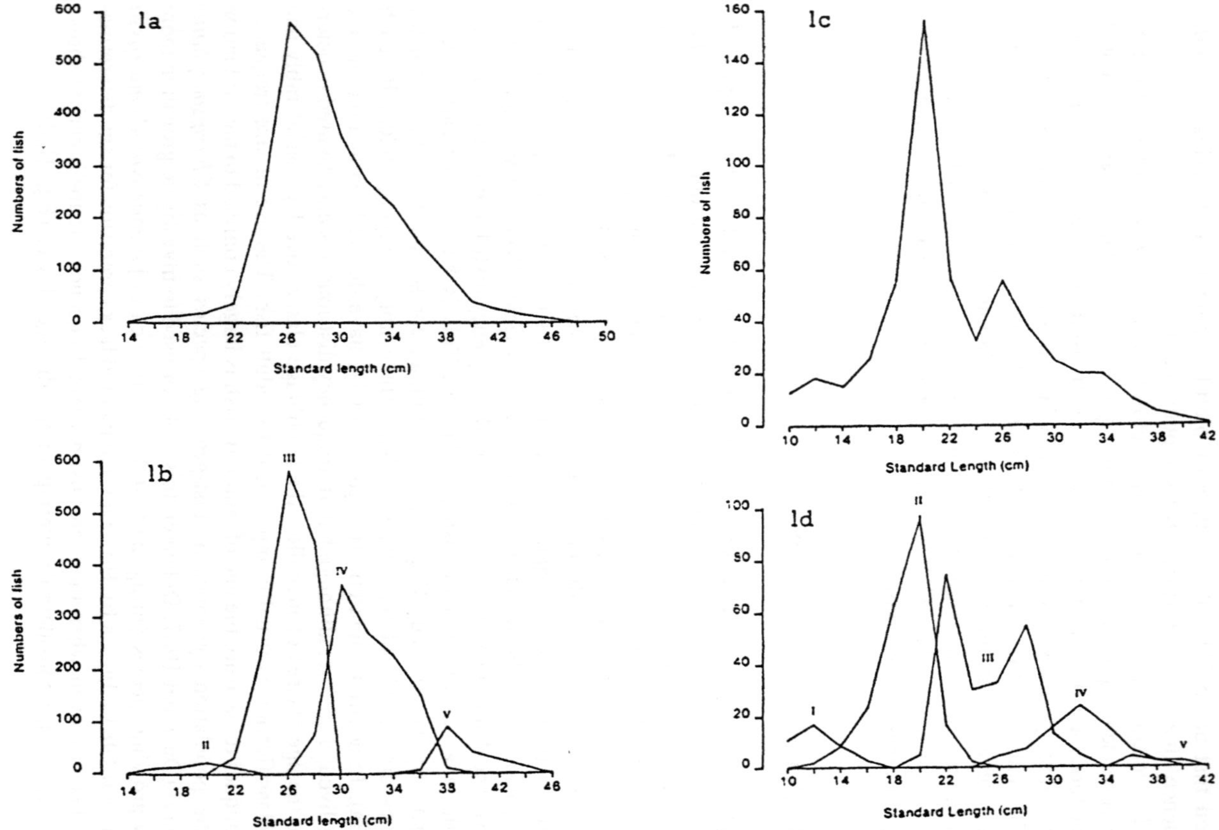

Fig. 1. Length frequency and age frequency of juvenile *Notothenia rossii* from South Georgia. Age indicated in *Roman numerals*. (**a, b**) 1978 and 1979 data recalculated from Burchett (1983b). (**c, d**) 1987/88 data White and Coggan (BAS unpubl. data)

increments in antarctic fish still requires detailed evaluation and validation as has been done for fish occurring elsewhere.

A recognition of these technical difficulties, and the need to achieve agreed age-determination protocols, has stimulated a number of technical workshops both within the BIOMASS programme (Anonymous 1980, 1982; Everson 1980; North et al. 1980) and by CCAMLR (Anonymous 1986; Kock, in press). These workshops have demonstrated that the traditional techniques (Bagenal 1974) are generally practical for use in determining the age of antarctic fish, but that some specific difficulties have been identified. The otoliths of antarctic fish usually have a dense calcareous structure and are opaque. As a result they cannot be read without recourse to crack and char or sectioning techniques to first reveal the annuli. Antarctic fish from unexploited populations have life spans amounting to 2–4 decades, but because the otoliths are often small, the growth increments are closely packed and can be difficult to quantify with accuracy. The channichthyids lack scales, and so the opportunity to cross reference between readings from scales and otoliths from the same individual is not possible. When this is possible, however (for example, with the nototheniids; Fig. 2), it has been found that the results from the two types of structure give different absolute estimates of age as well as having a different range of values about the average (Coggan et al. 1990). Similar difficulties arise from age determinations derived from alternative structures such as vertebrae and fin rays (Hoffmann 1982) and age is usually underestimated when analysis is based on these structures. Growth curves calculated from these can be markedly different from those prepared from the use of scales or otoliths (Coggan et al. 1990). Otoliths should be used for precise determinations of age in fish because they contain a complete record of the growth from hatching to capture and can be validated by examination of the microstructure.

In common with studies on non-antarctic fish, a whole range of techniques have been used to examine the seasonal increments in growth structures to obtain estimates of age. These have included: whole otoliths (Kock 1981), cracked otoliths (Sosinski 1981), cracked and charred otoliths (Clasing et al. 1985), otolith thin sections (Kochkin 1980b; Barber et al. 1986), multiple thin sections (Burchett 1984; Coggan et al. 1990), otolith staining (Schneppenheim and Freytag 1980), scale circuli analysis (Crisp and Carrick 1975; Burchett et al. 1984; Duhamel and Ozouf-Costaz 1985; Hureau and Ozouf-Costaz 1980); skeletal bones (Freytag 1980b); fin-ray section (Gubsch 1980a; Hoffman 1982) and vertebrae section (Kochkin 1980a, 1985). All have been shown to reveal seasonal increments.

Thus far it has not been possible to determine which are the most effective methods for determining the age of individual species, because few validation experiments have been undertaken to confirm and compare procedures. In addition, the requirement for precision will be different depending on whether the data is to be used in a scientific experiment with a small number of fish, or used for a gross estimate of population age structure in a commercial fisheries survey. In the former case, otoliths are the preferred structure to be examined with validation undertaken by the use of micro-increment analysis. In the latter, estimates of age taken by evaluating scales might be more practical because a large numbers of fish could be sampled rapidly. However, estimates of age based on the interpretation of scales must be viewed with caution because these underestimate the true age (Wohlschlag

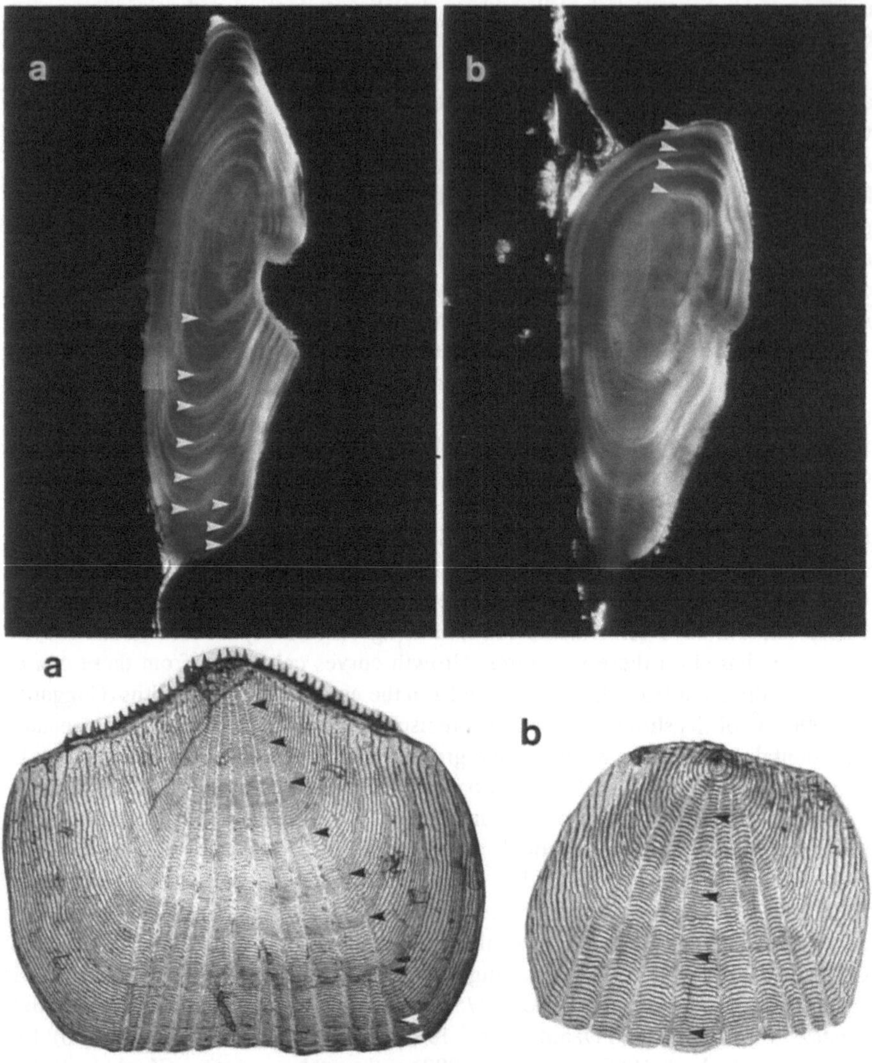

Fig. 2a,b. Scale and sectioned otolith from *Notothenia gibberifrons* from South Georgia. Estimated age, **a** 9 years; **b** 4 years. Annuli indicated by *arrows*

1962; Coggan et al. 1990). This is partially a consequence of the slow rates of development and delayed metamorphosis from larva to juvenile in antarctic fish, which results in a delay in the development of scales for several months after hatching. In addition, the complete growth history may also not be represented on scales, because they become eroded with age and the annulations are obscured by compression of the circuli in older fish. The underestimate in age is of the order of one year in *Notothenia gibberifrons*. The result of underestimating age by a year or two among the relatively long-lived antarctic fish might appear trivial; however, when the data are used to compute fisheries statistics such as yield-per-recruit

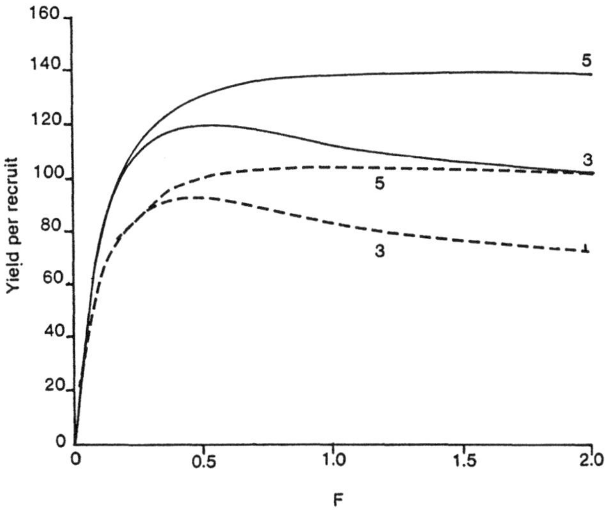

Fig. 3. Yield-per-recruit curves for *Notothenia gibberifrons* from South Georgia recalculated from Coggan et al. (1990). Calculations based on scales (*solid line*) and otoliths (*dashed line*)

curves (Coggan et al. 1990), the result using data from scales exaggerates the potential yield by 20-30% (Fig. 3).

Age derived from the examination of seasonal increments has now been examined in more than 20 antarctic species of fish, mostly species that have been subject to the commercial-scale fishery in the Southern Ocean. Table 1 indicates these species, the materials investigated and the source publications.

5 Microstructure of Otoliths

The presence of micro-increments in the otoliths of antarctic fish is now well documented (Townsend 1980; Radtke and Targett 1984; Linklowski 1987; Radtke 1987, 1990; Slosarczyk 1987; Radtke et al. 1989; Radtke and Hourigan 1990). These micro-increments are believed to be daily rings and this has been confirmed in three species, *Trematomus newnesi* (Radtke et al. 1989), *Nototheniops nudifrons* (Radtke and Hourigan 1990) and *Harpagifer antarcticus* (White, BAS unpubl. data), where experimental validation has been undertaken. The use of microstructures in otoliths to determine or validate the age of antarctic fish has not been widely used because the procedures are time consuming and also require relatively sophisticated equipment. For example, a scanning electron microscope is required to ensure that all the micro-increments are observed; the narrowest increments often cannot be satisfactorily resolved using a light microscope.

A detailed discussion of the use of micro-increments to investigate the age and the events in the life history of antarctic fish is to be found in Radtke and Kellermann (this Vol.). It is evident that the evaluation of the microstructure of otoliths is a most effective method for validating seasonal growth patterns, identi-

Table 1. Age determination by using seasonal growth techniques in antarctic fish species

Species	Structures	Source
Notothenia rossii	Scales	Crisp and Carrick (1975); Shcherbich (1975,1976); Shust and Pinskaya (1978); Linkowski and Zubowski (1980)
	Otoliths	Schneppenheim and Freytag (1980)
	Scales and otoliths	Burchett (1983a, 1984); Kock (in press); Olsen (1954)
	+ Bones	Freytag (1979, 1980a,b)
Notothenia neglecta	Scales	Linkowski and Zubowski (1980)
	Otoliths	Everson 1980
Notothenia gibberifrons	Fin rays	Hoffman (1982)
	Scales	Shust and Pinskaya (1978); Skora (1980); Boronin and Altman (1979); Kompowski (1983)
	Otoliths	Clasing et al. (1985) Boronin and
	Scales and otoliths	Frolkina (1976); Barrera-Oro (1989); Kock (in press); Coggan et al. (1990)
Notothenia squamifrons	Scales	Shust and Pinskaya (1978); Duhamel and Ozouf—Costaz (1985)
Notothenia tchizh	Scales	Gorelova and Shandikov (1988)
Nototheniops larseni	Scales	Shust and Pinskaya (1978); Radtke and Targett (1984)
Notothenia kempi	Scales	Shust and Pinskaya (1978)
Trematomus newnesi	Scales	Shust and Pinskaya (1978); Radtke et al. (1989)
Trematomus centronotus	Scales	Zur (1978)
Trematomus bernacchii	Scales	Wohlschlag (1962); Zur (1978)
Pagothenia hansoni	Scales	Shust and Pinskaya (1978)
Dissostichus mawsoni	Otoliths	Yukhov (1971); Burchett et al. (1984)
Dissostichus eleginoides	Otoliths	Yukhov (1971); Frolkina (1977)
	Scales	Hureau and Ozouf-Costaz (1980)
Pleuragramma antarcticum	Otoliths	Hubold and Tomo (1989); Kock (in press)
Chaenocephalus aceratus	Otoliths	Olsen (1955); Gubsch (1980ab,1982)
Pseudochaenichthys georgianus	Otoliths	Olsen (1955); Chojnacki and Palczewski (1981)
Champsocephalus gunnari	Otoliths	Kock (1980); Kochkin (1982, 1985); Sosinski (1981); Radtke (1990); Kock

Table 1. (*continued*)

Species	Structures	Source
	Bones	(in press) Kochkin (1980b)
Harpagifer antarcticus	Otoliths	Tomo and Cordero Funes (1972); Daniels (1983)
Micromesistius australis	Otoliths	Barrera-Oro (1989)
Electrona antarctica	Otoliths	Linkowski (1987)
Electrona carlsbergi	Otoliths	Linkowski (1987)
Gymnoscopelus nicholsi	Otoliths	Linkowski (1985)

fying the location and size of the first annulus and examining the development of larval fish (Fig. 4).

6 Validation

The validation of age derived from the examination of growth structures in antarctic fish falls into five main areas. These are:

1. The use of otoliths and scales from larval and juvenile fish to determine the dimensions of the nucleus (Slosarczyk 1987; North 1988).
2. The detailed comparison of readings derived from scales and otoliths sampled from the same individual fish (Freytag 1980b; Coggan et al. 1990).

Fig. 4. Daily increments in larval *Harpagifer antarcticus* otolith, 27 days after hatching. *Arrow* indicates presumed hatching mark

3. The examination of time series samples to demonstrate the period when the seasonal increments are deposited (North 1988; Barrero-Oro 1989).
4. The use of chemical markers such as tetracycline, acetozolamide or alizarin to time-mark otoliths and scales (Radtke 1989).
5. The exchange of materials used for determining age among specialists (Kock, in press).

Developments in all these fields are continuing and they will provide a means by which the age of antarctic fish can be estimated with confidence. However, a result of the exchange scheme initiated by CCAMLR was, that although the results demonstrated good agreement for determining the age of juvenile specimens of several species, there was poor agreement when materials from adult *Notothenia rossii, Notothenia gibberifrons, Champsocephalus gunnari* and *Pleuragramma antarcticum* were evaluated. This suggests that it remains a difficulty, even for experts, to interpret the age using otoliths and scales from longevous species. Some concern arises from the observation that, during this exchange scheme, there was little indication of better agreement among experienced readers than the less experienced investigators (Kock, in press). This is likely to be a result of the evaluations being made independently without any opportunity to confer once the initial estimates had been achieved. More successful results are achieved when samples are evaluated by two or more independent readers, who then resolve anomalous values by reexamination of the disputed results (Coggan et al. 1990)

Acknowledgments. I am indebted to Roger Coggan, Richard Price and Paul Burren who have made samples available for examination, and for constructive contributions, especially to Andrew Clarke who read and advised on earlier drafts.

References

Albrechtsen K (1968) A dyeing technique for otolith age reading. J Cons Perm Int Explor Mer 32:278–280

Andriashev AP (1987) A general review of the antarctic bottom fish fauna. In: Kullander SO, Fernholm B (eds) Proc V Congr Europ Ichthyol, Stockholm, 1985, pp 357–372

Anonymous (1980) Working party on antarctic fish biology. Rep 2nd Meeting, Dammarie-les-Lys, France, 27–31 May 1980. Biomass Rep Ser 12:1–47

Anonymous (1982) Working party on antarctic fish biology. Rep 2nd Workshop on the Ageing of Antarctic Fishes, Orono, Maine USA, 7–10 June 1982. Biomass Rep Ser 17:1–30

Anonymous (1986) Report of the workshop on antarctic fish age determination. SC-CAMLR-VII/BG/41

Bagenal TB (1974) The ageing of fish. Unwin Brothers, Old Woking, England

Barber DL, White MG, Coggan R (1986) Fish ageing; a method for decalcification and sectioning fish otoliths. In: Anonymous SC-CAMLR-VII/BG/41

Barrera-Oro ER (1989) Age determination of Notothenia gibberifrons from the South Shetland Islands, Antarctic Peninsula subarea (Subarea 48.1). SC-CAMLR Select Sci Pap 1988 2: 143–160

Barrera-Oro ER, Tomo AP (1988) New information on age and growth in length of *Micromesistius australis* Norman 1937 (Pisces, Gadidae) in the South-West Atlantic. Polar Biol 8:341–351

Bedford B (1983) A method for preparing sections of large numbers of otoliths embedded in black polyester resin. J Cons Int Explor Mer 41:4–12

Boronin AV, Altman YS (1979) Growth and natural mortality of green Notothenia. (in russ). Trudy AtlantNIRO 81:66–71

Boronin AV, Frolkina ZA (1976) Age determination of the green Notothenia (*Notothenia gibberifrons* Lönnb., fam. Nototheniidae) from the SW Atlantic. (in russ). Trudy AtlantNIRO 60:29–37

Burchett MS (1983a) Age and growth of the antarctic fish *Notothenia rossii* from South Georgia. Bull Br Antarct Surv 60:45-61

Burchett MS (1983b) The life cycle of *Notothenia rossii* from South Georgia. Bull Br Antarct Surv 61:71–73

Burchett MS (1984) Age determination of *Notothenia rossii* from South Georgia (Antarctica) using scales and otoliths. Cybium 8:19–28

Burchett MS, DeVries AL, Briggs AJ (1984) Age determination and growth of *Dissostichus mawsoni* (Norman 1937) (Pisces Nototheniidae) from McMurdo Sound (Antarctica). Cybium 8:27–31

Casselman JM (1983) Age and growth assessment of fish from their calcareous structures - techniques and tools. In: Prince ED, Pulos LM, (eds) Proc Int Workshop on Age Determination of Oceanic Pelagic Fishes: Tunas, Billfishes, and Sharks. NOAA Tech Rep NMFS 8, 211 pp

Chilton DE, Beamish RJ (1982) Age determination methods for fishes studied by the groundfish program at the Pacific Biological Station. Can Spec Publ Fish Aquat Sci 60, 102 pp

Chojnaki J, Palczewski P (1981) Age and growth rate of *Pseudochaenichthys georgianus* Norman 1937 (family Chaenichthyidae) of the South Georgia region. Pol Polar Res 2:145–157

Christensen JM (1964) Burning of otoliths, a technique for age determination of soles and other fish. J Cons Int Explor Mer 29:73–81

Clarke A (1988) Seasonality in the antarctic marine environment. Comp Biochem Physiol 90B:461–473

Clasing E, White MG, Moreno CA (1985) Estudio de la edad en una poblacion virginal de *Notothenia gibberifrons* (Pisces: Nototheniidae) de Bahia South, Antartica. Ser Cient INACH 32:75–90

Coggan R, White MG (1986) Techniques used to age antarctic fish. In: Anonymous SC-CAMLR-VII/BG/41

Coggan RA, Skora K, Murray A, White M G (1990) A comparison between age determination of the antarctic fish *Notothenia gibberifrons* Lönneberg using scales and otoliths. Cybium 14:43–55

Crisp DT, Carrick SM (1975) Some observations on the growth and length: weight relationship of the South Georgia Cod *Notothenia rossii marmorata* Fischer during the first four years of life. J Fish Biol 7:407–409

Croxall JP, North AW, Prince PA (1988) Fish prey of the wandering albatross *Diomedia exulans* at South Georgia. Polar Biol 9:9–11

Daniels RA (1983) Demographic characteristics of an antarctic plunderfish *Harpagifer bispinis antarcticus*. Mar Ecol Prog Ser 13:181–187

Dannevig EH (1956) Chemical composition of the zones in cod otoliths. J Cons Perm Int Explor Mer 21:156–159

Duhamel G (1987) Distribution and abundance of fish on the Kerguelen Islands shelf. In: Kullander SO, Fernholm B (eds) Proc V Congr Europ Ichthyol, Stockholm, 1985, pp 397–403

Duhamel G, Ozouf-Costaz C (1985) Age, growth and reproductive biology of *Notothenia squamifrons* Gunther 1880 from the Indian Sector of the Southern Ocean. Polar Biol 4:143–153

Everson I (1970) The population and energy budget of *Notothenia neglecta* Nybelin at Signy Island, South Orkney Islands. Bull Br Antarct Surv 23:25–50

Everson I (1980) Age determination methods of antarctic fishes. Biomass Handbook 8:1–24

Everson I (1984) Fish. In: Laws RM (ed) Antarctic ecology. Academic Press, London, pp 491–532

Fawell JK (1974) The use of image analysis in the ageing of fish. In: Bagenal TB (ed) The ageing of fish. Unwin Brothers, Old Woking, England

Freytag G (1979) Age determination in *Notothenia rossii marmorata*. ICES CM 1979 G:11:1–7

Freytag G (1980a) Length, age and growth of Notothenia rossii marmorata Fischer, 1885 in the West Antarctic waters. Arch FischWiss 30:39–66

Freytag G (1980b) Problems in ageing *Notothenia rossii marmorata*. Cybium 4:43–51

Frolkina Z A (1977) A method for age determination in Patagonian toothfish (in russ). Trudy AtlantNIRO 73:86-93

Gorelova TA, Shandikov GA (1988) On age related changes in the food composition and diurnal feeding rhythm of *Nototheniops tchizh tchizh* in the Lena Bank area (Indian Sector of the Southern Ocean). (in russ). Vopr Ikthiol 28:434–440

Gubsch G (1980a) Studies on age determination and growth of the ice fish *Chaenocephalus aceratus* (Lönnberg). ICES CM G 47:1–12

Gubsch G (1980b) Untersuchungen zur Altersbestimmung und zum Wachstum beim Eifisch *Chaenocephalus aceratus* (Lönnberg). Fisch-Forsch 18:7–10

Gubsch G (1982) Zur Verbreitung und Biologie der Eisfische (Chaenichthyidae) im atlantischen Sektor der Antarktis. Fisch-Forsch 20:39–47

Hecht T (1987) A guide to the otoliths of Southern Ocean fishes. S Afr J Antarct Res 17:1–87

Hecht T, Cooper J (1986) Length/mass relationships, energetic content and the otoliths of antarctic cod *Paranotothenia magellanica* (Nototheniidae: Pisces) at sub-Antarctic Marion Island. S Afr J Zool 21:294–296

Hoffmann U (1982) Zur Altersbestimmung und zum Wachstum von *Notothenia gibberifrons* Lönnberg bei South Georgia. Fisch-Forsch 20:49–53

Hubold KG, Tomo AP (1989) Age and growth of antarctic silverfish *Pleuragramma antarcticum* Boulenger, 1902, from the southern Weddell Sea and Antarctic Peninsula. Polar Biol 9:205–212

Hureau J-C (1962) Etude descriptive de l'otolithe (Sagitta) de quelques téléostéens antarctiques. Bull Soc Zool France 87:533–546

Hureau J-C Ozouf-Costaz C (1980) Age determination and growth of *Dissostichus eleginoides* Smitt 1898 from Kerguelen and Crozet Islands. Cybium 3(8):23–32

Kear AM (1989) Feeding mechanisms and diet in cephalopods: special reference to antarctic mesopelagic squid. Ph. D. Thesis, Univ Aberdeen. 290 pp

Kobayashi S, Yuki R, Furui T, Kosugiyama T (1964) Calcification in fish and shell-fish I. Tetracycline labelling patterns on scale, centrum and otolith in young goldfish. Bull Jp Soc Sci Fish 30:6–13

Kochkin PN (1980a) How to prepare fish vertebrae for age determination (in russ.). Vopr Ikthiol 20:941–944

Kochkin PN (1980b) On methods of ageing icefish as demonstrated in the pike glassfish (*Champsocephalus gunnari* Lönnberg) (in russ). VNIRO Moscow 29B:1–32

Kochkin PN (1982) A study of the age and growth of *Champsocephalus gunnari* Lönnberg (Channichthyidae) (in russ). Vopr Ikthiol 22:390–400

Kochkin PN (1985) Analysis of registering structures and linear growth of the pike glassfish *Champsocephalus gunnari* Lönnberg (Chaenichthyidae) (in russ). Vopr Ikthiol 25:834–842

Kochkin KH (1980) Graphical analysis of length frequency of *Champsocephalus gunnari* Lönnberg (Channichthyidae) from South Georgia. Cybium 8:33–42

Kock KH (1981) Fischereibiologische Untersuchungen an drei antarktischen Fischarten: *Champsocephalus gunnari* Lönnberg 1905, *Chaenocephalus aceratus* (Lönnberg 1906) und *Pseudochaenichthys georgianus* Norman 1937 (Notothenioidei, Channichthyidae). Mitt Inst Seefisch Hamburg 32:1–226

Kock KH (1985) Antarctic fish. In: Bonner WN, Walton DWH (eds) Key environments. Pergamon Press, Oxford, pp 173–192

Kock KH (in press) Results of the CCAMLR antarctic fish otoliths/scales/bones system. SC-CAMLR

Kock KH, Duhamel G, Hureau JC (1985) Biology and status of exploited antarctic fish stocks: a review. Biomass Sci Ser 6:1–143

Kompowski A (1983) Studies on growth rate of yellow notothenia *Notothenia gibberifrons,* Lönnberg 1905 off South Georgia. Acta Ichthyol Piscat 13:3–24

Linkowski TB (1985) Population biology of the myctophid fish *Gymnoscopelus nicholsi* (Gilbert 1911) from the western South Atlantic. J Fish Biol 27:683–698

Linkowski TB (1987) Age and growth of four species of *Electrona* (Teleostei, Myctophidae). In: Kullander SO, Fernholm B (eds) V Congr Europ Ichthyol, Stockholm, 1985 pp 435–442

Linkowski TB, Zukowski C (1980) Observations on the growth of *Notothenia coriiceps neglecta* Nybelin and *Notothenia rossii marmorata* Fischer in Admiralty Bay (King George Island, South Shetland Islands). Pol Polar Res 1:155–162

Mason JE (1974) A semi-automatic machine for counting and measuring circuli on fish scales. In: Bagenal TB (ed) The ageing of fish. Unwin Brothers, Old Woking, England, pp 87–102

Moksness E, Campana SE (1990) The intercalibration exercise on otolith microstructure. ICES CM (in press)

Mugiya Y (1977) Effect of acetazolamide on the otolith growth of goldfish. Bull Jpn Soc Sci Fish 43:1053–1058

North AW (1988) Age of antarctic fish: validation of the timing of annuli formation in otoliths and scales. Cybium 12:107–114

North AW, White MG, Burchett MS (1980) Age determination of antarctic fish. Cybium 8:7–11

North AW, Croxall JP, Doidge DW (1983) Fish prey of the antarctic fur seal *Arctocephalus gazellae* at South Georgia. Bull Br Antarct Surv 61:27–37

North AW, Burchett MS, Gilbert CJ, White MG (1984) Identification of fish from the Southern Ocean by means of otoliths. Bull Br Antarct Surv 62:83–94

Olsen S (1954) South Georgian cod, *Notothenia rossii marmorata* Fischer. Norsk Hvalfangsttid 43:373–382

Olsen S (1955) A contribution to the systematic and biology of chaenichthyid fishes from South Georgia. Nytt Mag Zool Oslo 3:79–93

Pannella G (1971) Fish otoliths: daily growth layers and periodic patterns. Science 173:1124–1127

Pannella G (1974) Otolith growth patterns: an aid in age determination in temperate and tropical fishes. In: Bagenal TB (ed) The ageing of fishes. Unwin Brothers, Old Working, England, pp 28–39

Pentilla J, Dery LM (1988) Age determination methods for northwest Atlantic species. NOAA Tech Rep NMFS 72:1–135

Prince ED, Pulos LM (1983) Proceedings of the International Workshop on Age Determination of Oceanic Pelagic Fishes: Tunas, Billfishes, and Sharks. NOAA Tech Rep NMFS 8, 211 pp

Radtke RL (1987) Information incorporated in antarctic fish otoliths. In: Kullander SO. Fernholm B (eds) V Congr Europ Ichthyol, Stockholm, 1985, pp 421–425

Radtke RL (1990) Age determination of the antarctic fishes *Champsocephalus gunnari* and *Notothenia rossii* from South Georgia. Polar Biol 10:321–327

Radtke RL, Hourigan TF (1990) Age and growth of the antarctic fish *Nototheniops nudifrons*. US Fish Bull 88:557–571

Radtke RL, Targett TE (1984) Rhythmic structural and chemical patterns in otoliths of the antarctic fish *Notothenia larseni*: their application to age determination. Polar Biol 3:203–210

Radtke RL, Targett TE, Kellermann A, Bell JL, Hill KT (1989) Antarctic fish growth: profile of *Trematomus newnesi*. Mar Ecol Progr Ser 57:103–117

Rauck G (1975) A new technique for sawing otoliths. ICES CM 1975/F:23

Richter H, McDermott JG (1990) The staining of fish otoliths for age determination. J Fish Biol 36:773–780

Schneppenheim R, Freytag G (1980) Age determination by staining otoliths of *Notothenia rossii marmorata* with ninhydrin. Cybium 4:13–15

Shcherbich LV (1975) On the age reading methods and the onset of maturity in marbled notothenia *Notothenia rossii marmorata*. (in russ). Vopr Ikthiol 15:94–100

Shcherbich LV (1976) Length weight composition and growth rate of marbled notothenia. (in russ). Trudy AtlantNIRO 65:151–159

Shust KV, Pinskaya IA (1978) Age and growth of six species of nototheniid fish (family Nototheniidae). (in russ). Vopr Ikthiol 18:837–843

Skora KE (1980) Changes in the composition of *Notothenia gibberifrons* Lönnberg population of the shelf of South Georgia in the years 1977–1979. Pol Polar Res 1:155–162

Slosarczyk W (1987) Contribution to the early life history of Channichthyidae from the Bransfield Strait and South Georgia (Antarctica). In: Kullander SO, Fernholm B (eds) V Congr Europ Ichthyol, Stockholm, 1985 pp 427–433

Sosinski J (1981) Biologia porownawcza Kerguelen (*Champsocephalus gunnari*) z rejonow Antarktyki. Bull Morsk Inst Ryb Gdynia 48:1–91

Targett TE, Radtke RL (1984) Growth and feeding ecology studies on coastal antarctic fishes. Antarct J US 19:147–149

Tesch FW (1971) Age and growth. In: Ricker WE (ed) Methods for assessment of fish populations in fresh waters. Int Biol Handbook 3, Blackwell, Oxford, pp 98–130

Tomo AP, Cordero Funes JR (1972) Edad y crecimiento en largo del *Harpagifer bispinis antarcticus*. Contrib Cient Inst Antart Argent 96:1–19

Townsend DW (1980) Microstructural growth increments in some antarctic fish otoliths. Cybium 8:17–22

Tsukamoto K, Kuwada H, Hirokawa J, Oya M, Sekiya H, Fujimoto H, Imaizumi K (1989) Size-dependent mortality of red sea bream, *Pagrus major*, juveniles released with fluorescent otolith-tags in News Bay, Japan. J Fish Biol 35 Suppl A:59–69

White MG (1977) Ecological adaptations by antarctic poikilotherms to the polar marine environment. In: Llano GA (ed) Adaptations within antarctic ecosystems. Gulf, Houston, pp 197–208

Williams T, Bedford BC (1974) The use of otoliths for age determination. In: Bagenal TB (ed) The ageing of fish. Unwin Brothers, Old Woking, England, pp 114–123

Wohlschlag DE (1962) Antarctic fish growth and metabolic differences related to sex. Ecology 43:589–597

Yukhov VL (1971) Otolith structure in the antarctic and patagonian blennies *Dissostichus mawsoni* and *D. eleginoides* family Nototheniidae of the Antarctic Ocean. (in russ). Vopr Ikthiol 11:587–594

Zur R (1978) The age and growth of *Trematomus bernacchii* and *Trematomus centronotus* at White Island, Antarctica. B.Sc. Thesis Univ Canterbury, Christchurch New Zealand, pp 1–110

Microstructural Analysis of Growth Patterns in the Early Life History of Antartic Fishes

R.L. Radtke and A. Kellermann

1 Introduction

Fundamental to responsible programs of biological resource management is a thorough understanding of the life-history parameters that characterize the population. Age-structured models of fish stocks are widely utilized in the analyses of fisheries management options. These models can include age-specific natural mortality rates, fecundity, growth, and vulnerability to fishing and stock-recruitment relationships. Such age-structured models have been used in management schemes to provide strategies for optimal harvesting. Management of fisheries often need to explore the relationship between the rate of harvesting and age-specific vulnerability. Age and growth are directly related; once ages are determined, average growth rates can be calculated. From growth rates, it is then possible to utilize age structure models to estimate cohort survival rates, i.e., mortality. Since fish stocks and recruitment are related in their overall population parameters, any advances in knowledge in these problem areas will enhance the accuracy of predicted yields and improve management of fish stocks.

Although the amount of data on population dynamics has increased during recent years, limited information exists on the life-history dynamics of many antarctic fish species. In their review of the biology and status of exploited antarctic fish stocks Kock et al. (1985) recognized that there are significant deficiencies in the information necessary for stock assessment and fisheries management. This refers especially to the data base relevant to stock assessment, e.g., to age structure of harvested fish populations (Kock et al. 1985).

The influence of environmental change on the growth and mortality of larval fish, and thus on year-class strength, is one of the more intriguing and prominent problems in fishery science. The larval period is considered to be the most critical stage in the survival of most fish (Hjort 1914), with numerous factors affecting survival and growth. Some of these factors are food (Cushing 1975), predation (Murphy 1967), temperature (Laurence 1973; Houde 1974) and ocean currents (Sette 1943). The first two of these factors are density dependent and tend to regulate population size. Temperature may have a pronounced effect on larval growth (Laurence 1973; Houde 1974). It can affect larval survival directly and indirectly, e.g., by reducing growth (smaller size at age), which can affect predation

School of Ocean and Earth Science and Technology, Hawaii Institute of Geophysics, University of Hawaii at Manoa, 2525 Correa Road, Honolulu, Hawaii 96822, USA vulnerability. In the Antarctic seas, however, temperature fluctuates little and has a potentially small influence on larval growth rates. Transport by ocean currents can carry eggs and larvae to or from areas favorable to survival. In antarctic fish, biotic and physical environments of the early life history are known for only a few species, and studies on the larval drift and juvenile migrations are needed especially for the commercially harvested species (Kellermann 1990a).

One of the foremost reasons for investigating the effects of environmental variations on larval fish survival is to understand what has happened and to use this knowledge as the basis for predicting what may happen to fish populations. To accomplish this task, it is necessary to determine the age structure of the larval fish populations and to follow their growth related to the environmental variations to which the larvae have been exposed.

Otoliths appear to be the most suitable structure from which to determine the age, as in larval fish they are the only hard parts with recognizable growth patterns and they are not susceptible to resorption in adult fish, unlike scales (Mugiya and Watabe 1977). Also, they undergo little alteration once formed, and they can be used to estimate the age of fish species which lack scales, such as channichthyids.

In antarctic fish, a variety of hard body parts have been used in age-estimation studies. These include annual marks on ossified structures such as vertebrae (Kochkin 1986), the spine of the pelvic fin (Gubsch 1980; Hoffmann 1982), scales (Duhamel and Ozouf-Costaz 1985; Freytag 1980), and annual structures of otoliths (e.g., Olsen 1954, 1955; Kock 1981; Clasing et al. 1985). Microstructures of otoliths and elemental ratios have been used to determine the age of adult antarctic fish (Townsend 1980; Radtke and Targett 1984). In some cases, a combination of different structures has been used (Kochkin 1986), but there is a need of more efforts to compare the different methods of estimating the age of Antarctic fish (Kock et al. 1985).

In addition to permitting age determination, otoliths can provide further insights in respect to Antarctic larval fish biology and ecology. Otoliths have been employed to discover environmental histories (Radtke 1989; Radtke et al. 1989, 1990) and migration patterns (Radtke et al. 1988; Radtke and Morales-Nin 1989). In the present investigation, otolith microstructures are analyzed for life-history changes in larval and juvenile antarctic fish, and increment width determined.

2 The Structure and Function of Fish Otoliths

Otoliths are gravity receptors, which are calcium carbonate concretions situated in the membraneous labyrinth of teleost fish (Lowenstein 1971; Popper and Coombs 1982). There are three otoliths, the sagitta, the asteriscus, and the lapillus, on each side of the brain area, but only the largest otolith, usually the sagitta, has been utilized for most studies. The other two otoliths (the asteriscus and the lapillus) may also contain valuable ecological information (e.g., Radtke 1987), but need more investigation. The otoliths of teleost fish are complex polycrystalline bodies which serve as balance organs in the inner ear (Carlstrom 1963). They are composed chiefly of crystallized calcium carbonate in the form of aragonite and of a collagenous, fibrous protein (Degens et al. 1969). Otolith growth occurs through the accretion of new material over its surface. This growth is cyclic and is a function of daily calcium metabolism rhythms and amino acid synthesis cycles geared by neuroprotein secretion that result in the formation of daily growth units or increments (Gauldie and Nelson 1988; Mugiya 1987; Tanaka et al. 1981). Each unit is made up of a continuous or incremental zone, deposited during the active period of the calcium metabolism, and a discontinuous zone (Pannella 1974). The incremental zone is mainly composed of aragonite microcrystals laid down transversally to the surface of the otolith, embedded in the organic matrix. The discontinuous unit is principally composed of the organic matrix laid down in concentric layers over the surface of the otolith. The size of the crystals and the width of the incremental units are linked to the stage of growth. Broad incremental units are laid down during active growth; during periods of slow growth, the increments are thinner and with compact microcrystals (Moralen-Nin 1987). Frequently two or more subunits are present in the increments, probably in relation to feeding periodicity, temperature changes, migrations, etc. (Panella 1980; Jones 1985a).

Otolith growth and body growth may be closely correlated, and the thickness of the otolith microincrements decreases with age following the general trend of a reduced growth rate in length of the organism. In addition, metabolic rate, stress, spawning, etc. may affect the growth rate and thus may be reflected in the increments laid down in the otoliths.

The daily information of the increments appears to be linked to an endocrine-driven, endogenous circadian rhythm. This circadian rhythm is entrained at an early age by photoperiod or other daily environmental clues (Campana and Neilson 1985). The entrainment to a clue requires that the clue periodicity does not differ by more than 2-4 h in 24 h, or that it consists of a cycle of multiple harmonics of 24 h. Only one environmental variable may act as the entraining factor, although others may act to mask an endogenous rhythm. Lunar and semi lunar deposition patterns have been observed in otoliths in a variety of species, including antarctic fish (e.g. Panella 1974; Kochkin 1986; Fig. 1). These periodicities are poorly understood, however.

Many authors have used increments for age determination (Brothers et al. 1976; Campana and Neilson 1985) in larvae and juvenile fish. Other authors have applied the increments to study daily growth changes (Gutiérrez and Morales-Nin 1987), detect life transitions (Brothers and McFarland 1981), estimate recruitment and mortality (Methot 1983), and in taxonomic studies (Brothers 1984).

3 Internal Otolith Morphology of Antarctic Fish Larvae

The otoliths are usually the first calcified bodies which appear during early ontogeny. The largest otoliths, the sagitta and lapillus, are present at hatching in most teleost fish, while the asteriscus appears later (Brothers 1984). In laboratory reared *Notothenia neglecta*, the sagitta and lapillus were found in embryos shortly before hatching of their siblings. They were of similar size and shape and could only be discriminated by their position in the sacculi. The central area of the sagitta may contain a variable number of primordia which form during embryogenesis. These primordia may fuse and become the starting point of growth of the nucleus, or

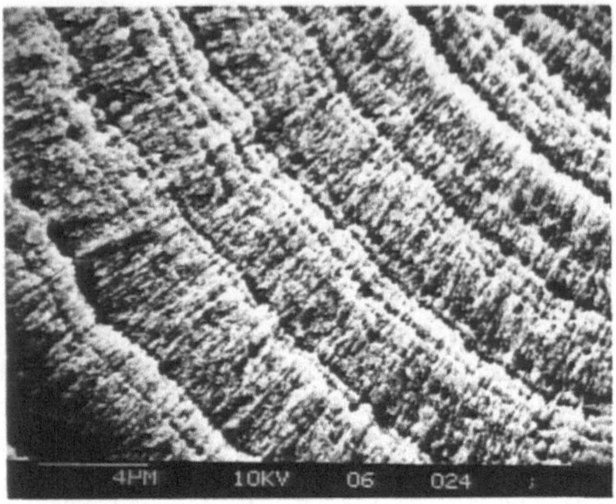

Fig. 1. Scanning electron micrograph of microincrements in the sagitta of a *Nototheniops kempi* larva, 32.2 mm (fresh) standard length. Semilunar periods are apparent in the deposition of daily growth rings, and indicated by the **light rings**, separated by 7 increments each

single primordia may give rise to several nuclei (Brothers 1984). In Antarctic fish, multiple nuclei have not been observed to date, and sagittal otoliths show a single, well-defined central core area. However, SEM micrographs of sagittal otoliths of *Chionodraco rastrospinosus* larvae revealed numerous small primordia of 1-2 microns in size (Fig. 2). The size of the central core area or nucleus varies among species (Table 1). The low intraspecific variation of the nucleus diameter suggests that it is species specific (Table 1), and it is therefore assumed that the nucleus edge

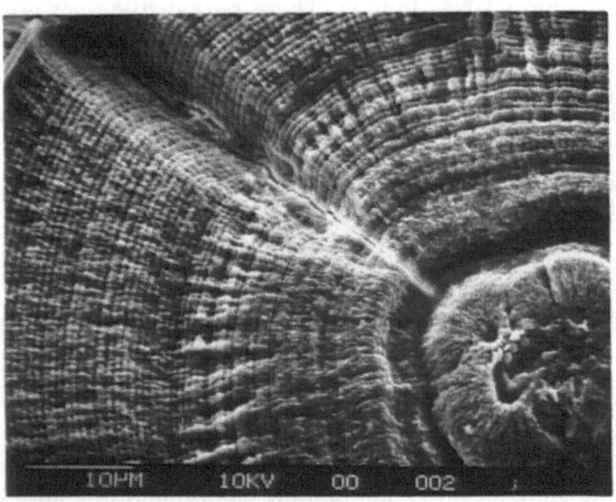

Fig. 2. Partial view of a sagittal otolith of a larval *Chionodraco rastrospinosus*, 27.2 mm (fresh) standard length. The SEM picture shows a distinct nucleus, in the central area of which numberous sphaerical primordia are visible. Radiating outward from the core are narrow microincrements of less than 1 μm in width

Table 1. Measurements of otolith structures and counts of microincrements (± SD) revealed by scanning electron and light microscope techniques from sagittae of four notothenioid fish[a]

Species	Diameter of nucleus (μm)	Radius to 2nd check (μm)	Increments to 2nd check	N[c]	Larval lengths (mm)
Trematomus newnesi	43–45	100	20	6	9.9–16.5
Nototheniops nudifrons[b]	106 ± 4	279	24[b]	18	7–38
Pseudochaenichthys georgianus	53 ± 5	109 ± 6	48 ± 4	20	17–23
Pagetopsis macropterus	53 ± 6	94 ± 6	62 ± 10	8	36–41

[a]Larval lengths are fresh standard lengths.
[b]All fish were caught in the field, except the *Nototheniops nudifrons*.which were reared from eggs spawned in captivity.
[c]N is the number of larvae studied.

marks the time of hatching. Hence the concentric rings radiating outward from the nucleus were laid down by the free-swimming larva.

Increment deposition in the larvae of teleost fish has been observed to occur in the egg, but in most species investigated, initial daily rings formed some time after hatching (Jones 1985a). Analysis of the sagitta of laboratory reared *Nototheniops nudifrons* showed that initial daily increment formation took place the day after hatching, although the core area contained eight to ten faint rings apparently deposited by the embryo (Hourigan and Radtke 1989). Otoliths taken from embryos of reared *Notothenia neglecta* showed four to six wide rings around a central area which were laid down during incubation of the egg (Kellermann and Radtke, unpub. data).

In addition to the distinct check as the benchmark of hatching, a second check can be regularly observed in the otoliths of all notothenioid fish larvae investigated. This check appears as a narrow, concentric dark band some distance from the center when viewing the sample with transmitted light (Fig. 3). It occurs both in the sagitta and in the lapillus. SEM techniques reveal the check as a well-etched grove (Fig. 4), which indicates a low calcium carbonate content in the increment, with adjacent rings richer in protein. The origin of this check is not yet fully understood. Significant events during the larval life such as yolk absorption, first feeding, or changes in habitat have a potential influence on the deposition pattern of daily and subdaily growth increments. The impact of external factors such as poor feeding, conditions or other stress directly related to the environment (Jones 1985a) seems unlikely, as the number of increments between the nucleus and the second check shows little intraspecific variation (Table 1).

Considerable differences between species in the position of this second check were noted: in the two nototheniid species investigated, 10-20 daily rings were counted between the nucleus and the second check, whereas the number of increments was substantially higher in the channichthyids (Table 1). Yolk absorption as the trigger can be ruled out (except for *Pagetosis macropterus*), as the observed absorption times did not correspond to increment counts, or because the sagittae

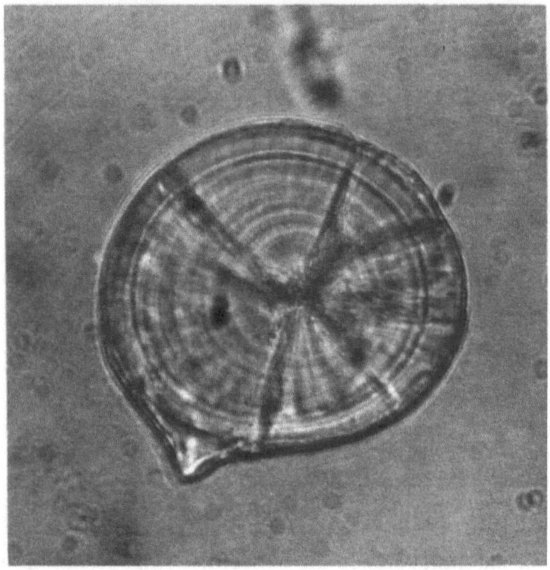

Fig. 3. Light micrograph of a sagittal otolith of a larval *Trematomus newnesi* (12 mm SL). Two *dark bands* are visible: the first surrounds the central core area as the benchmark of hatch, the second appears close to the margin (Radtke et al. 1989)

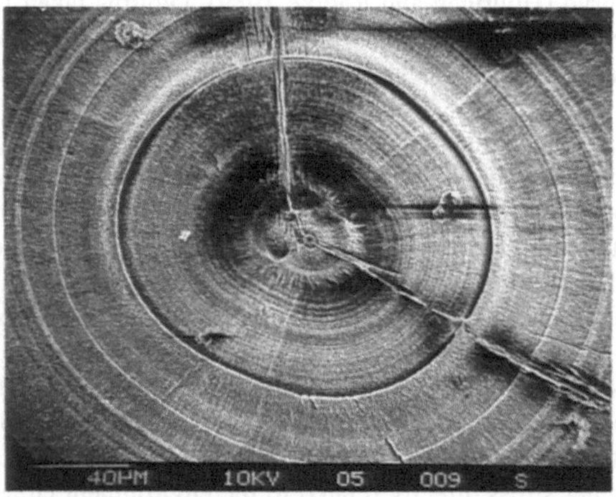

Fig. 4. Scanning electron micrograph of the central part of the sagitta of a transforming larva of *Pagetopsis macropterus* of 38.2 mm (fresh) standard length. The central core area is the nucleus of 49 μm in diameter, followed by a sequence of narrow, regular increments. The distinct, concentric second check is visible, followed by increments radiating towards the outer edge of the otolith

were dissected from the yolk-sac larvae, as in *Pseudochaenichthys georgianus*. Besides first feeding, other, intrinsic factors may be found in physiological changes that will occur during early ontogeny. It is speculated that the ossification of fin rays or skeletal bones may lead to changes of the calcium metabolism that affect otolith increment deposition. Fin ray formation was observed to occur relatively early in the genera *Pagetopsis* and *Pseudochaenichthys* (Kellermann 1990a), while pectoral or skull ossification may be responsible in the nototheniids. Little is known of the physiological processes responsible for check information. Thus, it is not clear when checks would be expected to form (Jones 1985a). However, the low intraspecific variability argues strongly in favor of intrinsic factors responsible for this check formation in notothenioid larvae.

4 Temporal Information Derived from Otolith Microincrements in Antarctic Fish Larvae

The hatching periods in the different species of Antarctic fish are spread over most of the year. Based on findings of young larvae at certain periods of the year, this has been observed in the subantarctic waters of South Georgia (North and White 1987), and in the zone of seasonal pack-ice cover of the Antarctic Peninsula (Kellerman 1986; 1989a). Through the use of microincrement techniques, it is now feasible to obtain birthday distributions of individual species from the known ages of larvae and adult fish. Otolith studies of larval and juvenile *Trematomus newnesi* have corroborated the previous assumption that hatching of this species occurs in spring (Kellermann 1986; Radtke et al. 1989). Back calculations of hatching dates from the ages of adult *Nototheniops nudifrons* resulted in an apparently long spawing and hatching period in this species, confirmed by concomitant gonad studies (Hourigan and Radtke 1989). Slosarczyk (1987) used light microscopy with otoliths of young *Chionodraco rastrospinosus* to demonstrate from the lack of an annual check that they belonged to age group 0. Kochkin (1986) counted daily increments in the otoliths of *Champsocephalus gunnari* early juveniles and concluded that they were less than one year old.

Microincrement techniques provide a more precise estimate of the hatching periods of antarctic fish than both length frequency analysis or the occurrence of smallest larvae, although the perceptions which have resulted from previous field studies are likely to be essentially correct. However, due to the greater depth of the Antarctic shelf areas, newly hatched larvae of some deep and demersal spawning fishes (e.g. *Chionodraco rastrospinosus*; Gubsch 1982) may not be recorded in plankton-tows fishing in surface waters (Kellermann 1986). Further, as a peculiar aspect of the antarctic environment, sampling of the early stages of winter hatching species may be prevented by the ice cover, although there are occasional records of these young larvae from the field studies (Regan 1916; Kellermann and Schadwinkel 1990). In such species, e.g., *Chaenodraco wilsoni*, microincrement analysis provides a readily available means to reliably estimate the age and growth of the early stages. An alternative, however incidental approach has recently been used for *Trematomus eulepidotus*, the eggs of which were collected in the field and reared in the laboratory, suggesting that hatching occurs in winter (Ekau 1989).

5 Growth Patterns in Antartic Fish Larvae

Growth curves of larvae have been established for a few nototheniid and channichtyid fish using the progression of mean or median length over the austral summer season (Keller 1983; Slosarczyk 1987; see also North, this Vol.). Typically, growth in length was curvilinear, that is growth rates were initially low and increased as the larvae develop. Growth rates varied between species with a long pelagic larval phase (e.g. in *Nototheniops larseni*, Kellermann 1986; *Trematomus scotti*, Loeb, in press), and faster growth in some icefishes (e.g., in *Chionodraco rastrospinosus*, Kellerman 1986; *Chaenocephalus àceratus*, Slosarczyk 1987). However, there are other nototheniids, e.g., *Notothenios nudifrons* and *Trematomus newnesi*, which conclude their pelagic development by the end of their first summer. Length frequency data from *Trematomus newnesi* suggest that the larvae grew fast during the first two months and then reached a plateau by the time they transformed to the juvenile stage (Radtke et al. 1989). Microincrement analysis of the larval sagittae in these species confirmed the high initial growth rates which are comparable with those of advanced channichthyid larvae (Hourigan and Radtke 1989; Radtke et al. 1989). It seems therefore that the fast growth of early larvae and the short pelagic phase are correlated, thereby reducing that part of life history that is most vulnerable to pelagic predators.

Rapid growth during the first months of life may be critical to survival. There is evidence to support this hypothesis (Brothers et al. 1983), and to contradict it (Methot 1983). Jones (1985b) found that differences in growth rate between herring larvae hatched early in the season, and those hatched later could be demonstrated through the use of otolith increments. Precise determination of growth is essential for reliable estimates of larval mortality, and to develop an understanding of the factors governing growth and survival during the various larval and post larval stages. Larval survival is obviously related to growth, as larvae become less vulnerable to predation, and increase the volume that they can search for prey as they grow. Hence, they become capable of preying upon larger and scarcer food. It has been suggested that large recruitment fluctuations can be caused by relatively small variations of growth rates and the related durations of the larval stages (Houde 1987). This would argue in favor of the growth mortality hypothesis to explain recruitment variability in fish stocks (Ware 1975).

Often the length frequencies observed are not the reflections of the age and growth of the larval fish (Radtke and Dean 1982; Radtke and Scherer 1982). Consequently, it is important that ageing techniques be as accurate as possible. The use of otolith microincrements makes it feasible to determine the ages of larval Antarctic fish, basing on the premise that sagittal growth rings are deposited daily, as has been demonstrated for two nototheniid species (Hourigan and Radtke 1989; Radtke et al. 1989). We have applied SEM and light microscopy techniques to sagittal otoliths of larvae of the channichthyids *Pagetopsis macropterus* and *Pseudochaenichthys georgianus* (Radtke and Kellermann, unpubl. data). These were collected during the first leg of the EPOS expedition with RV "Polarstern" off the South Orkney Islands and in the northwestern Weddell Sea in October/November 1988 (Hempel 1989; Kellermann 1989b).

Fig. 5. Close-up views of the same otolith as in Fig. 4. Note the regular, narrow width of increments in the upper part of the micrograph. Large radial aragonite crystals are visible next to the nucleus

Sagittal otoliths in both species revealed a well-defined central core area, followed by a series of narrow, regular increments (Fig. 5). Sagittae of *P. macropterus* were about twice as large as those of *P. georgianus*, and microincrements counts revealed that *P. macropterus* larvae had hatched between mid April and mid May 1988, i.e., late fall/early winter. Their increments were similar inside and outside the second check (see Fig. 4),and increment width varied between 0.9 and 1.4 microns. Estimated ages from sagittae of *P. georgianus* larvae obtained from SEM (Fig. 6) and light microscopy suggested hatching dates between late August and mid-September 1988, i.e., late winter. Compared with records of smallest larvae at South Georgia (Efremenko 1979; North and Ward 1989), hatching at the South Orkney Islands appears to occur later than at South Georgia, which is consistent with the observed shift in the spawning season between these areas (Kock and Kellermann 1991).

Microincrements of *P. georgianus* larvae were deposited at variable intervals radiating from the nucleus to the second check, and increment width varied between 0.8 and 2.0 microns (Fig. 7). Beyond the check, mean increment width increased markedly to 1.8-2.6 microns (Fig. 7). Changes in increment width may be related to external factors such as feeding conditions (Campana and Neilson 1985). Hence the obvious check may well be correlated with first feeding of larvae, especially since most larvae had food in their guts. It is attractive to assume that the wider increments beyond the check might reflect enhanced larval growth, once exogenous feeding is initiated. However, the time between hatching and the second check was up to two months in *P. georgianus* (Table 1), and it seems unlikely that first feeding commences so late in larval life.

Exceptionally dark and thick discontinuous zones, called species transition zones, are the primary landmarks present in otoliths and have been utilized extensively in dating events such as growth disturbances, injuries, incidents of poor feeding, or changes in environmental conditions. If these transition zones represent

Fig. 6. SEM micrograph of microincrements in a sagittal otolith of a yolk-sac larva of *Pseudochaenichthys georgianus*, 19.5 mm (fresh) standard length. The magnified, lower portion shows the increments between the nucleus and the second check. The aragonite was removed by EDTA etching, leaving the protein ridges (*light rings*) behind

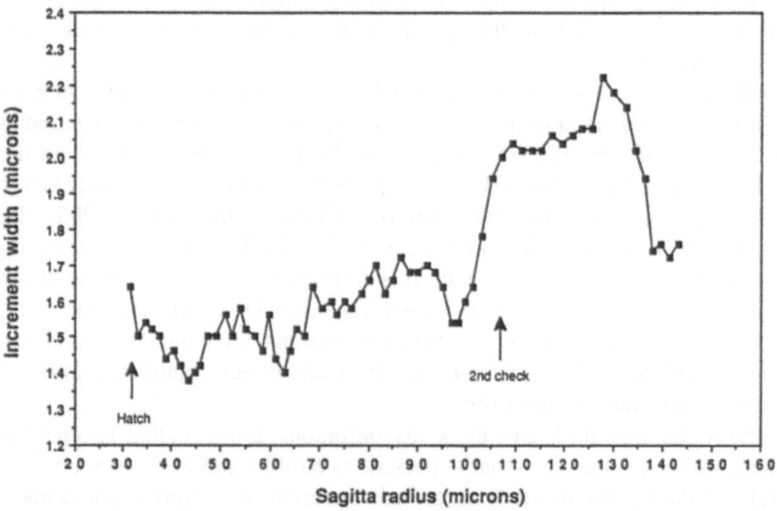

a fixed period during the day wherein calcium deposition into the organic matrix slows or ceases, as has been postulated by Watabe et al. (1982), then a transition zone is an indicator of a disturbance in the daily metabolism of the larval fish.

The highest density of organic matrix occurs when deposition is resumed after a resting phase, such as during the growth phase, which occurs once the light period has begun (Tanaka et al. 1981). This layer also corresponds to the interlamellar organic matrix as shown by Dunkelberger et al. (1980), Tanaka et al. (1981), and Radtke and Dean (1982). These authors demonstrated that discontinuous zones in otoliths contain more organic matrix than incremental layers.

From the estimated ages at larval lengths, average daily growth rates were calculated, assuming the length at hatch to be 14 mm in both species (Regan 1916; North and Ward 1989). For *P. macropterus*, the daily growth in length averaged over the winter months was 0.10–0.14 mm. For the *P. georgianus* larvae hatched in the late winter, 0.04-0.12 mm per day were estimated over a two-month period. These data have to be treated with some caution, as they were averaged over a considerable period of time. Microincrement width probably reflects instanteneous, i.e., daily, growth better (Fig. 7). The data available for the two species suggest that growth in the overwintering larvae of *P. macropterus* may be slower than in the late winter hatching *P. georgianus*.

How do these data compare with estimates derived from the length frequency analysis? The data given by Kellermann (1986) show that daily growth in the nototheniid *Nototheniops larseni* was 0.04 mm in spring and summer (November-January), and 0.12 mm in late summer (January-March). Similar rates were obtained for this species and for *Trematomus scotti* by Loeb (in press). Icefish larvae *Chionodraco rastrospinosus* grew faster at 0.07 mm per day in spring and summer, and 0.21 and 0.37mm in late summer and fall, respectively (Kellermann 1986). Similar rates were estimated for the late summer and fall growth for this and other icefishes by Slosarczyk (1987). The data obtained from otolith microincrements corroborate the initially low growth rates of larval icefish, but they are well above the estimates for early growth of nototheniid larvae.

The interspecific differences of larval growth apparent in nototheniids and channichthyids may well be related to their different habits. Larvae of *Nototheniops larseni* and *Trematomus scotti* feed on small particles (Balbontin et al. 1986; Kellermann 1990a). Food size increased rapidly in early feeding larvae of *N. larseni* due to their initially low feeding success, but median particle size in larger larvae remained virtually unchanged (Kellermann 1990b). In contrast, larval *Chionodraco rastrospinosus* commence feeding on large particle sizes such as krill furciliae and larval nototheniids, and prey size increased linearly with larval length (Kellermann 1986). Hunter (1980) concluded from prey size and volume considerations that the relatively large and rare prey in a larva's diet probably contributes most to the fish's growth, despite the fact that more small prey items are eaten. The faster growth of

Fig. 7. Increment widths (smoothed data) recorded along the sagittal radius of a larval *Pseudochaenichthys georgianus*, 20.5 mm (fresh) standard length. The arrows point to the radial distance of the second check at 107 μm, and the sagitta radius at hatch. Forty-four increments were counted from the core to the second check. The total number of increments in the sagitta of this larva was 63 channichthyids may thus be due to the larger prey consumed, but growth rates may be limited by the availability of food.

In summary, the emergence of structural techniques of otolith analysis in Antarctic larval fish now make it possible to examine the age and thus provide information about the developmental processes and ecology of the larvae. The data presented interpret the increments found in the microstructure of the otoliths as daily in occurrence and thus make it possible to obtain accurate age estimates. It is the authors' perception that daily increments can be used·accurately for age and growth determination. Otoliths are composed of calcium carbonate in an aragonite crystal form within a protein-rich matrix. They contain rhythmic incremental desposition. The information recorded in otoliths would be invaluable to population, ecological and evolutionary studies, and would supply a new level of knowledge to biologists who study fish. In paricular, these techniques hold great promise for allowing us to evaluate the origins and past environments of fish. It is possible that Antarctic larval fish otoliths can fuction as data storage units and thus provide a chronicle of a fish's physiological and ecological past. This information will become accessible with the proper techniques.

Acknowledgments. We thank J.J. Ruzicka and B. Shepherd for their help in dissections, sample preparation, and in preparing the manuscript. Some of the data presented here were collected during the European "Polastern" Study (EPOS) sponsored by the European Science Fondation and the Alfred Wegener Institute for Polar and Marine Research. The study was funded by the National Science Foundation of the U.S. grant DPP 88-16521.

References

Balbontin F, Garreton M, Neuling J (1986) Composicion del alimento y tamano de las presas en larvas de peces del estrecho Bransfield (SIBEX Fase II-Cile). Ser Cient INACH 35: 125–144

Brothers EB (1984) Otolith studies. Fish Bull US 85: 50–57

Brothers EB, McFarland WN (1981) Correlations between otolith microstructure, growth and life history transitions in newly recruited french grunts [*Haemulon flavolineatum* (Desmarest) Haemullidae]. Rapp P-V Reun Cons Int Explor Mer 178: 369–374

Brothers EB, Mathews CP, Lasker R (1976) Daily growth increments in otoliths from larval and adult fishes. Fish Bull Us 74:1–8

Brothers EB, Prince ED, Lee DW (1983) Age and growth of young-of-the-year bluefin tuna, *Thunnus thynnus*, from otolith microstructure. In: Prince ED, Pulos LM (eds) Proc Int Workshop on Age determination of oceanic pelagic fishes: tunas, bullfishes, and sharks. US Dep Commer, NOAA Tech Rep NMFS 8: 49–49

Campana SE, Neilson JD (1985) Microstructure of fish otoliths. Can J Fish Aquat Sci 42: 1014–1032

Carlstrom D (1963) A crystallographic study of vertebrate otoliths. Biol Bull 125:441–463

Clasing E, White MG, Moreno CA (1985) Estudio de la edad en una poblacion virginal de *Notothenia gibberifrons* (Pisces: Nototheniidae) de Bahia South, Antarctica. Ser Cient INACH 32: 75–90

Cushing DH (1975) Marine ecology and fisheries. Univ Press, Cambridge

Degens ET, Deuser WG, Haedrich RL (1969) Molecular structure and composition of fish otoliths. Mar Biol 2: 105–113

Duhamel G, Ozouf-Costaz C (1985) Age, growth and reproductive biology of *Notothenia squamifrons* Gunther, 1880 from the Indian Sector of the Southern Ocean. Polar Biol 4: 143–153

Dunkelberger DG, Dean JM, Watabe N (1980) The ultrastructure of the otolithic membrane and otolith in the juvenile mummichog, *Fundulus heteroclitus*. J Morphol 163: 367–377

Efremenko VN (1979) Description of the larvae of six species of the family Chaenichthyidae from the Scotia Sea. Vopr Ikhtiol 19:65–75
Ekau W (1989) Egg development of *Trematomus eulepidotus* Regan, 1914 (Nototheniidae, Pisces) from the Weddell Sea, Antarctica. Cybium 13: 213–219
Freytag G (1980) Length, age and growth of *Notothenia rosii marmorata* Fischer 1885 in the West Antarctic waters. Arc Fishwiss 30:39–66
Gauldie RW, Nelson DGA (1988) Aragonite twinning and neuroprotein secretion are the cause of daily growth rings in fish otoliths. Comp Biochem Physiol 90A:501–509
Gubsch G (1980) Untersuschungen zur Altersbestimmung und zum Wachstum beim Eisfisch *Chaenocephalus aceratus* Lönnberg. Fischerei-Forsch 20:39–47
Gubsch G (1982) Zur Verbreitung und zur Biologie der Eisfische (Chaenichthyiidae) im atlantischen Sektor der Antarktis. Fischerei-Forsch 20:39–47
Gutierrez E, Morales-Nin B (1987) Time series analysis of daily growth in *Dicentrarchus labrax* L. otoliths. J Exp Mar Biol Ecol 103:163–179
Hempel I (ed) (1989) The expedition Antarktis VII/1 and 2 (EPOS I) of RV "Polastern" in 1988/1989. Ber Polarforsch 62:185
Hjort J (1914) Fluctuations in the great fisheries of Northern Europe reviewed in the light of biological research. Rapp P-V Const Int Explor Mer 20:1–228
Hoffmann U (1982) Zur Altersbestimmung und zum Wachstum von *Notothenia gibberifrons* Lonnberg bei South Georgia. Fischerei-Forsch 20:49–53
Houde ED (1974) Effects of temperature and delayed feeding on growth and survival of larvae of three species of subtropical marine fishes. Mar Biol 26:271–285
Houde ED (1987) Fish early life dynamics and recruitment variability. Am Fish Soc Symp 2:17–29
Hourigan TF, Radtke RL (1989) Reproduction of the antarctic fish *Nototheniops nudifrons*. Mar Biol 100:277–283
Hunter JR (1980) Feeding ecology and predation of marine fish larvae. In: Lasker R (ed) Marine fish larvae. Morphology, ecology, and relation to fisheries. Washington Sea Grant Programs pp 33–77
Jones C (1985a) Detrmining age of larval fish with the otolith increment technique. Fish Bull US 84: 91–103
Jones C (1985b) Within-season differences in growth of larval Atlantic herring, *Clupea harengus harengus*.Fish Bull US 83:289–298
Keller R (1983) Contributions to the early life history stages of *Pleuragramma antarcticum* Boul 1902 (Pisces, Nototheniidae) in the Weddell Sea. Meeresforsch 30:10–29
Kellermann A (1986) Zur Biologie der Jugendstadien der Notothennioidei (Pisces) an der antarktischen Halbinsel. Ber Polarforsch 31:149
Kellermann A (1989a) The larval fish community in the zone of seasonal ice cover and its seasonal and interannual variability. Arch Fischwiss 39:89–109
Kellermann A (1989b) Fish and eggs and larvae. In Hempel I (ed) The expedition Antarktis VII/1 and 2 (EPOS I) of RV "Polastern" in 1988/1989. Ber Polarforsch pp 44–136
Kellermann A (1990a) Catalogue of early stages of antarctic nototheniod fishes. Ber Polarforsch 67: 44–136 (also published in Biomass Sci Ser 10(1989):44–136
Kellermann A (1990b) Food and feeding dynamics of larval Antarctic fish *Nototheniops larseni*. Mar Biol 105:00–00
Kellermann A, Schadwinkel S (1990) Winter aspect of the ichthyoplankton community in Antarctic Peninsula water. Polar Biol 11:117–127
Kochkin PN (1986) Analysis of the age sensitive structures and linear growth in the pike glassfish, *Champsocephalus gunnari* Channichthyidae). Vopr Ikhtiol 25:110–119
Kock K-H (1981) Fischereibiologische Untersuchungen an drei antarktischen Fischarten: *Champsocephal aceratus* (Lonnberg 1906) und *Pseudochaenichthys georgianus* Norman 1937 (Nothothenioidel, Channichthyidae). Mitt Inst Seefisch Hamburg 32:1–226
Kock K-H, Kellermann A (1991) Reproduction in antarctic fish. A review. Ant Sci 3:125–150
Kock K-H, Duhamel G, Hureau J-C (1985) Biology and status of exploited antarctic fish stock: a review. Biomass Sci Ser 6:1–143
Laurence GC (1973) Influence of temperature on energy utilization of embryonic and prolarval tautog, *Tautoga onitis*. J Fish Res Board Can 30:435–442
Loeb VJ (1990) Distribution and abundance of larval fishes collected in the western Bransfields Strait region, 1986–87. Deep Sea Res (in press)

Lowenstam HA (1963) Sr/Ca ratio of skeletal aragonites from recent marine biota at Palau and from fossil gastropods. In: Craig H, Miller SL, Wasserburg GC (eds) Isotopic and cosmic chemistry. North Holland, Amsterdam, pp 114–132

Lowenstein O (1971) The labyrinth. In: Hoar WS, Randall DJ (eds) Fish physiology, vol 5. Academic Press, New York pp 207–240

Methot RD (1983) Seasonal variation in survival of larval northern anchovy, *Engraulis mordax*, estimated from the age distribution of juveniles. Fish Bull US 81:741–750

Morales-Nin B (1987) Ultrastructure of the organic and inorganic constituents of the otoliths of the sea bass. In: Summerfelt RC, Hall GE (eds) Age and growth of fish. Iowa State Univ Press, Ames Iowa, pp 331–343

Mugiya Y (1987) Phase difference between calcification and organic matrix formation in the diurnal growth of otoliths in the rainbow trout, *Salmo gairdneri*. Fish Bull US 85:395–401

Mugiya Y, Watabe N (1977) Studies on fish scale formation and resorption II. Effect of estradiol on calcium homeotasis and skeletal tissue resorption in the goldfish, *Carassius auratus*, and the killfish, *Fundulus heteroclitus*. Comp Biochem Physiol 57A:395–401

Mugiya Y, Watabe N, Yamada J, Dean JM, Dunkelberg DG, Shimuzu M (1981) Diurnal rhythm in otolith formation in the goldfish, *Carassius auratus*. Comp Biochem Physiol 68A:659–662

Murphy GI (1967) Vital statistics of the Pacific sardine (*Sadinops sagax*) and the population consequences. Ecology 68:731–736

North AW, Ward P (1989) Initial feeding by antarctic fish larvae during winter at South Georgia. Cybium 13:357–364

North AW, White MG (1987) Reproductive stategies of antarctic fish. In: Proc V Congr Europ Ichthyol, Stockholm 1985, pp 381–390

Olsen S (1954) South Georgia cod (*Notothenia rossii marmorata* Fischer). Norsk Hvalfangstid 7, 43 Arg: 373–382

Olsen S (1955) A contribution to the systematic and biology of chaenichthyid fishes from South Georgia. Nytt Mag Zool 3:79–93

Pannella G (1971) Fish otoliths: daily growth layers and periodical patterns. Science (Wash DC) 173:1124–1127

Panella G (1974) Otolith growth patterns: an aid in age determination in temperate and tropical fishes. In: Bagenal TB (ed) Proc Symp on the Ageing of Fish. Unwin Broth Surrey UK, pp 28–39

Pannella G (1980) Growth patterns in fish sagittae. In: Rhoads DC, Lutz RA (eds) Skeletal growth of aquatic organisms: biological records of environmental change. Plenum Press New York, pp 519–560

Popper AN, Coombs S (1982) Structure and function of the auditory system in clown knifefish, *Notopterus chitala* J Exp Biol 97:225–239

Radtke RL (1984) Formation and structural composition of larval striped mullet otoliths. Trans Am Fish Soc 113:186–191

Radtke RL (1987) Age and growth information available from the otoliths of the Hawaiian snapper, *Pristipomoides filamentosus*. Coral Reefs 6:19–25

Radtke RL (1989) Strontium-calcium concentration ratios in fish otoliths as environmental indicators. Comp Biochem Physiol 92A:189–193

Radtke RL, Dean M (1982) Increment formation in the otoliths of embryos, larvae, and juveniles of the mummichog, *Fundulus heteroclitus*. Fish Bull US 80:201–215

Radtke RL, Morales-Nin B (1989) Mediterranean juvenile bluefin tuna: life history patterns. J Fish Biol 35:485–496

Radtke RL, Scherer M (1982) Daily growth of winter flounder (*Pseudopleuronectes americanus*) larvae in the Plymouth Harbor estuary. In: Bryan CF, Connors JV, Truesdale FM (eds) Fifth Annual Larval Fish Conference. La Coop Fish Res Unit, LSU Baton Rouge pp 1–5

Radtke RL, Kinzie RM III, Folsom SD (1988) Age at recruitment of Hawaiian freshwater gobies. Envirn Biol Fish 23:205–213

Radtke RL, Targett TE, Kellermann A, Bell JL, Hill KT (1989) Antarctic fish growth: profile of *Trematomus newnesi*. Mar Ecol Prog Ser 57:103–117

Radtke RL, Townsend DW, Folsom SW Morrison MA (1990) Stronium:calcium concentration ratios in larval herring otoliths as indicators of environmental histories. Envirn Biol Fish 27:51–61

Regan CT (1916) Larval and postlarval fishes. I. Antarctic and subantarctic fishes. Nat Hist Rep British Antarctica "Terra Nova" Expedition 1910, Zool I:125–156

Sette DE (1943) Biology of the Atlantic Mackerel (*Scomber scombrus*) of North America. Part I. Early life history, including growth, drift and mortality of the egg and larval populations. Fish Bull 50:149–237

Slosarczyk W (1987) Contribution to the early life history of Channichthyidae from the Bransfield Strait and South Georgia (Antarctica). Proc V Congr Europ Ichthyol, Stockholm 1985, pp 427–433

Tanaka K, Mugiya Y, Yamada J (1981) Effects of photoperiod and feeding on daily growth patterns in otoliths of juvenile *Tilapia nilotica*. Fish Bull US 79:459–466

Townsend DW (1980) Microstructural growth increments in some Antarctic fish otoliths. Cybium 3e Ser 8:17–22

Ware DM (1975) Relation between egg size, growth and natural mortality of larval fish. J Fish Res Board Can 32:2503–2512

Watabe NK, Tanaka J, Yamada J, Dean JM (1982) Scanning electron microscope observations of the organic matrix in the otolith of the teleost fish *Fundulus heteroclitus* (Linnaeus) and *Tilapia nilotica* (Linneaus). J Exp Mar Biol Ecol 58:127–134

The Fossil and Modern Fish Faunas of Antarctica: Evolution and Diversity

J.T. Eastman

1 Introduction

The paleontological exploration of Antartica is constrained by unusual physiographic conditions. For example, the average thickness of the ice cap is 2160 m and only 2.4% of the continent's rock surface is exposed above the ice at various coastal and mountainous sites (Drewry 1983). Antarctica has also experienced extreme tectonic displacement and accompanying paleoenvironmental change during geologic time. Nevertheless, the existing fossil record indicates that fish have inhabited waters in and around Antarctica at various intervals during the past several hundred million years. These faunas are separated by large gaps in time and are not ancestral to each other. Collectively, however, they indicate that freshwater and marine biotas diversified in the antarctic component of Gondwana at specific intervals during the Paleozoic, Mesozoic and Cenozoic eras.

The isolation and endemicity of the modern antarctic fish fauna might lead to the expectation that the preceding faunas exhibited similar characteristics. These faunas were, however, much more widely distributed than are the modern notothenioids. Only during the last few tens of millions of years have the modern antarctic fish separated from other elements of the Southern Hemisphere fauna and acquired the fascinating array of physiological and biochemical adaptations enabling life at low water temperatures. Many of the chapters in this volume deal with these aspects of the biology of antarctic fish, especially notothenioids.

As this volume is broadly focused on the biology of antarctic fish, I will adopt a similar approach in considering the evolution and diversity of antarctic fish. "Antarctic fish" will therefore denote the various faunas that have inhabited Antarctica during its geologic history. The specific objectives of this chapter are to: (1) review and update work on fossil fish faunas from the Paleozoic, Mesozoic and Cenozoic; and (2) comment on the evolution and diversification of the modern notothenioid fauna.

2 Geologic History of Gondwana and Antarctica

The vast southern continent of Gondwana provided a suitable habitat for fish at various geologic intervals before and after its fragmentation. The breakup of

---

Department of Zoological and Biomedical Sciences, Ohio University, Athens, Ohio, 45701–2979, USA

Gondwana influenced the composition and distribution of the various fish faunas occupying the waters in and around the antarctic plate. The following summary of paleogeographic events (Craddock 1982; Kennett 1982; Woodburne and Zinsmeister 1984; Zinsmeister 1987) provides a background for understanding the paleoenvironment of the fish faunas from Antarctica.

Gondwana existed at least since the late Precambrian (590 Ma) and was near the South Pole for much of this time (Parrish 1990). It remained intact for approximately 375 Ma, through the Paleozoic (Fig. 1) and into the Mesozoic (Jurassic) (Fig. 2). Antarctica, occupying a central position in Gondwana, attained a south polar position by the Cretaceous. During the Late Cretaceous (Fig. 2), South America, Antarctica, New Zealand and Australia were still continuous, although there were prominent breaks between all crustal blocks. Separation of Australia from East Antarctica may have begun by the Late Jurassic, although northward drift of Australia did not create an ocean between the two continents until about 53 Ma. Final separation took place between Late Eocene and Early Oligocene (38 Ma; Fig. 3). Separation of West Antarctica from South America occurred during the Early Tertiary, however deep-sea conditions have prevailed in the Drake Passage only since the Oligocene/Miocene boundary, approximately 25 Ma. Once Antarctica was fully isolated by seafloor spreading, the unrestricted Circum-Antarctic Current began to reach full development. By decoupling warm subtropical gyres from the continent, this current served as a barrier to heat flow and thermally isolated Antarctica. While sea ice may have formed at high latitudes as early as 38 Ma, sparse precipitation prevented major ice sheet formation until the Middle Miocene (14 Ma) (Kennett 1978, 1980, 1982).

The latest work on the Cenozoic history of the antarctic ice sheets challenges the notion that the ice cap has been static for the last 14 Ma. Harwood, Webb and colleagues (Webb et al. 1984; Harwood 1986, 1987; Webb and Harwood 1987; Pickard et al. 1988; Webb 1990; Harwood, pers. comm.) have evidence from reworked planktonic diatom assemblages and from nonfossilized wood fragments (stems and roots of *Nothofagus* sp.) indicating repeated growth and decline of ice masses on a 1-3 Ma scale over the last 14 Ma. More specifically, Harwood and Webb suggest that during periods of ice sheets retreat, large marine embayments reached deep (85° S) into the interior of the continent. These seaways moderated inland temperatures and provided an ice-free marine environment for part of the year. These relatively warm (2-6 °C) bodies of water extended across part of East Antarctica as recently as 3.1-2.5 Ma (Pliocene). Reasonably diverse coniferous vegetation developed inland during these periods. Harwood and Webb suspect that the latest expansion of the ice cap commenced 2.5 Ma during a period of gradual marine and terrestrial cooling, and that the present antarctic climate is as severe as it has ever been. Although a post-Eocene fish fauna is not represented in the fossil record, the work of Harwood and Webb implies that, in addition to coastal habitat, there was extensive inland marine habitat (> 100 m deep) available for such a fauna, if indeed such a fauna existed. It is even conceivable that modern species occupied this habitat previous to the latest advance of the ice cap.

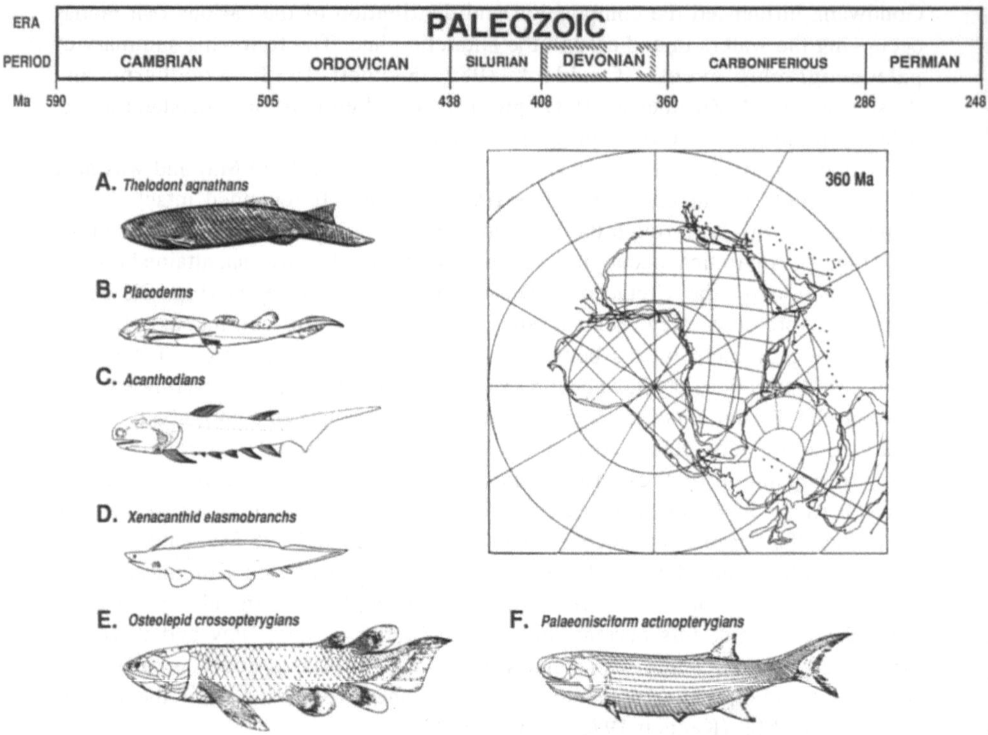

Fig. 1. Paleozoic fish from Antarctica. Map from Smith et al. (1981) shows the antarctic plate within Gondwana during the Late Devonian. *Hatched areas* in the time line indicate approximate ages for known fossils in this and subsequent figures. Most fossil fish from Antarctica are represented by incomplete specimens. Hence most of the illustrations in Figs. 1–3 are not reconstructions of Antarctic material but depict representatives of the same fossil or modern taxa from elsewhere in the world. Most of the Paleozoic fish in this figure are representative of the late Middle Devonian Aztec fauna from Southern Victoria Land (Young 1989). **A, B** and **C** from Kuhn-Schnyder and Rieber (1986); **D** from Nelson (1984); and **E** and **F** from Moy-Thomas and Miles (1971)

3 Fossil Fish Faunas

Like the fossil record for most other vertebrate groups, that for fish in Antarctica is incomplete. There is not a continuous record of evolutionary succession, and the known fossil record affords only brief glimpses of separate and unrelated faunas during the three geologic eras of the Phanerozoic (Figs. 1–3). They probably represent only a small fraction of the fish inhabiting Antarctica during its long geologic history. This brief review of fossil fish faunas is intended to update the work of Grande and Eastman (1986) and Eastman and Grande (1989).

Following Carroll (1988), a brief synopsis of the higher level taxonomy of the Agnatha and Gnathostomata may be helpful in interpreting the nomenclature in the following sections. Three of the seven extant vertebrate classes are fish: the Agnatha (jawless fish); the Chondrichthyes (cartilagenous fish); and the Osteichthyes (bony fish). The Placodermi, a class of heavily armored, jawed fish,

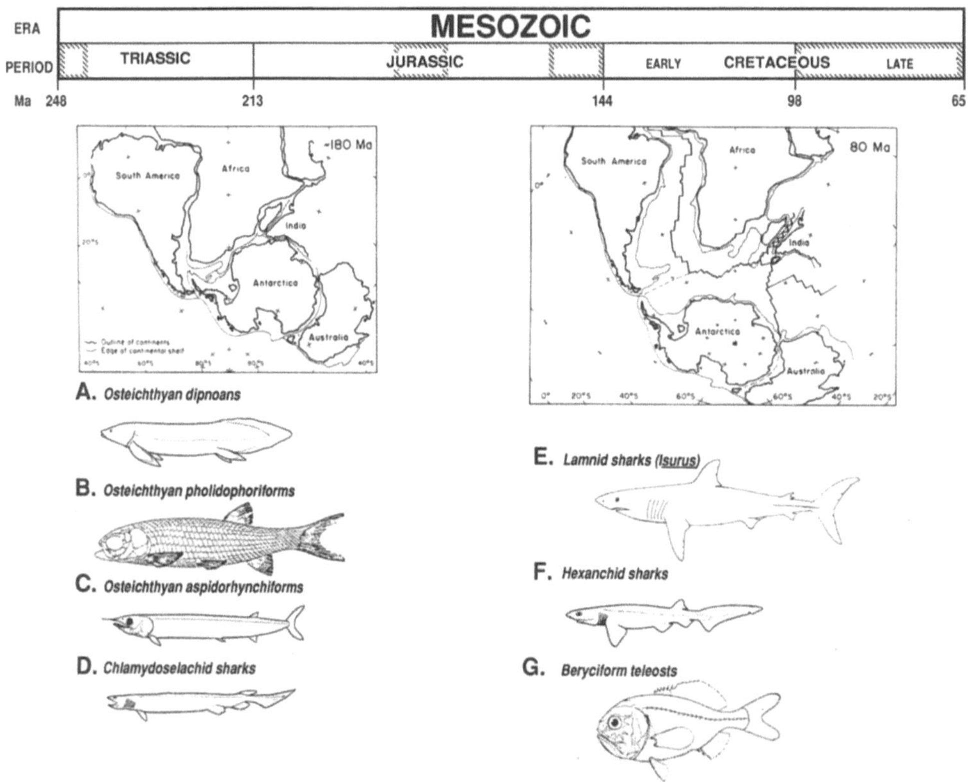

Fig. 2. Mesozoic fish from Antarctica. Middle Jurassic and Late Cretaceous maps from Elliot (1985). Represented are **A** an Early Triassic lungfish; **B** a Middle Jurassic pholidophoriform; **C** a Late Jurassic aspidorhynchiform; and **D-F** several Late Cretaceous sharks and **G** a teleost. **A, C, D** and **F** from Nelson (1984); **B** from Schaeffer (1972); **E** from Leim and Scott (1966) and **G** from Ayling (1982)

appeared, radiated and became extinct almost entirely within the Devonian. Another extinct class of jawed fish, the Acanthodii, are known from the Silurian to the end of the Early Permian. The one agnathan and four gnathostome classes first appeared during the Paleozoic and all are represented in the fossil record of Antarctica during this era.

3.1 Paleozoic

Two distinct faunas are recognizable in the Paleozoic (Fig. 1). The oldest is an Early Devonian (408-387 Ma) fauna from the Horlick Formation of the Ohio Range in West Antarctica (Young 1986). This fragmentary material consists of spines of the acanthodian *Machaeracanthus* and plates from actinolepid arthrodires, a group of placoderms. Associated brachiopods indicate that this fauna occupied a marine habitat, probably a shallow epicontinental sea. There was abundant moisture during

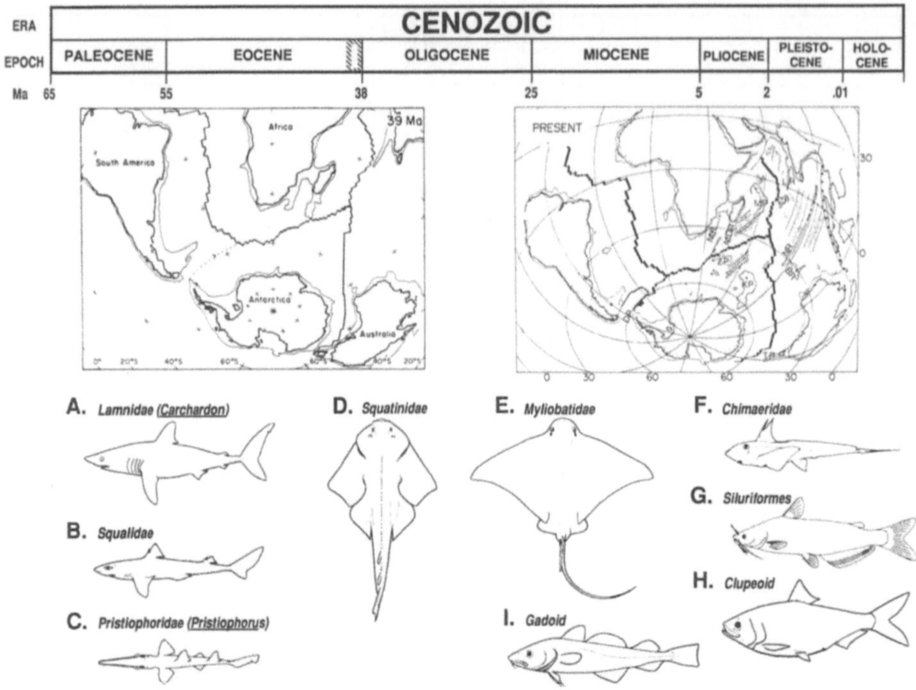

Fig. 3. Cenozoic fish from Antarctica. Late Eocene and present-day maps from Elliot (1985) and Norton (1982), respectively, show increased isolation of Antarctica over the last 39 Ma. Represented are a variety of chondrichthyans (**A-F**) and teleosts (**G-I**) from the Late Eocene La Meseta formation on Seymour Island, Antarctica Peninsula. Examples of the modern Antarctica fauna are not illustrated. **A, D, H** and **I** from Nelson (1984); **B, E** and **F** from Last et al. (1983); **C** from Smith and Heemstra (1986); **G** from Becker (1983)

this time and glaciers were probably confined to mountainous areas during the Early Devonian (Parrish 1990). While the fish specimens are too incomplete to provide significant biogeographic information, Young indicates that the brachiopods exhibit South American and South African affinities.

Young (1989) recently summarized his work on an extensive fauna containing over 30 taxa from the late Middle Devonian (380-374 Ma). Known as the Aztec fauna, these fossils were collected in freshwater deposits in Southern Victoria Land, East Antarctica. This fauna is preserved as isolated skeletal pieces like spines, plates, teeth, and scales. While there is geologic evidence for episodes of glaciation during the Paleozoic (Craddock 1982), this Middle Devonian fauna lived during an ice-free period (Tasch 1977).

Young (1989) identified these components of the Aztec fauna (Fig. 1): thelodont agnathans (*Turinia*); xenacanthid elasmobranchs (*Xenacanthus* and *Antarctilamna*); three taxa of acanthodians (*Gyracanthides*, *Antarctonchus* and *Culmacanthus*), and various osteichthyans (lungfishes, rhipidistian crossopterygians, and palaeoniscoid actinopterygians). He found that the remains of placoderms were more numerous and diverse than those of any other fish group in the Aztec fauna. Especially common were the antiarch *Bothriolepis* and the arthrodire *Groenlandaspis*, with species of *Bothriolepis* accounting for about one-third of the taxonomic diversity in the fauna. The phyllolepid *Phyllolepis* was far less common.

Although 20 endemic species are contained in the Aztec fauna, many of these are shared with eastern Australia. Young's (1989) work revealed the existence of a distinct East Gondwana vertebrate faunal province involving Australia and East Antarctica. This province persisted until at least the late Middle Devonian. Essential to interpreting the significance of the Aztec fauna in the East Gondwana Province are Young's observations that the fauna contains several taxa which are more primitive and/or of earlier occurrence in the fossil record than in Euramerican sequences. Furthermore, there are notable differences in the composition of the nonmarine vertebrate faunas of the Australian-Antarctic region versus the Euramerican region during the Early and Middle Devonian. By the Late Devonian, however, there are close similarities between the two faunas. Young suggests that this pattern may be explained by a biotic dispersal episode between Gondwana and Euramerica during the Late Devonian. This was presumably made possible by the disappearance of the ocean between Gondwana and Euramerica.

In conclusion, Young (1989) emphasizes that the fossil record does not suggest that Antarctica was the site of the initial origin and radiation of higher taxa of jawed vertebrates during the Paleozoic. At lower taxonomic levels, however, the East Gondwana province may have been a center of origin and evolution for some chondrichthyan (*Antarctilamna*) and placoderm (*Bothriolepis*) genera. Evidence for this view is based on the occurrence of the most primitive representatives of these groups in Antarctica or East Gondwana.

3.2 Mesozoic

Both freshwater and marine fossil fish are represented in the Mesozoic (Fig. 2). World climates were generally warm during this era and oceans lacked pronounced temperature gradients. There was little biotic provincialization among marine faunas (Kennett 1982).

Triassic. Colbert (1982) summarized a collection of fossil amphibians and reptiles (including *Lystrosaurus*) from the Early Triassic (248-243 Ma) Fremouw Formation of the Shackleton and Beardmore Glacier areas, Transantarctic Mountains. Included in the fossils was the detached tooth plate of a ceratodontid lungfish, possibly *Ceratodus* (Dziewa 1980), which probably lived in a seasonally hypoxic or arid habitat like extant lungfish. This fossil is significant because it indicates that lungfish had a wide distribution, including all six major continents, in the Triassic. Furthermore, it is the only known example of an osteichthyan occurring in sediments containing *Lystrosaurus*.

Jurassic. Several complete individuals of the osteichthyan pholidophoriform *Oreochima ellioti* (Fig. 2B) are known from sedimentary beds within the Kirkpatrick Basalt, Transantarctic Mountains, Victoria Land (Schaeffer 1972). Originally considered early Jurassic, the deposits bearing these fish are now thought to be early Middle Jurassic (180 ±5 Ma) according to D. H. Elliot (pers. comm.) of The Ohio State University. These fish lived in freshwater under ice-free conditions (Tasch 1977). A member of the family Archaeomaenidae, *O. elliot*i has Australian affinities, as all other species in this family are confined to that continent. Antarctica and Australia were still in contact during the Early Jurassic.

Richter and Thomson (1989) recently described *Aspidorhynchus antarcticus*, the oldest known fossil teleost from Antarctica. The specimen, consisting of pelvic and caudal regions, was collected from the late Jurassic (156-144 Ma) Nordenskjold Formation, James Ross Island, Antarctic Peninsula. Aspidorhynchids were pelagic marine fish previously unknown from deposits outside Europe. Richter and Thomson note that, although the area of James Ross Island was situated at 60° S latitude during the Late Jurassic (versus 64° S today), the climate was temperate or cool temperate.

Cretaceous. Seymour and James Ross Islands, near the tip of the Antarctic Peninsula, are the only known localities for Late Cretaceous (98-65 Ma) fish in Antarctica. The Late Cretaceous climate at this location was warm temperate, with marked seasons (Elliot 1985). Most of the fossils are elasmobranch teeth and undetermined vertebral centra. Based on 78–65-Ma teeth from Seymour Island, Grande and Eastman (1986) recognized the lamnid shark *Isurus* sp., and Grande and Chatterjee (1987) and Cione and Medina (1987) reported hexanchiform sharks. Richter and Ward (1990) recently described the oldest known frilled shark (family Chlamydoselachidae) from an 87-53-Ma deposit on James Ross Island. To date the only known teleost from the Late Cretaceous of Seymour Island is *Antarctiberyx seymouri* (Grande and Chatterjee 1987), a member of the beryciform family Trachichthyidae. Beryciforms are an extant group largely restricted to deep-sea habitats; however, Late-Cretaceous beryciforms may have been present in inshore waters, the ecological equivalents of modern perciforms (Moyle and Cech 1988).

Cretaceous paleoenvironment. As mentioned previously, parts of Gondwana have occupied a polar position throughout the Cretaceous. Therefore the fish faunas of these areas were subjected to extreme seasonality in both the light regime and in productivity. Often overlooked is the fact that pre-Cenozoic fish faunas lived under conditions that are usually considered unique to the modern antarctic marine environment. It is unlikely that these conditions restricted diversity or caused extinction. For example, Rich et al. (1988) found a diverse terrestrial biota of 150 taxa, including freshwater fish and dinosaurs, preserved as fossils along the southeast coast of Australia. Occupying a latitude of 70-80 °S, this Early Cretaceous (130-105 Ma) fauna experienced mean annual temperatures of 5 °C and marked seasonality, with several months of winter darkness.

3.3 Cenozoic (Fig. 3)

The Late Eocene La Meseta Formation (40 Ma) on Seymour Island is the only known locality for Cenozoic fossil fish in Antarctica. This fauna lived in an inshore marine habitat under temperate conditions (Woodburne and Zinsmeister 1984). Included in the fossils are a diverse array of chondrichthyans, including sharks, saw sharks, rays, and ratfish (Welton and Zinsmeister 1980; Grande and Eastman 1986; Eastman and Grande 1989). Douglas J. Long (pers. comm.) of the University of California, Berkeley, is recognizing 14 taxa in his revision of the shark fauna from the La Meseta Formation. Some Eocene sharks are identical to living genera and then, as now, sharks were most numerous and diverse in warm seas (Compagno 1988).

Artifactual preservation may be responsible for the greater perceived diversity of chondrichthyans relative to osteichthyans in the Late Cretaceous/Early Tertiary fauna from Seymour Island (Grande and Chatterjee 1987). This is because most of the fish fossils from Seymour Island are preserved as fragments, including teeth. Isolated teeth are diagnostic for both fossil and living chondrichthyans, but not for most teleosts. Therefore, the chondrichthyan fauna appears more diverse, as isolated teeth are identifiable, whereas those of teleost are assigned to indeterminate species.

At the time of the review of Grande and Eastman (1986), known teleost fossils from La Meseta Formation included only a siluriform pectoral spine, a variety of unidentifiable vertebral centra, and unidentifiable jaw bones from large and small individuals. Recent fieldwork on Seymour Island has contributed substantially to the known diversity of Late Eocene teleosts, all first occurrences for the continent. Doktor et al. (1988) provide a photograph of two nearly complete clupeoids. These specimens are being formally described by A. Jerzmanska of Wroclaw University. Jerzmanska (1988) has also described isolated gadiform vertebrae from La Meseta Formation. If this diagnosis is correct, Seymour Island is the only known locality for nonotolithic gadiform fossils in the Southern Hemisphere.

Eastman and Grande (1991) described the well-preserved dorsal aspect of a gadiform skull from another site in the La Meseta Formation. This specimen, the most completely preserved teleost cranium yet described from Antarctica, possesses a mosaic of merlucciid (hake) and gadid (cod) characters. A variety of additional teleost material from the La Meseta Formation is being described by D. J. Long (pers. comm.). Included among these fossils are at least three families of perciform teleosts previously unknown from Antarctica and also unrepresented in the modern fauna.

3.4 Do Late Cretaceous/Early Tertiary Fossil Fish Exhibit Heterochroneity?

Zinsmeister and Feldmann (1984) found that some mollusks, starfish, and crabs first appeared in Seymour Island strata that were as much as 40 Ma older than the rocks containing the descendant taxa in lower latitudes. They formulated a hypothesis accounting for the temporal disparity of these fossils from Seymour Island compared with those from lower latitudes in the Southern Hemisphere. Heterochroneity, their explanation for the disparity in time of appearance, has two central tenants. First, high-latitude (> 60° S) areas served as centers of origin for various groups. Second, these areas were also evolutionary incubators for new taxa which remained isolated until climatic conditions favored their dispersal to lower latitudes.

Recently, the Antarctic fossil record for other taxa has contributed additional support for the concept of heterochroneity. Late Cretaceous/Early Tertiary antarctic groups with descendants that subsequently dispersed to lower latitudes include some taxa of plants (Askin 1989; Dettmann 1989), lobsters (Feldmann and Tshudy 1989), amphipods (Watling and Thurston 1989), and marsupials (Case 1989).

The known fossil record for teleosts in Antarctica does not supplement the argument for heterochroneity; it is simply too incomplete for any conclusions to be drawn. There are no "first occurrences in Antarctica" among the known fossil record for teleosts. It should be noted, however, that most diagnoses to date are of higher level taxa known from partial specimens. As more complete specimens are discovered, it will be possible to make more precise diagnoses at lower taxonomic levels. Thus more biogeographic information should become available.

Seymour Island is the southernmost locality in the world with outcropping rocks spanning the Cretaceous/Tertiary boundary (Askin 1988). The Late Cretaceous and Late Eocene faunas from this site occupy a critical time in percomorph diversification and they deserve additional study.

3.5 Conclusions from the Fossil Record

The freshwater fossil fish faunas from Antarctica indicate that the continent was an integral part of Gondwana during the Paleozoic and Mesozoic. The Late Cretaceous and Late Eocene marine faunas are reasonably cosmopolitan. The modern fauna, however, is highly endemic, and unlike the Late Eocene Seymour Island fauna, the only fauna known to precede it in geologic time.

4 Modern Antarctic Fish

4.1 Absence of a Fossil Record for Notothenioids

The modern antarctic fish fauna is exclusively marine, and is less diverse than might be expected given the considerable age and large area of the ecosystem. This fauna is also markedly different in composition from the fauna that preceded it in geologic time. This is not true for other southern continents. The Tertiary marine fauna of Australia, for example, is similar to the modern fish fauna (Long 1982).

Although there have been reports of reputed notothenioid fossils since early in this century, the dominant group of modern antarctic fish are not represented in the known fossil record. Hence we can say nothing about the suspected center of origin or prior distribution of this group. In fact, there is no post-Eocene fossil record for any antarctic fish group, leaving a gap of 38 Ma between the Eocene/Oligocene boundary and the present (Fig. 3). Evidence for a transition fauna during this period has not yet come to light.

Woodward (1908) attributed isolated vertebral centra from Eocene deposits on Seymour Island to the Nototheniidae. Since centra are not diagnostic for most nototheniids, it is doubtful that his identification is correct. Stinton's (1957) report of *Notothenia* from the Miocene of New Zealand is also a misidentification (Fordyce 1982). Unfortunately, Stinton's listing has been incorporated into the major English language textbooks of vertebrate paleontology (Romer 1966, p. 360; Carroll 1988, p. 610).

Since they have no fossil record, we cannot be certain how long notothenioids have been associated with Antarctica. However, knowing the fossil record for some other teleosts, we have approximate guidelines as to the time of appearance of notothenioids. For example, many of the modern percomorph families probably evolved rapidly during the Late Cretaceous (98–65 Ma) with many modern genera appearing during the Eocene (55–38 Ma) (Carroll 1988). Hence the supposition for most of this century has been that notothenioids evolved in situ on the margins of the Antarctic Continent, while gradually adapting to cooling conditions during the Tertiary (Regan 1914; Norman 1938; DeWitt 1971). With the exception of a time frame for Southern Ocean paleotemperatures (Kennett 1978, 1982; see Clarke 1983 for review) and historical biogeographic scenarios based on plate tectonics (Andersen 1984; Miller 1987), no new information has been brought to bear on the evolution of notothenioids.

4.2 Composition of the Fauna

In summarizing information concerning the distribution and endemism of modern antarctic fish, DeWitt (1971) notes that the fauna comprises 120 species and is dominated by about 80 species of the perciform suborder Notothenioidei. Since 1971, many new taxa have been described. Over 200 species of bottom fish are now recognized (with about 100 notothenioids in the antarctic region), and 28 families are represented (Andriashev 1987). There are an additional 75 mesopelagic species. Both DeWitt and Andriashev also note that endemism is high within the Notothenioidei; 97% for species and 85% for genera. In *Fishes of the Southern Ocean*, Gon and Heemstra (1990) recognize 72 species representing 49 families from waters south of the Antarctic Convergence.

Although the recent discovery of new nonnotothenioid species has reduced the numerical dominance of notothenioids in the fauna (Andriashev 1987), they still constitute about 53% of the species in the antarctic region. Thirty-four species of snail fish (Liparididae) and 19 species of eelpouts (Zoarcidae), families of North Pacific origin, are prominent among non-notothenioids (Andriashev 1987). Other typical boreal fish such as cods, herrings, salmons, smelts, sculpins, and flatfish are either absent or poorly represented in the antarctic region.

Eastman and Grande (1989) evaluated factors that might have contributed to the paucity of nonnotothenioid osteichthyans in the modern fauna. In agreement with Clarke (1983), they concluded that low water temperature was not paramount, and that factors in the realm of ecological constraints were probably at least as important in restricting diversity. They cited limited shallow water habitat on the continental shelf and seasonal oscillation in the food supply as plausible ecological factors.

4.3 Why so Few Chondrichthyans in the Modern Fauna?

From Late Cretaceous and Late Eocene fossils at Seymour and James Ross Islands, it is obvious that many chondrichthyan genera were then present in essentially modern form in the temperate waters near the Antarctic Peninsula. However, the group did not become well established in the modern fauna and are today represented by two species of sharks and five species of skates (Fischer and Hureau 1985). There is no obvious physiological reason for their scarcity. The chondrichthyan osmoregulatory strategy employs high concentrations of urea and trimethylamine oxide in the tissue fluids. Chondrichthyans are therefore slightly hyperosmotic to their environment. The freezing point depression of the blood serum of *Squalus acanthias* is –1.95 °C (Watts and Watts 1974). Since seawater freezes at –1.86 °C, this would seem to confer adequate protection against freezing, provided there is no contact with ice. In addition, the melting point of squalene, a buoyancy agent in many sharks, is lower than –20 °C (Handbook of Chemistry and Physics, 69th edn., p. C-494), hence it is fluid at ambient temperatures in the Southern Ocean.

Perhaps the scarcity of chondrichthyans in the modern fauna is an ecological consequence of the unusual trophic conditions in the Southern Ocean (Grande and Eastman 1986). More than one-half of the 800 species of modern chondrichthyans are batoids (Compagno 1988). These are generally most abundant in water less than 1000 m deep (Moyle and Cech 1988). Shallow continental shelf habitat is limited in Antarctica, and the benthos consists largely of inedible filter feeding invertebrates (Hedgpeth 1969). Mollusks, an important food group for many rajids, exhibit reduced species richness in the Antarctic compared to the tropics or even to New Zealand (Dell 1972; Clarke and Crame 1989). Similarly, the reduced diversity of teleost fish in the antarctic midwaters may have restricted the entry of sharks into the ecosystem.

4.4 Diversification of Notothenioids

The antarctic continental shelf is 400–500 m deep at the edge and also contains innershelf depressions 1000 m deep (Andriashev 1965). When compared to the depth distribution of temperate fish, species diversity among notothenioids is greatest at 300–600 m rather than at 100–200 m (Andriashev 1965, 1987; DeWitt 1971). This pattern of distribution, known as glacial submergence, may be attributable to destruction of bottom habitat by continental ice sheets and ice shelves (Andriashev 1987). As these sheets were ground on the shelf during the times of maximum glaciation, habitat and faunal destruction could have caused the local extinction of most of the Tertiary fish fauna. The resulting ecological void was filled by a notothenioid fauna tolerant of deep-water conditions.

Most notothenioids are bottom fish confined to water less than 1000 m deep, although the depth range of individual species may be considerable (DeWitt 1971). They lack swim bladders, are usually denser than seawater (Eastman and DeVries 1982), and commonly feed and reproduce on the substrate. There is, however, a documented trend toward the diversification of a few neutrally buoyant species that are permanent members of the midwater community (Eastman and DeVries 1981; Eastman 1985).

In an ecological sense, the Southern Ocean is probably underutilized by fishes and could theoretically support more species. The waters south of the Antarctic Convergence are productive during the summer, but contain relatively few non-notothenioid fish. Thus reduced competition and the isolation of Antarctica have provided the opportunity for speciation within this group. Notothenioids fill ecological roles normally occupied by taxonomically diverse fish in temperate waters (Eastman 1991).

5 Concluding Remarks

Like the continent itself, the fish faunas of Antarctica have changed remarkably through geologic time. Continuing work in paleoichthyology, geology and sytematic biology will expand our fragmentary conceptions concerning the origins, relationships and diversity of both fossil and modern faunas. Current information indicates that the various fossil faunas are: (1) different in taxonomic composition and paleoecology; (2) separated by large gaps in geologic time; and (3) not ancestral to each other. Paleozoic fish are represented by most higher taxonomic groups in existence at that time. Mesozoic fish are not known from extensive faunal associations, but rather as single taxa from sites separated by distance and geologic time. A reasonably diverse chondrichthyan fauna inhabited temperate waters near the Antarctic Peninsula during the Late Cretaceous and Late Eocene, but there are only about seven species in the modern fauna. Continuing work on Seymour Island fossils is expanding the known diversity of the late Eocene teleostean fauna. However, the known Seymour Island fauna does not share any taxa with the modern fauna.

Unlike the situation in the shelf waters of the other continents, the modern Antarctic fauna is dominated by a single endemic group: the notothenioids. Notothenioids probably first appeared in the early Tertiary and began to diversify in the middle Tertiary while the climate was cooling. Unfortunately, there is no paleontologic and insufficient systematic data to allow more precise refinement of the time frame for the evolution of this group. The radiation of notothenioids probably indicates that the Southern Ocean is, in an ecological sense, underutilized by fish and theoretically capable of supporting more species.

Acknowledgments. I am grateful to the Italian National Programme for Antarctic Research (PNRA) for a travel grant allowing my attendance at the Ravello meeting. I also thank Guido di Prisco, Bruno Maresca and Bruno Tota for their efforts in organizing a successful conference. This work was supported by funds from the Ohio University College of Osteopathic Medicine. Danette Pratt prepared Figs. 1-3. Linda Watkins typed the manuscript.

References

Andersen NC (1984) Genera and subfamilies of the family Nototheniidae (Pisces, Perciformes) from the Antarctic and Subantarctic. Steenstrupia 10:1–34

Andriashev AP (1965) A general review of the antarctic fish fauna. In: van Oye P, van Mieghem J (eds) Biogeography and ecology in Antarctica, Monographiae Biologicae, vol XV. Junk, The Hague, pp 491–550

Andriashev AP (1987) A general review of the antarctic bottom fish fauna. In: Kullander SO, Fernholm B (eds) Proc V Congr Europ Ichthyol, Stockholm 1985. Swed Mus Nat Hist Stockholm, pp 357–372

Askin RA (1988) The palynological record across the Cretaceous/Tertiary transition on Seymour Island, Antarctica. In: Feldmann RM, Woodburne MO (eds) Geology and paleontology of Seymour Island, Antarctic Peninsula. Geol Soc Am Mem 169. Geol Soc Am, Boulder, Colorado, pp 155–162

Askin RA (1989) Endemism and heterochroneity in the late Cretaceous (Campanian) to Paleocene palynofloras of Seymour Island, Antarctica: implications for origins, dispersal and palaeoclimates of southern floras. In: Crame JA (ed) Origins and evolution of the antarctic biota. Geol Soc Spec Publ 47, Geol Soc Lond, pp 107–119

Ayling T (1982) Collins guide to the sea fishes of New Zealand. Collins Auckland, 343 pp

Becker GC (1983) Fishes of Wisconsin. Univ Wisconsin Press, Madison, 1052 pp

Carroll RL (1988) Vertebrate paleontology and evolution. Freeman New York, 698 pp

Case JA (1989) Antarctica: the effect of high latitude heterochroneity on the origin of Australian marsupials. In: Crame JA (ed) Origins and evolution of the Antarctic biota. Geol Soc Spec Publ 47, Geol Soc Lond, pp 217–226

Cione AL, Medina FA (1987) A record of *Notidanodon pectinatus* (Chondrichthyes, Hexanchiformes) in the Upper Cretaceous of the Antarctic Peninsula. Mesozoic Res 1:79–88

Clarke A (1983) Life in cold water: the physiological ecology of polar marine ectotherms. Oceanogr Mar Biol Ann Rev 21:341–453

Clarke A, Crame JA (1989) The origin of the Southern Ocean marine fauna. In: Crame JA (ed) Origins and evolution of the antarctic biota. Geol Soc Spec Publ 47, Geol Soc Lond pp 253–268

Colbert EH (1982) Triassic vertebrates in the Transantarctic Mountains. In: Turner MD, Splettstoesser JE (eds) Antarctic research series, vol 36. Geol Central Transantarctic Mountains. Am Geophys Union, Washington, pp 11–35

Compagno LJV (1988) Sharks of the order Carcharhiniformes. Univ Press, Princeton, 486 pp

Craddock C (1982) Antarctica and Gondwanaland. In: Craddock C (ed) Antarctic geoscience. Int Union Geol Sci Ser B, No. 4 Univ Wisconsin Press, Madison, pp 3–13

Dell RK (1972) Antarctic benthos. In: Russell FS, Yonge M (eds) Advances in marine biology, vol 10. Academic Press, London, pp 1–216

Dettmann ME (1989) Antarctica: Cretaceous cradle of austral temperate rainforests? In: Crame JA (ed) Origins and evolution of the antarctic biota. Geol Soc Spec Publ 47, Geol Soc Lond, pp 89–105

DeWitt HH (1971) Coastal and deep-water benthic fishes of the Antarctic. Antarct Map Folio Ser 15:1–10

Doktor M, Gazdzicki A, Marenssi SA, Porebski SJ, Santillana N, Vrba AV (1988) Argentine-Polish geological investigations on Seymour (Marambio) Island, Antarctica, 1988. Pol Polar Res 9:521–541

Drewry DJ (1983) Antarctica: glaciological and geophysical folio. Scott Polar Res Inst, Cambridge, England, 9 sheets

Dziewa TJ (1980) Note on a dipnoan fish from the Triassic of Antarctica. J Paleontol 54:488–490

Eastman JT (1985) The evolution of neutrally buoyant notothenioid fishes: their specializations and potential interactions in the antarctic marine food web. In: Siegfried WR, Condy PR, Laws RM (eds) Antarctic nutrient cycles and food webs. Springer Berlin Heidelberg New York Tokyo pp 430–436

Eastman JT (1991) Evolution and diversification of antarctic notothenioid fishes Am Zool 31:93–109

Eastman JT, DeVries AL (1981) Buoyancy adaptations in a swim-bladderless antarctic fish. J Morphol 167:91–102

Eastman JT, DeVries AL (1982) Buoyancy studies of notothenioid fishes in McMurdo Sound, Antarctica. Copeia 2:385–393

Eastman JT, Grande L (1989) Evolution of the antarctic fish fauna with emphasis on the recent notothenioids. In: Crame JA (ed) Origins and evolution of the antarctic biota. Geol Soc Spec Publ 47, Geol Soc Lond, pp 241–252

Eastman JT, Grande L (1991) A late Eocene gadiform (Teleostei) skull from Seymour Island, Antarctic Peninsula. Antarct Sci 3:87–95

Elliot DH (1985) Physical geography-geological evolution. In: Bonner WN, Walton DWH (eds) Key environments: Antartica. Pergamon Press, Oxford, pp 39–61

Feldmann RM, Tshudy DM (1989) Evolutionary patterns in macruronus decapod crustaceans from Cretaceous to early Cenozoic rocks of the James Ross Island region, Antarctica. In: Crame JA (ed) Origins and evolution of the antarctic biota. Geol Soc Spec Publ 47, Geol Soc Lond, pp 183–195

Fischer W, Hureau JC (eds) (1985) FAO species identification sheets for fishery purposes, Southern Ocean, vol 1. FAO, Rome, 232 pp

Fordyce E (1982) The fossil vertebrate record of New Zealand. In: Rich PV, Thompson EM (eds) The fossil vertebrate record of Australasia. Monash Univ Offset Printing Unit, Clayton, pp 629–698

Gon O, Heemstra PC (eds) (1990) Fishes of the Southern Ocean. JLB Smith Inst Ichthyol, Grahamstown, South Africa, 462 pp

Grande L, Chatterjee S (1987) New Cretaceous fish fossils from Seymour Island, Antarctic Peninsula. Paleontology 30:829–837

Grande L, Eastman JT (1986) A review of antarctic ichthyofaunas in the light of new fossil discoveries. Paleontology 29:113–137

Harwood DM (1986) Recycled siliceous microfossils from the Sirius Formation. Antarct J US 21(5):101-103

Harwood DM (1987) Diatom biostratigraphy and paleoecology with a Cenozoic history of antarctic ice sheets. Diss Abstr Int 47B:3276–3277

Hedgpeth JW (1969) Introduction to antarctic zoogeography. In: Bushnell VC, Hedgpeth JA (eds) Distribution of selected groups of marine invertebrates in waters south of 35° S latitude, Antarctic Map Folio Ser, Folio 11. Am Geog Soc, New York, pp 1–9

Jerzmanska A (1988) Isolated vertebrae of teleostean fishes from the Paleogene of Antarctica. Pol Polar Res 9:421–435

Kennett JP (1978) The development of planktonic biogeography in the Southern Ocean during the Cenozoic. Mar Micropaleontol 3:301–345

Kennett JP (1980) Paleoceanographic and biogeographic evolution of the Southern Ocean during the Cenozoic, and Cenozoic microfossil datums. Palaeogeogr Palaeoclim Palaeoecol 31:123–152

Kennett JP (1982) Marine geology. Prentice Hall, Englewood Cliffs, New Jersey, 813 pp

Kuhn-Schnyder E, Rieber H (1986) Handbook of paleozoology. Hopkins Univ Press, Baltimore, 394 pp

Last PR, Scott EOG, Talbot FH (1983) Fishes of Tasmania. Tasmanian Fish Develop Authority, Hobart, 563 pp

Leim AH, Scott WB (1966) Fishes at the Atlantic coast of Canada. Fish Res Board Can, Bull 155, Ottawa, 485 pp

Long J (1982) The history of fishes on the Australian continent. In: Rich PV, Thompson EM (eds) The fossil vertebrate record of Australasia. Monash Univ Offset Printing Unit, Clayton, pp 53–85

Miller RG (1987) Origins and pathways possible for the fishes of the Antarctic Ocean. In: Kullander SO, Fernholm B (eds) Proc V Congr Europ Ichthyol, Stockholm 1985. Swed Mus Nat Hist, Stockholm, pp 373–380

Moy-Thomas JA, Miles RS (1971) Palaeozoic fishes, 2nd edn. Saunders, Philadelphia, 259 pp

Moyle PB, Cech JJ Jr (1988) Fishes: an introduction to ichthyology, 2nd edn. Prentice Hall, Englewood Cliffs, New Jersey, 559 pp

Nelson JS (1984) Fishes of the world, 2nd edn. Wiley, New York, 523 pp

Norman JR (1938) Coast fishes. Part III. The Antarctic zone. Discovery Rep 18:1–104

Norton IO (1982) Paleomotion between Africa, South America, and Antarctica, and implications for the Antarctic Peninsula. In: Craddock C (ed) Antarctic geoscience. International Union of Geological Sciences, Ser B, No 4. Univ Wisconsin Press, Madison, pp 99–106

Parrish JT (1990) Gondwanan paleogeography and paleoclimatology. In: Taylor TN, Taylor EL (eds) Antarctic paleobiology: its role in the reconstruction of Gondwana. Springer Berin Heidelberg New York Tokyo, pp 15–26

Pickard J, Adamson DA, Harwood DM, Miller GH, Quilty PG, Dell RK (1988) Early Pliocene marine sediments, coastline, and climate of East Antarctica. Geology 16:158–161

Regan CT (1914) Fishes. Br Antarct ("Terra Nova") Exped 1910, Nat Hist Rep, Zool 1:1–54

Rich PV, Rich TH, Wagstaff BE, McEwen Mason J, Douthitt CB, Gregory RT, Felton EA (1988) Evidence for low temperatures and biologic diversity in Cretaceous high latitudes of Australia. Science 242:1403–1406

Richter M, Thomson MRA (1989) First Aspidorhynchidae (Pisces: Teleostei) from Antarctica. Antarct Sci 1:57–64

Richter M, Ward DJ (1990) Fish remains from the Santa Marta Formation (Late Cretaceous) of James Ross Island, Antarctica. Antarct Sci 2:67–76

Romer AS (1966) Vertebrate paleontology, 3rd edn. Univ Chicago Press, Chicago, 468 pp

Schaeffer B (1972) A Jurassic fish from Antarctica. Am Mus Novit 2495:1–17

Smith AG, Hurley AM , Briden JC (1981) Phanerozoic paleocontinental world maps. Univ Press, Cambridge, 102 pp

Smith MM, Heemstra PC (eds) (1986) Smith's sea fishes. Macmillan South Africa Johannesburg, 1047 pp
Stinton FC (1957) Teleostean otoliths from the Tertiary of New Zealand. Trans R Soc New Zeal 84:513–517
Tasch P (1977) Ancient antarctic freshwater ecosystems. In: Llano GA (ed) Adaptations within antarctic ecosystems. Smithsonian Inst, Washington, pp 1077–1089
Watling L, Thurston MH (1989) Antarctica as an evolutionary incubator: evidence from the cladistic biogeography of the amphipod Family Iphimediidae. In: Crame JA (ed) Origins and evoution of the antarctic biota. Geo Soc Spec Publ 47, Geol Soc Lond, pp 297–313
Watts RL, Watts DC (1974) Nitrogen metabolism in fishes. In: Florkin M, Scheer BT (eds) Chemical zoology, vol 8. Academic Press New York, pp 369–446
Webb PN (1990) The Cenozoic history of Antarctica and its global impact. Antarct Sci 2:3–21
Webb PN, Harwood DM (1987) Terrestrial flora of the Sirius Formation: its significance for late Cenozoic glacial history. Antarct J US 22(4):7–11
Webb PN, Harwood DM, McKelvey BC, Mercer JH, Stott LD (1984) Cenozoic marine sedimentation and ice-volume variation on the East Antarctic craton. Geology 12:287–291
Welton BJ, Zinsmeister WJ (1980) Eocene neoselachians from the La Meseta Formation, Seymour Island, Antarctic Peninsula. Contrib Sci Nat Hist Mus Los Angeles Co 329:1–10
Woodburne MO, Zinsmeister WJ (1984) The first land mammal from Antarctic and its biogeographic implications. J Paleontol 58:913–948
Woodward AS (1908) On fossil fish-remains from Snow Hill and Seymour Islands. Wiss Ergeb Schwed Südpolar-Exped 1901–1903 3:1–4
Young GC (1989) The Aztec fish fauna (Devonian) of Southern Victoria Land: Evolutionary and biogeographic significance. In: Crame JA (ed) Origins and evolution of the antarctic biota. Geol Soc Spec Publ 47, Geol Soc Lond, pp 43–62
Young VT (1986) Early Devonian fish material from the Horlick Formation, Ohio Range, Antarctica. Alcheringa 10:35–44
Zinsmeister WJ (1987) Cretaceous paleogeography of Antarctica. Paleogeogr Paleoclim Paleoecol 59:197–206
Zinsmeister WJ, Feldmann RM (1984) Cenozoic high latitude heterochroneity of Southern Hemisphere marine faunas. Science 224:281–283

The Contribution of the BIOMASS Program to Antarctic Marine Ecosystem Research

S.Z. El-Sayed

1 Introduction

The BIOMASS Colloquium, to be held in Bremerhaven, FRG, in 1991, marks the end of a glorious scientific era in the study of the Southern Ocean. BIOMASS is an acronym for Biological Investigations of Marine Antarctic Systems and Stocks. It is therefore fitting to examine the accomplishments of this program and to discuss its impact on the study of the ecosystem of the Southern Ocean and on the conservation of the marine living resources of Antarctica in general, and future biological research in the Southern Ocean in particular. In order to do so, it is useful to know the circumstances that led to the development of this international cooperative venture (see El-Sayed 1977 for details).

2 History of the BIOMASS Program

In the early 1970s, as the world seemed poised to begin harvesting the rich antarctic marine living resources, concern over the need for proper management and conservation of these resources was voiced by members of the scientific community and national and international agencies and organizations. The concern of scientists for the conservation of antarctic marine resources, and in particular the shrimplike organism, krill (*Euphausia superba*), stemmed from the fact that despite past research activities, notably those of the *Discovery* Investigations in the late 1920s and 1930s, there was no adequate information concerning the biology, population dynamics, and stocks of these resources. Yet, effective conservation and management depended upon a continuing flow of data and information regarding these stocks.

Realizing that unwise and unregulated past exploitation had decimated the southern fur seal and baleen whale populations, and recognizing krill's key position in the Southern Ocean food web and its impending exploitation, the Scientific Committee on Antarctic Research (SCAR) of the International Council of Scientific Unions (ICSU) foresaw a need for substantial expansion of scientific research on the antarctic marine ecosystem. In August 1972, SCAR established a group of experts to address this need. This group was later reorganized as the SCAR Group of Specialists on Southern Ocean Ecosystems and Their Living Resources. This

Department of Oceanography, Texas A&M University, College Station, Texas, USA group is cosponsored by the Scientific Committee on Oceanic Research (SCOR), the International Association for Biological Oceanography (IABO), and the Advisory Committee on Marine Resources Research (ACMRR) of the Food and Agriculture Organization (FAO).

In the summer of 1976, the United States hosted the First International Conference on Living Resources of the Southern Ocean at the U.S. National Academy of Sciences' Summer Studies Center in Woods Hole, Massachusetts. The chief objective of the conference was to review present knowledge of the living resources of the Southern Ocean and to develop a proposal for future cooperative studies in the area. The proposal became known by its acronym BIOMASS.

The primary goal of the BIOMASS program is to build a sound scientific foundation on which to base future management decisions, thereby ensuring rational exploitation and the continued maintenance of important ecological relationships between exploited and unexploited stocks. Because of the pivotal role of krill in the antarctic food web and because of its potentially significant contribution to world protein supplies, krill studies have played a key role in the BIOMASS program.

The austral summer of 1980–81 was chosen for the First International BIOMASS Experiment (FIBEX) in which 13 ships from 11 nations participated in the largest biological oceanographic expedition ever mounted in the Southern Ocean. The Second International BIOMASS Experiment (SIBEX), Phase I (1983–84) and Phase II (1984–85) were the final collaborative field effort of this ambitious ten-year program. The primary objective of SIBEX was to enhance our understanding of the dynamics of the krill-dominated part of the Antarctic marine ecosystem, and to obtain a broad picture of krill-swarm distribution in relation to the prevailing meso -scale enviromental features, particularly the advancing and retreating of the sea ice. The main focus of SIBEX was on the seasonal and annual variations in the concentration and production of krill in three relatively small areas which are noted for their high krill concentrations: Bransfield Strait/Elephant Island (Atlantic Sector); Prydz Bay (Indian Sector); and 60° E (Pacific Sector).

2.1 Accomplishments of the BIOMASS Program

Although a careful appraisal of the accomplishment of the BIOMASS program must await the publication of the results of ongoing and future workshops, and in particular the synthesis of the results that will take place at the BIOMASS Colloquium in Bremerhaven, FRG in 1991, all indicators are that the program has been a great success. There are several yardsticks by which one can measure the success of the program, but these fall into two basic categories: (1) scientific accomplishments; (2) international cooperation. With regard to the former, we will discuss some of the program's significant results. And since krill studies were central to the program, special emphasis will be placed on the results of krill investigations.

3 BIOMASS Krill Research

3.1 Estimation of Krill Standing Stocks

One of the most vexing problems that faced krill investigators was the determination of the size of krill standing stocks. Prior to BIOMASS, krill stock estimates were subject to inspired guesses. Past attempts to estimate total krill standing stocks indirectly (from the decline in large baleen whale stocks) or directly (by plankton net sampling) were proven unsatisfactory. The great discontinuities in the distribution and swarming behavior of krill, the relatively small areas in which sampling was conducted, and the diverse methods used in estimating standing stocks were responsible for the great variance in krill stock estimates. According to these estimates, krill stocks ranged between 200 million to 3.5 billion metric tons! The exercise in determining krill stocks has been tenuous and less useful than was expected.

In recent years, the introduction of quantitative acoustic techniques for stock assessment has shown great promise, although not without some disadvantages (Fig. 1). Between January and March 1981, joint krill hydroacoustic surveys during FIBEX were carried out in four sectors of the Southern Ocean (Fig. 2). Eleven ships from 10 nations surveyed these sectors quasisynoptically. Joint analysis of the data, using a data base specifically designed for the purpose, roughly estimated the biomass of krill at 250–600 million metric tons. However, estimates extrapolated from hydroacoustic surveys are difficult to make, particularly since krill distribution is patchy. In addition, krill may also be scattered over a wide part of the Antarctic Ocean outside the dense concentrations. Other independent estimates of

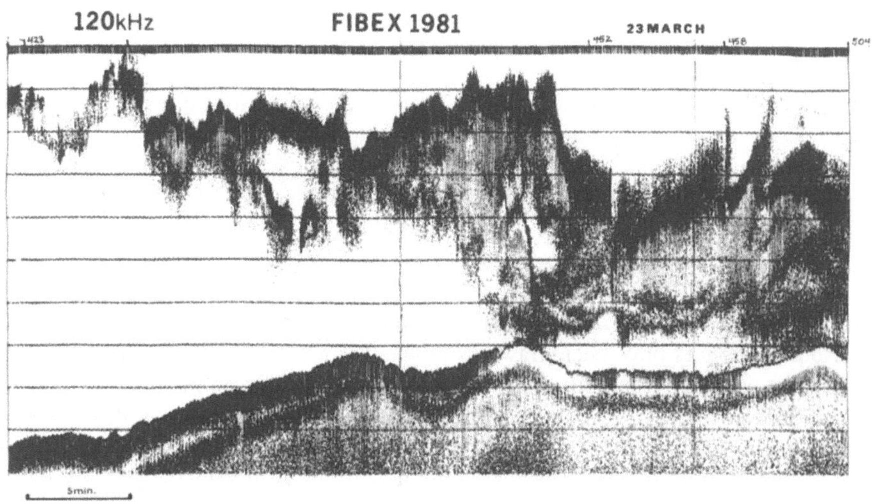

Fig. 1. A patch of krill swarm (located to the northwest of Elephant Islands), as shown on the 120 kHz system. Depth marks are at 20-m intervals: time scale at *lower left* indicates ships' speed (After Macanlay and Mathiesen 1981)

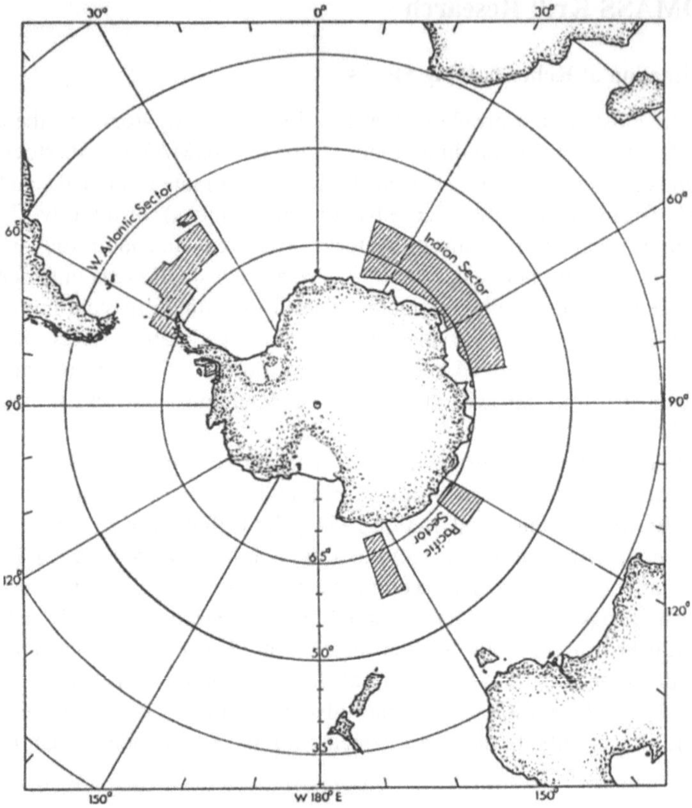

Fig. 2. The four study sites (*shaded areas*) where krill hydroacoustic surveys were carried out during FIBEX

overall krill biomass amounted to 500–700 million metric tons or even more. The FIBEX survey represents the most concerted attempt by far to assess krill acoustically, and gives a good insight into the methodological problems involved in the collection and joint analysis of the acoustic data (Miller and Hampton 1989).

3.2 Krill Stock Identification

As a result of the observations on krill distribution made during FIBEX and SIBEX, various inferences have been drawn as to whether local krill concentrations essentially are isolated from one another or whether substantial intermixing occurs. These inferences have profound implications, as they will ultimately determine to what extent local and/or regional concentrations can be treated as separate stocks for management purposes. The conventional methods of tagging and of relying on morphometric measurements, successfully used by fishery biologists in identifying other discrete management stocks, are not applicable to krill. Realizing this, investigators have resorted to electrophoretic analysis of the variations in the structure of enzymatic protein as an alternative approach to the problem of separating krill stocks.

Early electrophoretic analysis suggests the existence of at least two discrete krill populations in the Antarctic Peninsula region. More recently, samples of krill collected from the Weddell Sea, Scotia Sea, the Antarctic Peninsula area, and near Prydz Bay (in the Indian Sector of the Southern Ocean) indicate that they were all from a single breeding population and do not materially affect conclusions concerning population structure (Fevolden 1986). Further, it has been suggested that physical barriers may induce genetic subdivision of krill populations. Despite the considerable progress made in recent years, the successful separation of individual krill stocks (by genetic or other means) remains elusive and requires further research.

3.3 Krill Age Determination

Almost all aspects of krill biology, physiology, biochemistry, and ecology have fared well under the BIOMASS program. The progress made in these fields cannot be detailed in this chapter; however, two impotant aspects must be mentioned: krill age determination and krill's relation to its food base, the phytoplankton (discussed in Sect. 3.4). As to the former, it is now well established that the traditional method of determining krill age (by examining the length frequency distribution of catches, regarding the peaks in the histogram as year-classes) is fraught with error. This is due largely to the observations made by the late Mary Alice McWhinnie, who showed that mature krill may shrink in body size as an overwintering strategy. This could result in the overlapping of successive year-classes of mature krill. It has also been suggested that there is a regression of external sexual characteristics during winter. Drawing on his background in entomology where for years he has been studying aging in the fleshfly (*Sarcophaga bullata*), George Ettershank of Monash University (Australia) succeeded in using lipofuscin (also called age pigment) to estimate krill age (Ettershank 1983). Lipofuscin accumulates as a consequence of metabolic activity, and its assay is thus a measure of the cumulative metabolic activity of an organism. In other words, it is a measure of the "physiological" rather than "chronological" age. However, due to the uniformity of the temperature regime in the regions where krill live, one could expect reasonable agreement between physiological and chronological age. Although results of the lipofuscin assay independently confirm that krill may live as long as seven years, the technique, although more reliable than traditional methods, still requires refinement.

3.4 Food Chain Studies

The relation of krill to its food base has occupied much of the attention of krill biologists. Until the early 1970s, the herbivorous nature and food selectivity of krill seemed well established. Later, it was demonstrated that krill are omnivorous and, using electron micrographs of krill filtering appendages, researchers showed that the krill filtering basket is capable of retaining nanoplankton (organisms between 2 and 20 μm) with 30–40% efficiency. This changed the concept of *E. superba* as an omnivore feeding mainly on large diatoms, and substantially expands the food resources we know to be available to krill. Recently, German scientists have shed some light on another aspect of krill's food base, krill feeding during winter. During the Weddell Sea Winter Expedition in 1986, Hans-Peter Marschall observed that krill were feeding on ice algae. He noted that their guts were full although virtually no phytoplankton existed in the water column. The krill were feeding by grazing on algae adhering to the underside of the pack ice. According to Marschall, the flexibility of the krill to change behavior and habitat from pelagic life in the summer to "demersal" life in winter, thus gaining access to a valuable food source at times of low primary production, seems to be a major factor contributing to its success in the antarctic marine ecosystem.

Interest in the role played by the nanoplankton in krill feeding stimulated research in another even more important direction: assessing the role of the nanoplankton and the picoplankton (2 μm) in the antarctic food web. In contrast to net phytoplankton (organisms > 35 μm), which in the past have recieved considerable attention and form the basis of the classic food chain (diatom → krill → whale), the contribution of the nanoplankton and picoplankton to the standing crop and primary production have, until recently, been overlooked. BIOMASS investigators have convincingly demonstrated that these nanoplankton and picoplankton could be responsible for 70% of the standing crop and about 90% of the primary production of the antarctic waters. Moreover, it has been postulated that protozoans feeding on antarctic nanoplankton and picoplankton form a major link between the primary producers (phytoplankton) and thier consumers (zooplankton, including krill). This finding could have significant implications for the overall community structure and trophic organization of the antarctic marine ecosystem. Thus, in addition to the classic food chain, a complex food web consisting of pico-, nano-, and microplankton-sized producers and consumers, and bacterial secondary producers, at least in the region of the sea ice, is now emerging as the new paradigm (El-Sayed 1984; Fig. 3).

3.5 Fish-Related Investigations

The contributions of the BIOMASS program to our knowledge of organisms at higher tropic levels (in particular, fish and birds) are no less impressive than those of the krill. Considerable gains have been made in the fields of ichthyoplankton, fish biology, ecology, and physiology. During FIBEX most icythyoplankton research was carried out on Ichthyoplankton and little had been done on adult fish. Research showed that large quantities of juvenile fish are being caught during krill fishing with commercial pelagic trawls. Conceivably, krill fishing could have a strong impact on fish stock recruitment. During SIBEX, ichthyoplankton surveys were directed toward target species (fish) in order to use the distribution and abundance of eggs and early larval stages to obtain estimates of the biomass of adult spawning populations. This is particularly important within several areas where active commercial fishing is taking place (e.g., South Georgia, Scotia Sea, and Kerguelen Islands).

In Post-SIBEX years, fish-stock assessment surveys have been conducted either by individual nations or in joint operations between two or three nations (e.g.,

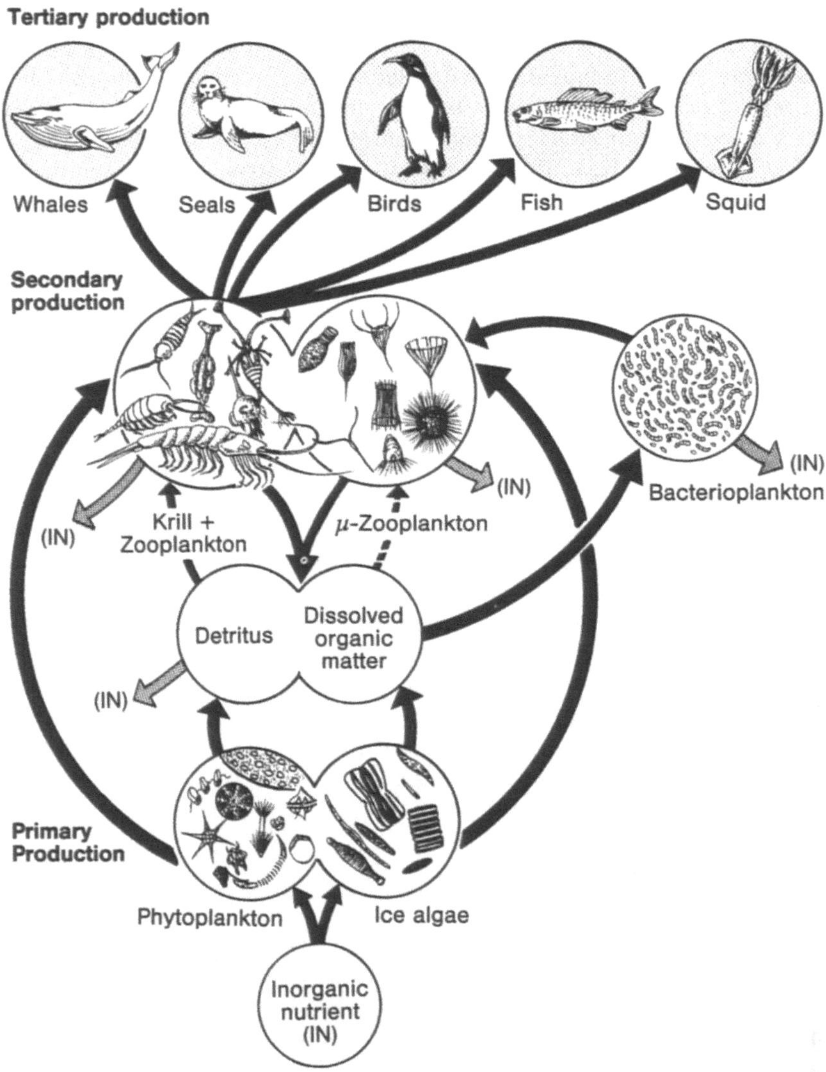

Fig. 3. Paradigm of energy flow within the antarctic marine ecosystem (After El-Sayed 1984)

between France and the USSR in the Kerguelen region; between Spain, Poland, and the USA in the South Georgia area). The results of these surveys indicate that most of the commercial fish stocks are being depleted and that strong conservation measures need to be adopted soon.

3.6 The BIOMASS Data Centre

Perhaps one of the crowning achievements of the BIOMASS program is the establishment, in June 1984, of the BIOMASS Data Centre, housed within the British Antartic Survey in Cambridge, UK. The Centre was created to support the BIO-

MASS program. It uses data from 14 countries to model ecosystem dynamics in the Southern Ocean, with a particular emphasis on krill ecology. The main goals of the Data Centre are (1) to provide a central information location; (2) to assist in the exchange of marine biological data between countries; and (3) to enable the BIOMASS community to produce detailed analyses of biological stocks using the BIOMASS computer base. This data base allows data from several scientific disciplines to be interrelated in order to examine causes and effects on different parts of the marine systems. This makes the BIOMASS Data Centre unique among other data centers, and the experience developed as a result is now available to other data centers who wish to emulate/adopt the Cambridge experience.

3.7 BIOMASS Publications

In an effort to standardize the physical (including acoustics), chemical, and biological data collected during FIBEX and SIBEX, the BIOMASS planners foresaw the need to develop and publish a handbook series to be used in field exercises and in laboratory experiments. As a result, a total of 25 authoritatively written handbooks, covering the gamut of research, from measurement of primary production to calibration of hydroacoustic instruments, penguin census methods, and an atlas of fish larvae of the Southern Ocean, have been published by BIOMASS. Other BIOMASS publications include the BIOMASS Scientific Series, dealing comprehensively with such topics as biology and status of exploited antarctic fish stocks and review of the biology and ecology of *Euphausia superba*; the BIOMASS Report Series, detailing the meetings and workshops of the specialized BIOMASS groups; and the BIOMASS Newsletter, published twice a year.

4 International Cooperation

So far we have discussed some of the scientific accomplishments of BIOMASS. This section focuses on the role the program has played, and still plays, in fostering international scientific cooperation. Within its short life span, BIOMASS has rekindled great international interest in Antarctic affairs reminiscent of the International Geophysical Year (IGY) in 1956-57. The era of individual national expeditions (e.g., *Belgica, Pourquoi Pas?, Gauss, Meteor, Discovery, Ob,* and *Eltanin*) has been replaced by well-coordinated, multiship, multinational expeditions. This is attested to by the mounting of the largest biological expedition ever undertaken in the Southern Ocean, and by the numerous scientific expeditions to the antarctic during the past few years by such newcomer nations as India, China, Italy, the Republic of Korea, and most recently, Spain, Peru, and Uruguay. Further, the worshops, symposia, and colloquia held in recent years in such places as Cambridge (UK), Tokyo (Japan), Bremerhaven (FRG), Bariloche (Argentina), Wilmington (North Carolina, USA), Rio de Janeiro (Brazil), Ravello (Italy), and College Station (Texas, USA), where BIOMASS data and results have been discussed by scientists from the BIOMASS community, underscore the high degree of scientific cooperation and camaraderie that has developed among the antarctic community as a result of the BIOMASS program. Perhaps the height of this cooperative spirit is best exemplified by the highly successful data analysis workshops (15 so far) where, in the best tradition of scientific cooperation, scientists from varied political persuasions and backgrounds have agreed to pool their unpublished data and subject them to communal analysis, and to collaborate later in joint publications.

It was in this atmosphere of international collaboration, and with the growing interest in the conservation of the Southern Ocean ecosystem and its living resources, that the Convention for the Conservation of Antarctic Marine Living Resources (CCAMLR), of the Antarctic Treaty, was ratified in 1982. BIOMASS and CCAMLR have much in common, as both are rooted in the concern over the development of a sound ecological strategy for the exploitation of living marine resources and for the conservation of the antarctic marine ecosystem. Wise use and conservation of the living resources of the Southern Ocean have been always key considerations of the BIOMASS program. Early on BIOMASS planners realized that in developing successful fishery research programs, the entire ecosystem would need to be considered. The "species-by-species approach" has proven inadequate and faulty. Thus, any attempt at harvesting a single species without regard to possible effects on other components in the system may set up irreversible reactions that will result in major changes in the ecosystem. This approach struck a responsive cord in CCAMLR and was put into immediate use by the drafters of the CCAMLR convention, who foresaw that harvesting levels should take into account the effect on the whole ecosystem, not only on the large but also on the dependent species.

5 Future Biological Studies of the Southern Ocean

As mentioned before, the BIOMASS program was the first major international collarobative effort to study the antarctic marine ecosystem and to provide the necessary information for the wise management of antarctic marine living resources. With the successful completion of FIBEX and SIBEX and the establishment of the BIOMASS Data Centre, the program has now entered a new phase of data analysis and data interpretation. To ensure that full advantage is taken of the enormous amount of data collected, a series of workshops to process the information and report on findings are being held. Among the workshops to be held in the near future are: SIBEX Acoustic Data Validation, Catch per Unit Effort as an Estimator of Krill, and Antarctic Fish Ecology.

Although the BIOMASS program has been criticized as too ambitious, too diffuse, and too krill-oriented, there is no denying that it has contributed and still contributes to a better understanding of the antarctic marine ecosystem. However, much remains to be done, not only in terms of further evaluation of the existing data but also in the initiation of new programs at sea, ashore, and in laboratories regarding the quantitative assessment of standing stock, production, and variability of krill, squid, fish, and top predators. The following are some of those new programs (El-Sayed 1986);

1. *Toward a "more-focused" post-BIOMASS program.* A review of recent FIBEX and SIBEX results has shown that despite the large quantities of data collected, much remains to be done. At present, some of the deficiencies are being addressed by ongoing or proposed programs. These include:
   a) *International target strength study.* Definitive target strength experiments of *Euphausia superba* are recognized as being of utmost importance for making better use of both present and future acoustic data.
   b) *nternational krill physiology project.* Physiological studies of krill growth, energy requirements, and metabolic rates are extremely important for understanding the ecological relationships in the Southern Ocean. Future research plans need to focus on the standardization of methodologies to improve the current low degree of comparatibility of results from different research groups.
   c) *Winter studies.* The objective of these studies is to provide relevant information on the physical-chemical components of the marine ecosystem in winter, in particular:
      i) To describe the physical-chemical factors governing the distribution, abundance, productivity, and behavior of the marine organisms, especially phytoplankton, zooplankton (including krill), and fish;
      ii) To describe the short- and long-term environmental fluctuations governing the biology of the organisms under the ice cover; and
      iii) To describe the behavior of the penguins and seals in the park ice.
2. *Other "Spin-off" research programs from the BIOMASS program* include:
   a) Krill biology and its relation to the environment. Krill biology is still a research field of high priority. Key topics that need to be addressed include:
      i) Growth and mortality in different areas;
      ii) Spawning success and recruitment mechanisms;
      iii) Stock separation and migration patterns;
      iv) Krill survival.

   All these topics should be viewed in relations to the biotic and abiotic environment of krill.
   b) *Methodology for abundance estimates of krill.* Solving the great uncertainties regarding target strength and net avoidance calls for concerted international action in areas of hydroacoustics and net sampling.
   c) *Research on other dominant components of the ecosystem.* These components include fish, squid, and key predators; this research has, to a large extent, been subordinated to krill research.
3. *Complementary research program with CCAMLR.* The Convention on the Conservation of Antarctic Marine Living Resources (CCAMLR) came into force in April 1982. The Convention provides the essential legal framework for management fo the living resources of the Southern Ocean. The Scientific Committee of CCAMLR has the responsibility of providing information and management advice about the Southern Ocean ecosystem and its living resources; however, the necessary research cannot be handled by CCAMLR

alone. There is a need for cooperation between SCAR and its subsidiary bodies and CCAMLR on several key research activities. It is assumed that this policy of cooperation and assistance is a mutual one, with BIOMASS and CCAMLR each having unique strengths and resources. The BIOMASS community has developed a level of competence and expertise capable of advancing basic scientific understanding of the antarctic marine ecosystem and can thus make valuable contributions to CCAMLR in its mandate to conserve the living resources of the Antarctic within the context of the ecosystem.

In planning future collaborative research activities between SCAR and CCAMLR, it is important to ensure that the research programs are not competing with each other but, rather, that they complement each other. Necessarily, CCAMLR must concentrate and maximize collaborative efforts. It would be desirable to have formal links between CCAMLR and SCAR.

4. *Cooperative research with other international programs.* Another option that Antarctic biologists have is to join forces with ongoing or planned programs in the Southern Ocean. These might include:
   a) *Antarctic sea-ice studies.* The seasonal advance and decay of sea ice around the Antarctic Continent, a unique physical feature of the region, plays a dominant role in determining the overall structure of this ecosystem. It is likely that interannual variability in this process has a significant impact on the recruitment success of antarctic krill and perhaps other species.

      The proposed ten-year program for an international collaborative study of the Antarctic Sea-Ice Zone includes a biological component. The expertise and competence of antarctic scientists gained during their recent winter studies should ensure that the new program will significantly contribute to a better understanding of the pack-ice ecosystem.
   b) SCAR Group of Specialists on Southern Ocean ecology. This newly established (1986) group will identify important fields of research on antarctic marine ecology and will propose cooperative studies, including multiship experiments. It will encourage and facilitate interdisciplinary studies in antarctic marine ecosystems, and will develop Southern Ocean ecosystem studies through workshops and other activities.

6 Who Will Inherit BIOMASS?

Although the SCAR/SCOR Group of Specialists on Southern Ocean Ecosystems and Their Living Resources was disbanded in 1985 (their responsibilities have since been transferred to the BIOMASS Executive), it became clear that some new international scientific group was required to continue to facilitate international cooperation in the Southern Ocean ecology and related fields. This led SCAR, with its extensive experience of fruitful coordination of such research, to establish (jointly with SCOR) in 1986 the SCAR Group of Specialists on Southern Ocean Ecology. The Group was charged with identifying important fields of research in antarctic marine ecology and proposing cooperative studies. Another SCAR group is the Group of Specialists on the Antarctic Sea-Ice Zone, whose ten-year program for an international collaborative study includes a biological component.

The third group is the governmental organization, CCAMLR, which has the mandate to conserve the living resources of the Antarctic within the context of the ecosystem. The fledging CCAMLR has already made good progress toward implementing a well-designed predator-based monitoring study and has adopted several conservation measures in connection with fish resources around South Georgia. It has reached agreement regarding the protection of immature fish and established a working group on krill to review and evaluate new research applications to krill abundance and distribution assessment. In view of the recently observed depletion of the ozone layer over Antarctica and its possible effect on life in the Southern Ocean, the United States has recommended that the Working Group on Ecosystem Monitoring develop an appropriate sampling protocol for measuring possible impacts of the stratospheric ozone decrease on the productivity of the Southern Ocean.

One looks forward with great anticipation to fruitful cooperation between these organizations in maintaining the high degree of scientific interest and international cooperation that has been the hallmark of the BIOMASS program.

References

El-Sayed SZ (ed) (1977) Biological Investigations Of Marine Antarctic Systems And Stocks (BIOMASS), vol I. Research proposals prepared by SCAR/SCOR Group of Specialists on Living Resources of the Southern Ocean. Scott Polar Res Inst, Cambridge, England, 79 pp

El-Sayed SZ (1984) Productivity of antarctic waters-a reappraisal. In: Holm-Hansen O, Bolis L, Giles R (eds) Marine phystoplankton and productivity. Springer, Berlin Heidelberg, New York. Tokyo, pp 19–34

El-Sayed SZ (1986) Biological studies of the Southern Ocean: research plans and international coordination. BIOMASS Occasional Paps published by SCAR/SCOR

Ettershank G (1983) Age structure and cyclical annual size change in the antarctic krill, *Euphausia superba* Dana. Polar Biol 2:189–193

Fevolden SE (1986) Genetic variation of *Euphausia superba* Dana. in the Antarctic Peninsula waters. Sarisa 71:169–175

Miller DGM, Hampton I (1989) Biology and ecology of the antarctic krill (*Euphausia superba* Dana): a review BIOMASS, vol 9. SCAR/SCOR, Cambridge, England

Part II

Physiology, Biochemistry and Molecular Biology

The Sensory Biology of Nototheniod Fish

J.A. Macdonald and J.C. Montgomery

1 Introduction

The antarctic marine environment is notable not only for its low temperature, but also for the light levels which occur under areas of ice cover, and during the polar winter. Within the confines of this environment has evolved a unique assemblage of fish.

This assemblage is characterized by a relative lack of diversity, with a total of 203 bottom species reported from antarctic waters, and a very high degree of endemism (88%). One hundred and four of these species, including the most common forms, belong to a single suborder of fish, the Notothenioidei (DeWitt 1971; Andriashev 1987). It appears that this suborder is derived from an ancestral species which at some point in its evolution overcame the difficulties of survival at extreme low temperature. Subsequent speciation and evolution of this group has resulted in the current species occupying diverse niches, providing an excellent example of a monophyletic radiation. Research has been directed at understanding the success of this group in the Antarctic and the relationship of form and function to niche specialization (Macdonald et al. 1987).

In this paper we review the available literature on the sensory biology of antarctic fish, placing particular emphasis on what appear to be specialized aspects of their sensory repertoire. However, it is also pertinent to enquire as to the extent to which antarctic fish are able to maintain sensory function at "normal levels" despite the extreme cold. Finally, we will attempt to highlight some of the areas where there are obvious gaps in our knowledge, or where particularly productive lines of research are evident.

2 Vision

In the aquatic habitat, vision is degraded by the optical properties of water; however, it is still a reasonable generalization to say that vision is the dominant sensory modality of teleost fish. The importance of vision is evident from the degree and range of specializations documented from deep-sea fish occupying a marginal visual environment (Munk 1966, 1984). It is not until the extreme abyssal depths

Department of Zoology, University of Auckland, Auckland, New Zealand

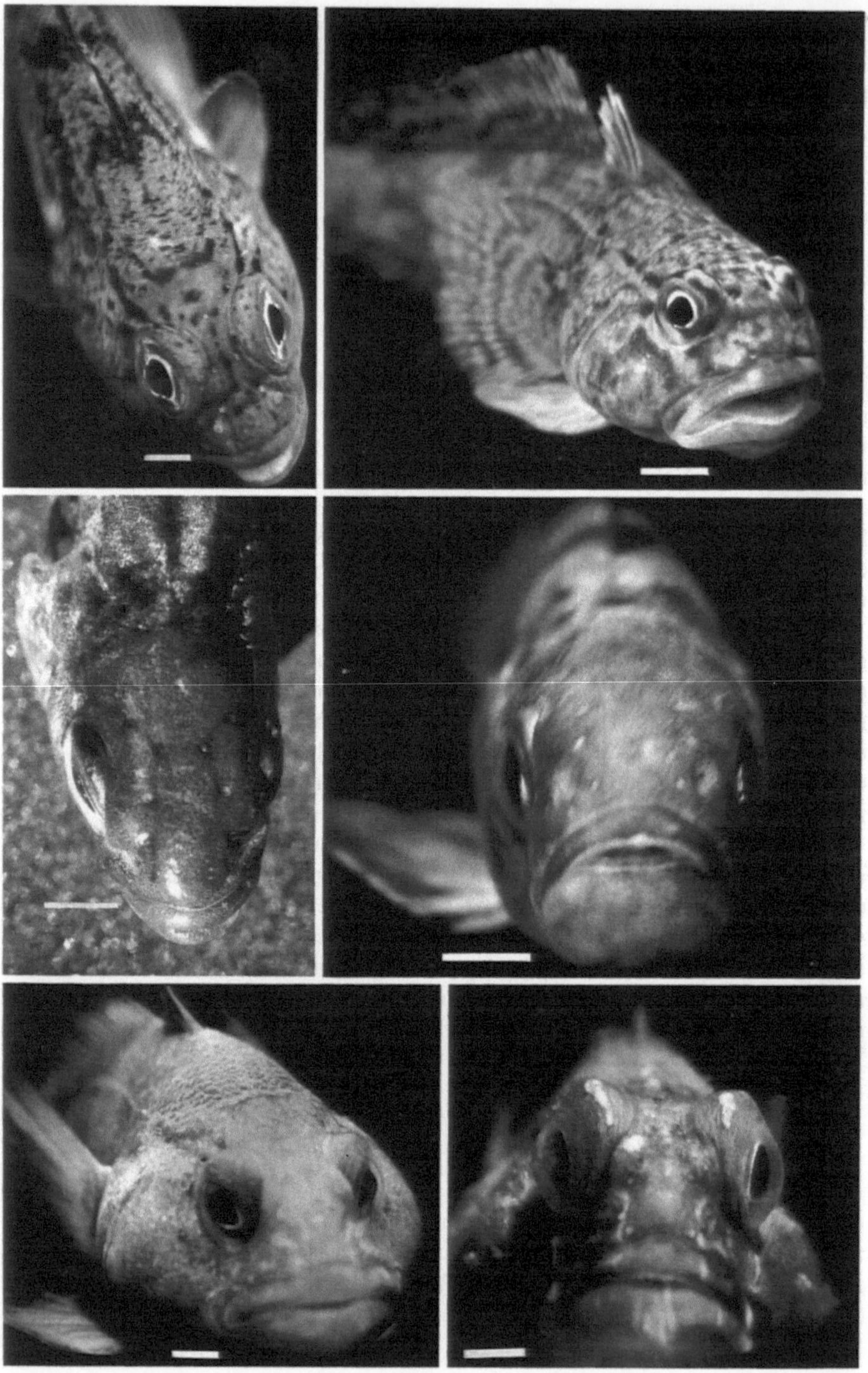

a b c d e f

that vision is abandoned and the eyes commonly become regressed. The antarctic marine environment is comparable to the deep sea in its low temperature and low wintertime light levels. Despite this, antarctic fish appear to show little in the way of comparable visual specializations. The common conclusion of the main structural (Meyer-Rochow and Klyne 1982; Eastman 1988; Pankhurst and Montgomery 1990) and functional studies (Pankhurst and Montgomery 1989) is that the eyes of antarctic fish are generally similar to those of coastal fish elsewhere.

2.1 Eye Morphology

The position and orientation of eyes varies somewhat within the notothenioids, and appears to be correlated with feeding vector (Pankhurst and Montgomery 1989). In some demersal species, the eyes are dorsolateral and directed forwards and upwards (e.g., *Trematomus nicolai*, *T. hansoni*, *Gymnodraco acuticeps*), a position and orientation appropriate to a feeding vector in the overlying water column. Other benthic forms have dorsal eyes oriented anterolaterally (*T. centronotus*, *T. bernacchii*), and feed rather unselectively on benthic invertebrates, while more active species frequenting the water column tend to have eyes situated and oriented laterally (*T. newnesi*, *Pagothenia borchgrevinki*, *Pleuragramma antarcticus*, *Chionodraco hamatus*; Fig. 1)

In many benthic species the pupil is assymmetric, with a pronounced anterior aphakic space, thought to enhance forward vision, suggesting that the greatest visual interest lies in that direction (Pankhurst and Montgomery 1989). Midwater species such as *P. borchgrevinki* lack an obvious aphakic space, and presumably have more uniform visual fields.

In some of the benthic feeders (*T. bernacchii*, *T. centronotus*, *Histiodraco velifer*) there is a distinct corneal iridescence, which is absent in those species feeding in the water column (e.g., *T. hansoni*, *P. borchgrevinki*, *Dissostichus mawsoni*). Corneal iridescence is thought to screen out bright downward light which would otherwise mask less intense, laterally incident light from the region of visual interest (Pankhurst and Montgomery 1989)

The relative eye size of the cryopelagic *Pagothenia borchgrevinki* is shown in Fig. 2. In this species, eye size ranges from 7–8% of the total length (Pankhurst and Montgomery 1990), which would be typical of the nototheniids as a group, and is

Fig. 1. Relative eye positions in nototheniid fish. **a** *Trematomus nicolai*: a sedentary benthic species resembling *T. bernacchii* in many respects. Vertical view of the head; note the relatively small interocular distance, with eyes directed upwards. **b** *Trematomus hansoni*: a benthopelagic species. Oblique view of head. The eyes are oriented more laterally than those of *T. nicolai*. **c***Pagothenia borchgrevinki*: a cryopelagic species. Vertical view of head; the eye orientation is mainly lateral. **d** *P. borchgrevinki*: frontal view of head. **e** *T. bernacchii*: a sedentary benthic species; oblique view of head. The eyes are directed laterally, downwards and slightly forwards. Note the pronounced aphakic space on the anterior edge of the pupil. **f** *Trematomus centronotus*: a sedentary benthic species, similar in habit to *T. bernacchii*; frontal view of head. The eyes are directed more ventrally than those of *T. bernacchii*. All of these species were collected in McMurdo Sound; **a, b, d, e, f** were photographed live in aquaria; **c** is a freshly killed specimen in which the anterior lateral line was injected with methylene blue. *Scale bar* for all specimens = 10 mm; photographs by the authors

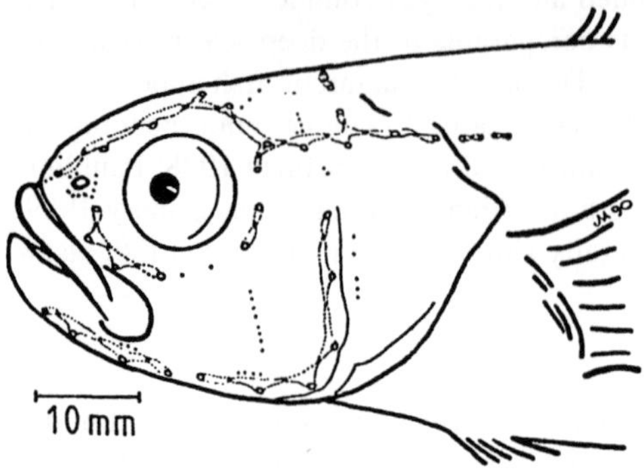

Fig. 2. The head of *Pagothenia borchgrevinki*, showing large eye, anterior lateral line canals and pores, and locations of neuromasts in pit rows

within the range of relative eye sizes reported for shallow-water marine teleosts (Pankhurst 1989). Relative eye size was highest among smaller individuals, and declined with increasing fish size (Pankhurst and Montgomery 1990).

The major structural features of the eye are shown in Fig. 3A. There is a well-developed spectacle over the surface of the eye (Eastman 1988), which in conjunction with the cornea prevents ocular freezing. The ocular fluids themselves are slightly hypotonic with respect to plasma and contain very low levels of serum proteins; unlike other body fluids, they are not protected from freezing by glycoprotein antifreeze. The ocular fluids must remain permanently supercooled, and can only do so because the spectacle and cornea provide a suitable barrier to prevent ice propagation into the eye (Turner et al. 1985). Both the spectacle and the cornea are avascular, and rely on oxygen and nutrient diffusion from the ocular margin (Eastman 1988).

The main structural component of the globe is the sclera, which is both fibrous and cartilaginous. The vascular and darkly pigmented choroidal layer lies between the sclera and retina. In some species (*Pagothenia borchgrevinki*), a reflective layer called the stratum argenteum lies between the choroid and the sclera. This is continued onto the anterior surface of the iris, effectively masking the black choroidal and iridal pigments, and providing for camouflage when viewed against the light background of the platelet ice (Eastman 1988). The pupil is fixed, and all species have a well-developed retractor lentis muscle attached midventrally to the inner surface of the iris, which moves the lens to provide for accommodation (Eastman 1988). Accommodation is presumably associated with the anterior aphakic space evident in most species (Pankhurst and Montgomery 1989).

A choroid body occurs in 14 of the 22 notothenioid species examined by Eastman (1988), with a distribution suggesting it to be a primitive notothenioid character. The choroid is a rete mirabile, or network of countercurrent blood vessels capable of producing elevated oxygen tensions in the choroid adjacent to the outer (sclerad) surface of the retina. The avascular retina, especially the visual cell layer, can then be supplied by diffusion. The significance of retention or loss of the choroid body amongst the different notothenioid groups is not understood, although

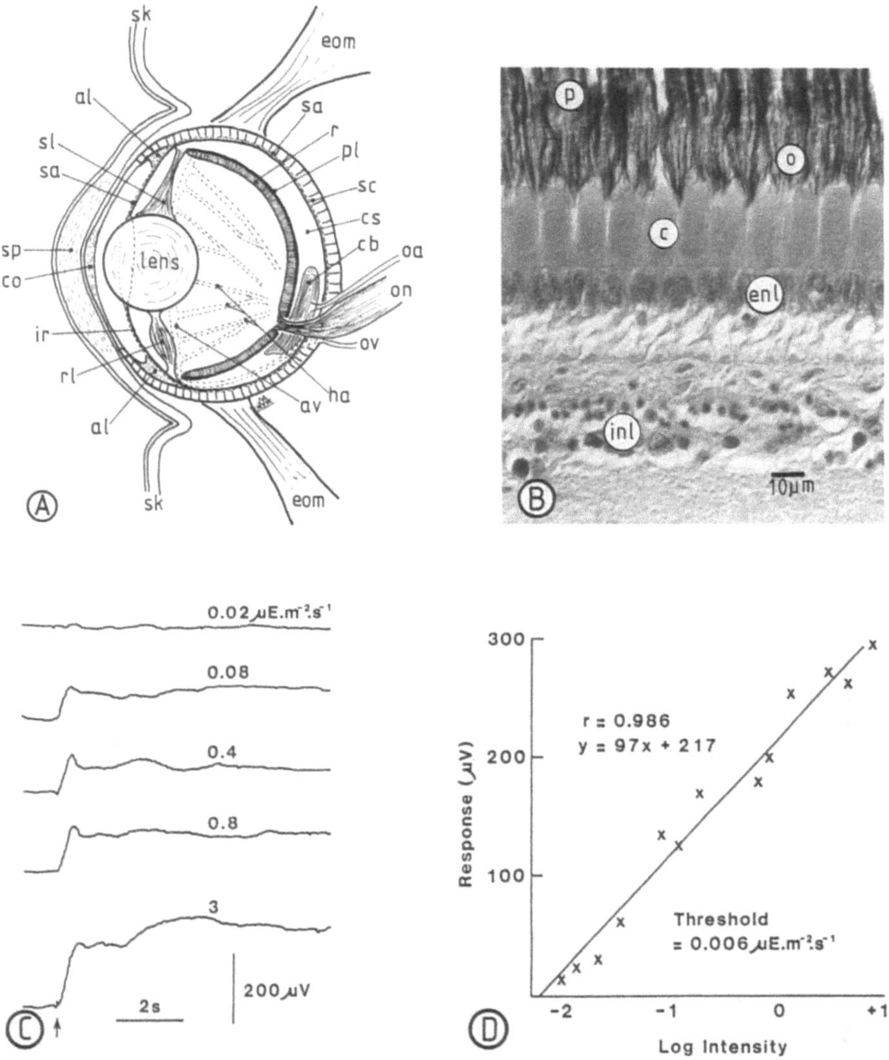

Fig. 3. The notothenioid eye. **A** Diagrammatic section through a generalized notothenioid eye. *al* Annular ligament; *av* annular vein; *cb* choroid body; *co* cornea; *cs* choroid space; *eom* extraocular muscle; *ha* hyaloid arteries; *ir* iris; *oa* ophthalmic artery; *on* optic nerve; *ov* ophthalmic vein; *pl* pigment layer of retina; *r* retina; *rl* reflector lentis; *sa* stratum argenteum; *sc* sclera; *sk* skin; *sl* suspensory ligament; *sp* spectacle. Note that the choroid body does not occur in some species, such as *Pagothenia borchgrevinki*, *Gymnodraco acuticeps* and the channichthyidae. **B** Transverse section through the light-adapted retina of a 206-mm specimen of *P. borchgrevinki*. *c* Cone ellipsoid; *enl* external nuclear layer; *inl* inner nuclear layer; *o* photoreceptor outer segments; *p* pigment epithelium. Haematoxylin and eosin stain; scale = 10 μm (Pankhurst and Montgomery 1990). **C** Electroretinogram (ERG) traces from *P. borchgrevinki*, showing the effects of increasing stimulus intensity. A small spikelike negative "A" potential is visible on some traces, followed by a clear "B" wave and slower "C" wave (Montgomery et al. 1989). **D** Semilogarithmic graph of ERG responses *vs* log stimulus intensity. The measured height of the "B" wave is plotted against the log of stimulus intensity, and threshold obtained by extrapolation of the regression line to the "0" response level (Montgomery et al. 1989)

it is intriguing that five of the eight species lacking choroid bodies are channichthyids. In those species lacking a choroid body, there is an arterial choriocapillaris, a capillary bed sclerad to the retina. Oxygen for the retina is also provided by the hyaloid arteries which ramify over the vitral surface of the retina. In the notothenioids, the hyaloid arteries show a more extensive and uniform distribution than do most other teleost species studied to date (Eastman 1988).

2.2 Eye Movement

Effective vision requires stabilization of the visual image on the retina. Thus the extraocular muscles form part of the eye structure and play an important role in visual function. During whole body movements, or head movements, vestibulo-ocular and optokinetic reflexes help stabilize the eye in space and maintain a fixed visual reference frame. Realignment of the visual reference frame is usually done by rapid eye movements called saccades. Both saccades and slow eye movements have been extensively studied as models for the organization of reflex behaviour and motor function.

In antarctic fish, extraocular muscles have been used to study motor performance and temperature compensations occurring within nerve and muscle (Macdonald and Montgomery 1982, 1986; Montgomery and Macdonald 1984, 1985; Montgomery 1984; Pockett and Macdonald 1986). The most significant results with respect to visual function are that in stationary fish, spontaneous eye movements are relatively infrequent, although this is species-dependent. Although spontaneous saccades are quite rare in *P. borchgrevinki*, they are much more common in demersal species such as *T. bernacchii* and *T. centronotus*. In swimming fish, anticipatory saccades precede turning movements, and the saccades themselves are about half the speed of those in temperate teleosts. Optokinetic movements may be elicited relatively easily in the benthic species *T. bernacchii*, but are much more difficult to observe in the cryopelagic *P. borchgrevinki*. The latter species has a tendency toward locomotory optokinesis, sometimes tracking a moving pattern, but often exhibiting a less coordinated general kinesis (Collins and Macdonald, unpubl. observ.). One further observation of eye movement is also pertinent to visual function: obvious vergence movements of the eyes occur when the large *Dissostichus mawsoni* approaches prey in captivity, indicating the use of binocular vision in this situation. (Montgomery and Macdonald, unpubl. observ.)

2.3 Retinal Structure

The retina contains visual cells and several layers of neuronal elements which process visual information before transmission to the brain. A substantial amount of visual function can be inferred from a structural study of the retina, and this approach has been adopted in a number of studies (Meyer-Rochow and Klyne 1982; Phan 1986; Eastman 1988; Pankhurst and Montgomery 1990).

A retinal pigment epithelium is present in all species, and retinomotor movement of the pigment has been demonstrated histologically in *P. borchgrevinki*

(Meyer-Rochow and Klyne 1982; Pankhurst and Montgomery 1989). Dark adaptation takes about 4 h, during which time the screening pigment is retracted from the rod outer segments. During light adaptation, full vitread pigment extension occurs after about 1 h.

All notothenioids have well-developed duplex retinae containing both rods and cones, with a mean thickness between 150–280 μm, which is near the low end of the 100–500 μm range reported for fish (Eastman 1988). Layering of the retina is typical for teleosts, with the different layers representing different retinal components (Fig. 3B). The highly organized nature of the retina allows one to recognize important cell types and calculate their packing densities.

Three types of photoreceptors have been recognized in notothenioids: (1) single cones; (2) equal twin cones; and (3) rods. Next in importance to the photoreceptors is the ganglion cell layer. This is the most vitread layer, and these cells send their axons to the brain as the optic nerve. The convergence ratio between photoreceptors and ganglion cells gives an indication as to whether the retina is specialized towards high spatial acuity (low convergence ratio) or high sensitivity (high convergence ratio). There are considerable interspecific differences in type of dominant photoreceptor present, density of photoreceptor packing and convergence ratios. Details of these differences for 22 species of notothenioids are given by Eastman (1988), who recognizes four groups. Non-antarctic notothenioids typically live in shallow water and have well-developed pigment layers, large numbers of visual cells (140–300/100 μm^2) with moderate cone:rod ratios (1:7–14) and moderate convergence ratios (17-28:1). Cryopelagic species such as *Pagothenia borchgrevinki* live under the sea ice and have low visual cell densities (*ca.* 50/100 μm^2), high cone:rod ratios (1:2–4) and low convergence (5–10:1) indicating good visual acuity. Pelagic and benthopelagic species are deeper dwelling and have large numbers of visual cells (100–200/100 μm^2), low ratios of cones:rods (1:16–57) and high convergence ratios (12–58:1). *Dissostichus* is the most extreme example of this latter group, with a rod-dominated retina (cone:rod ratio = 1:57) and a high degree of convergence (58:1), which increase sensitivity at the expense of acuity. The final group of six benthic species which live at moderate depths have intermediate retinas with moderate to high cone:rod ratios and moderate to low convergence ratios. In general terms, the retinae appear balanced for vision under dim but seasonally variable light conditions with a degree of acuity traded for sensitivity.

From a measurement of the cone spacing, and taking into account the optics of the eye, it is possible to calculate theoretical acuity. This has been done for a range of sizes of the cryopelagic *Pagothenia borchgrevinki* (Pankhurst and Montgomery 1990). The results of this study show that as these fish grow, relative eye size and cone density decrease. These changes are offset by the changing optics of the eyes to maintain a virtually constant theoretical acuity over a wide range of fish sizes. The actual value of their theoretical acuity (minimum separable angle of 25–40') falls towards the lower end of the values recorded for other teleosts and is a consequence of large cone diameters rather than wide cone spacing. This provides further evidence of an eye adapted to operate under low light conditions, since the large cones enhance sensitivity, but due to structural constraints, do so at the expense of acuity.

The pineal organ of *P. borchgrevinki* contains typical photoreceptor cells (Meyer-Rochow *et al.* 1986). Both rodlike cells, which react immunocytochemically to anti-opsin, and conelike cells, which react to anti-visinin, have been reported. The pineal extends through a layer of adipose tissue to reach the skull (Morita *et al.* 1990), which contains a pronounced pineal window (Meyer-Rochow and Morita, pers. comm.). These findings suggest that pineal photoreception may be of some importance to these fish, probably for the neuroendocrine regulation of reproductive cycles.

2.4 Visual Function

Some aspects of visual function have been determined from electroretinogram (ERG) and a behavioural studies (Pankhurst and Montgomery 1989). The ERG (Fig. 3C) is the compound electrical activity of the retina recorded during light flashes of known intensity and spectral composition. For dark-adapted *P. borchgrevinki* and three species of *Trematomus* mean ERG thresholds to white light were 1-3 $\times 10^{-3}$ µE m^{-2} s^{-1} (Fig. 3D; about 0.36–1.1 mlux for the light source used), which falls at the lower end of the range determined for a variety of dark-adapted shallow marine species, as one might expect from the retinal histology. Morita and Meyer-Rochow (1990) report very similar ERG threshold values (0.2 mlux) for *P. borchgrevinki.* The above measurements were made in summertime, and it is conceivable that rod outer segments may lengthen somewhat during the winter months, which could increase scotopic sensitivity.

In both studies, spectral sensitivity curves peaked at wavelengths around 500 nm in all four species, matching the spectral irradiance under the ice. Sub-ice phytoplankton growth in mid to late summer would shift irradiance to longer wavelengths, degrading visual ability.

ERGs can also be used to determine the flicker fusion frequency (FFF), which provides a measure of the temporal resolving power of the eye. The FFF was lower in *Trematomus* (1–5 Hz at the maximum ambient light level of 0.5 µE m^{-2} s^{-1}) species than in *P. borchgrevinki* (5–10 Hz) but was still surprisingly close to values reported for other shallow temperate marine species. FFF values of fish have been shown to be very temperature sensitive, which suggests that the antartic fish show considerable compensation in terms of temporal resolving power.

Light-adapted ERG thresholds in *P. borchgrevinki* were about ten times higher than scotopic thresholds (mean $19.3 \pm 11.5 \times 10^{-3}$ µE m^{-2} s^{-1}; Pankhurst and Montgomery 1989). The minimum light-adapted threshold of 6×10^{-3} µE m^{-2} s^{-1} closely matched the observed threshold for feeding in light-adapted *Pagothenia* (5×10^{-3} µE m^{-2} s^{-1}) and is thought to correspond to the threshold for cones. When the visual threshold for prey detection is compared with the light conditions and food availability under the ice, it becomes apparent that even under the most favourable conditions for vision, *Pagothenia* must be operating close to its photopic visual threshold (Fig. 4; Montgomery et al. 1989). It should be noted that prey items taken by *P. borchgrevinki* are quite small in size (Foster *et al.* 1987); species feeding on larger prey may detect it scotopically, and thus be visually competent at greater depths.

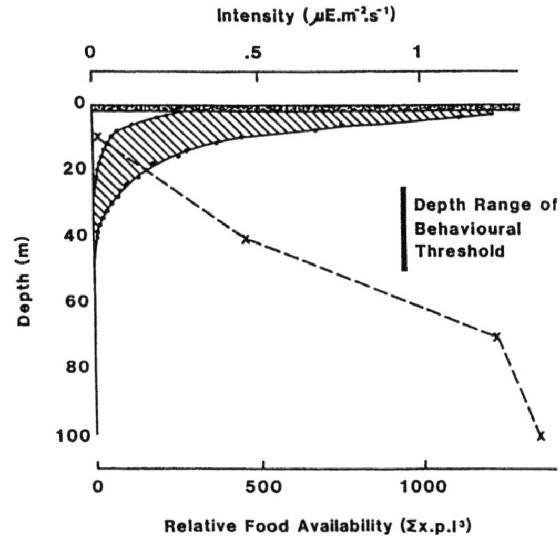

Fig. 4. Depth profiles of light intensity and food availability for *Pagothenia borchgrevinki* in McMurdo Sound. The *shaded layer* at the water surface represents sea ice cover. The *cross-hatched band* represents the range of midsummer light intensities between those determined for midday, with high sun angle (*lower* margin of band), and those determined for midnight, with low sun angle (*upper* margin of band). The depth range corresponding to the behavioural threshold for feeding is shown by the *heavy bar* to the *right* of the figure. The vertical profile of food availability is also shown, obtained by summing the availability for each prey species at each depth. Food availability for each prey species was calculated by multiplying the density of individuals (x) times the proportional representation of the species in the *P. borchgrevinki diet* (p) times the cube of the length l^3 of an average specimen (Montgomery et al. 1989)

2.5 Conclusions

The eyes of notothenioids do not show any extreme structural specializations for vision in light of low intensity, and visual function does not appear to have been substantially modified as a response to specific demands of the antarctic environment. There appears to be some increase in sensitivity which has been won at the expense of reduced acuity. Small prey items will only be detectable visually to depths of about 40 m under typical snow and ice conditions found in early summer. Larger objects can be detected with rod, or scotopic, vision which will extend to greater depths. Spectral sensitivity would appear to be appropriate for the ambient spectral irradiance occurring during only part of the antarctic summer. When conditions for vision deteriorate due to lower ambient light levels (increased snow and ice cover, sub-ice algal blooms, phytoplankton blooms, low sun angle, winter darkness or depth) it is expected that reliance on non-visual systems will increase.

3 Chemosensory Systems

Olfaction and gustation can both be considered chemosenses. There have been no functional studies of either sense in antarctic fish, but some structural information is available.

All five families of the suborder notothenioidei have a single external nostril on each side which is somewhat unusual amongst fish. The presence of a single nare is associated with an accessory nasal sac, which is thought to assist in the exchange of water at the sensory epithelium (Iwami 1986). Water exchange over the olfactory epithelium is tidal, driven by respiratory pressure fluctuations in the buccal chamber conveyed to the accessory sac through the soft palatal tissues (Janssen, pers. comm.).

Taste buds of three species have been examined with the scanning electron microscope (Meyer-Rochow 1981). Size-corrected densities of lingual sensory papillae are intermediate between those of trout (high density) and mesopelagic fishes (low density). Among the three nototheniids studied, receptor density increased slightly from *P. borchgrevinki* – *D. mawsoni* –*T. bernacchii*.

Behavioural observations of *P. borchgrevinki* provide clear evidence for the use of gustation in choice of available food items. The pteropod *Clione limacina* is a conspicuous and abundant element of the McMurdo Sound zooplankton, but is rare in *P. borchgrevinki* gut contents, although other smaller pteropods are common (Foster et al. 1987). Captive fish avoid *Clione*, but feed avidly on the planktonic amphipod *Hyperiella dilatata*. However, *Hyperiella* occasionally attaches to *Clione*, and the tandem pair of amphipod + pteropod are taken and rapidly spat out again (McClintock and Janssen, 1990).

4 Mechanosensory Systems

4.1 Tactile Sense

There have been no published studies of the tactile or somatosensory systems of notothenioid fish. However, preliminary behavioural studies with one of the plunderfish (*Histiodraco velifer*; Fig. 5) and with some of the other benthic species indicate that tactile inputs are likely to be important in the feeding responses of these fish. In the case of *Histiodraco*, gently grasping the extended mental barbel in a captive specimen was repeatedly observed to elicit a complex response, consisting of a sudden opening of the mouth, expansion of the entire buccal chamber, withdrawal of the barbel into the mouth, and a short forward lunge. As the *Histiodraco* barbel bears a fanciful resemblance to a caprellid amphipod (e.g., *Antarcturus* sp.), and is twitched or "jigged" while deployed, it is apparent that the barbel is being used both as fishing lure and as prey sensor. The barbel lacks taste buds, but is studded with what appear to be Pacinian corpuscles. (Janssen, pers. comm.).

4.2 Octavolateralis Sense Organs

The octavolateralis sense organs include the vestibular labyrinth and the anterior and posterior lateral lines (see Montgomery 1988 for a review of this system in elasmobranch fish). All these sense organs are associated with the eighth cranial nerve, and are based on mechanosensory hair cells. The anatomical structure around the hair cells determines the effective stimulus for each type of octavolateralis sense

Fig. 5. *Histiodraco velifer*. The mental barbel is used as a fishing lure, and contains mechanoreceptors which trigger a reflex feeding lunge. Note also the prominent tubular openings of the anterior lateral line. *Scale bar* = 10 mm (photo J.A. Macdonald 1989)

organ. The lateral line is a short-range, exteroceptive, mechanosensory system which responds to water currents and vibrations, whereas the vestibular senses are predominantly concerned with the animal's own movement in space, responding to angular rotations and linear accelerations (including gravity). Information from these senses is used in the control of eye and body movement and in the control of posture. The labyrinth also contains the exteroceptive sense of hearing.

4.3 The Vestibular Labyrinth

The horizontal vestibulo-ocular reflex has been adopted as a model system in which to examine aspects of temperature compensation in reflex behaviour (see Sec. 2.2). The sensory input to the reflex is from the horizontal semicircular canals of the labyrinth which respond to horizontal head rotations.

Electrophysiological recordings have been made from central vestibular neurons during horizontal head rotation (Montgomery and McVean 1987; Montgomery et al. 1990). A detailed quantitative comparison between goldfish (Allum and Graf 1977; Hartmann and Klinke 1980) and *Pagothenia borchgrevinki* shows that despite the 20 °C difference in temperature, vestibular function is very comparable in these two species. The mean gain of *P. borchgrevinki* central vestibular units is 2.6 nerve impulses/s per degree head rotation/s, which is in fact higher than that reported for temperate fish. Vestibular function is dependent on a number of processes which are temperate sensitive; their precise temperature dependence and the sites of temperature compensation require further study.

Chemical analyses confirm that the composition of endolymph from *P. borchgrevinki* and *Dissostichus mawsoni is* typical of that from other vertebrates, with high K^+, low Na^+ and very little protein. The viscosity of nototheniid endolymph at -2 °C (Macdonald and Wells, this Vol.) is very close to values extrapolated to that temperature from measurements of temperate fish endolymph (ten Kate and Kuiper 1970). Thus there is no indication that any compensatory changes in endolymph viscosity have been made to offset the effect of low temperature.

Published models of temperature effects on endolymph viscosity have been generally unsatisfactory in predicting viscosities for antarctic fish. Oman (1980) used a linear model, which predicts a viscosity less than that of pure water at -2 °C; the model is satisfactory in the 20–40 °C range, but was never intended for subzero temperatures. A model published by ten Kate and Kuiper (1970) is unworkable, probably due to typographical errors. For this paper, endolymph viscosity has been predicted using a modification of the polynomials published for pure water by Kestin *et al.* (1978; Fig. 6). In effect, the Kestin polynomials were offset to the right by a constant term (equivalent to 11.42887 °C) to give a viscosity of 1.200 at 23 °C, which is the viscosity of pike endolymph determined by ten Kate and Kuiper (1970). Equation (1) gives the predicted viscosity of fish endolymph:

$$\log_{10}\eta_T = [(R-T)/(T+107.42887)][1.2364 - 1.37 \times 10^{-3}(R-T) + 5.7 \times 10^{-6}(R-T)^2],$$
where R = 31.42887, (i.e., 20 °C plus displacement factor).

Endolymph density is about 1.0 at 20 °C (ten Kate et al. 1970); standard hydrographic tables (e.g., Riley and Skirrow 1975, Table 10) would predict an increase in density of about 0.2–0.3% at –2 °C. Because of the similarities in endolymph composition between antarctic fish and other vertebrates, we assume that endolymph density, like viscosity, is not temperature compensated.

However, in *P. borchgrevinki* there is a suggestion that the vestibular canals may be disproportionately large for the size of the fish, which would tend to offset the effects of low temperature. In *P. borchgrevinki* dissected immediately after death, the vestibular canals appear turgid; the walls are extremely tenuous, and the canals deflate to about half their in vivo diameter when punctured. Measurements of

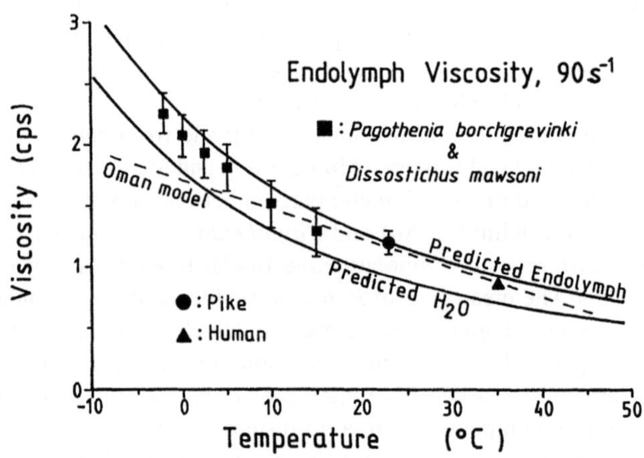

Fig. 6. The viscosity of endolymph and water vs temperature. Data have been combined for *Pagothenia borchgrevinki* and *Dissostichus mawsoni*; mean viscosities ± SE, measured at a shear rate of 90 s^{-1} are plotted over the temperature range –1.9 to 15 °C. The viscosity or pure water, calculated according to Kestin et al. (1978) is shown for comparison, as is a prediction for endolymph viscosity of pike derived from the Kestin model for water (see text). At this intermediate shear rate, nototheniid endolymph viscosities overlap the predicted values of pike. Oman's (1981) linear model for endolymph viscosity is also shown; this model is not valid beyond its intended temperature range of 20–40 °C

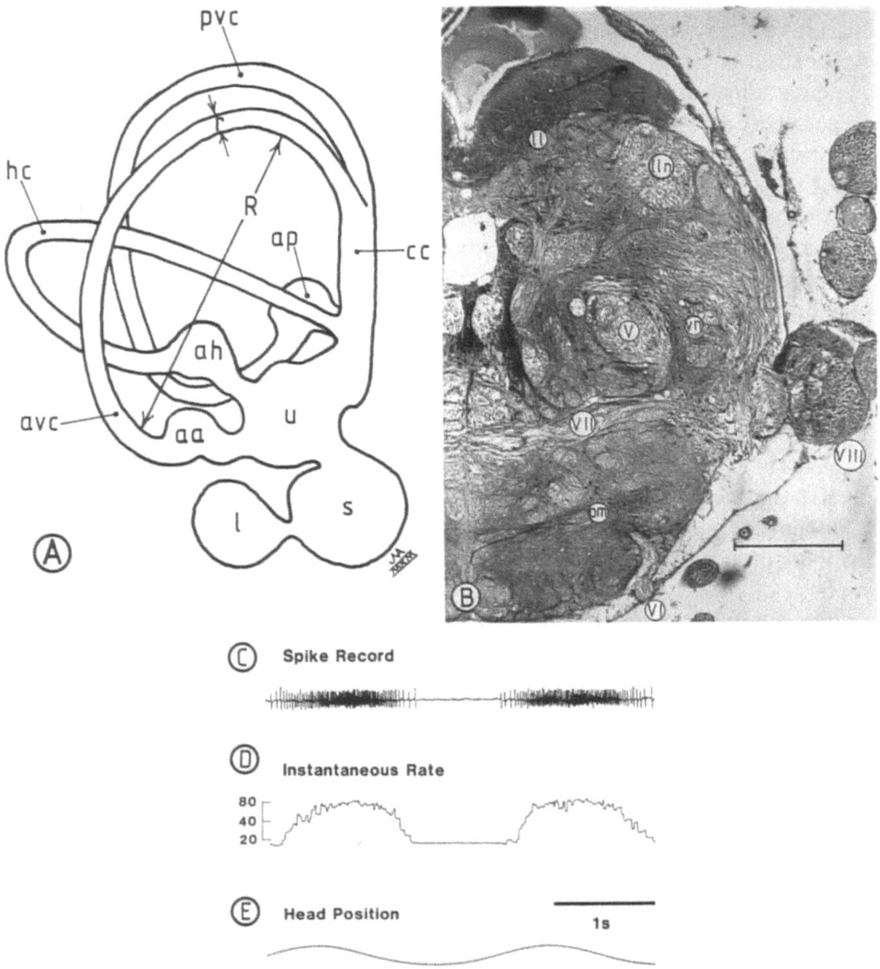

Fig. 7. The vestibular system. **A** Diagrammatic anterior view of right vestibular apparatus, showing dimensions critical for sensitivity. *aa* Ampulla of anterior vertical canal; *ah* ampulla of horizontal canal; *ap* ampulla of posterior verticall canal; *avc* anterior vertical canal; *cc crus commune*; *hc* horizontal canal; *l* lagena; *pvc* posterior vertical canal; r lumen diameter of semicircular canal; *R* diameter of toroid formed by a semicircular canal; *s* sacculus; *u* utriculus. **B** Section of hindbrain of *Pagothenia borchgrevinki* showing vestibular nucleus, from which were made the recordings illustrated in **C** and **D**. ll Lateral line lobe; lln tract of lateral line nerve; om abducens nucleus; *vn* 2° vestibular nucleus; *V* descending tract of 5th cranial nerve; *VI* root of 6th cranial nerve; *VII* motor tract of 7th cranial nerve; *VIII* branch of 8th cranial nerve. *Scale bar* = 500 μm. **C-E** Vestibular response to head rotation (Montgomery and McVean 1987). **C** Record of spike discharge of a central vestibular unit on the fishes' left side, responding with an increase in firing rate during ipsilateral head rotation (type I unit). This unit had a moderate rate of spontaneous activity (29 impulses/s), a high coefficient of variation (19), and a high gain (8.7 impulses/s/°/s). **D** Output of an instantaneous logarithmic rate meter; calibration in impulses/s. **E** Record of movement of the fish; head position to the *right* is upwards, and the peak to peak rotation is 5°. Time calibration is common to all records; each cycle of head movement is 2.5 s (0.4 Hz)

the anterior vertical canal in a single specimen of *P. borchgrevinki* (230 mm TL, 120 g) gave an external diameter of 890 µm. Subtracting the wall thickness that one would expect in canals of the same external dimensions from temperate fish leaves an internal lumen diameter of about 400 µm. The chord of the toroid formed by the anterior vertical canal measured 9–12 mm from the *crus commune* to near the ampulla, which would correspond roughly to the toroidal diameter (Macdonald, unpubl. observ.; Fig. 7).

Allometric relationships established by other authors (Jones and Spells 1963; ten Kate et al. 1970; Howland and Masci 1973) have been used to predict the dimensions of semicircular canals in a 230-mm specimen of *P. borchgrevinki*. All of the predictions seriously underestimated the dimensions of the *P. borchgrevinki* canals by factors of 30–50%. A temperate fish of the same size would have an external canal diameter of about 480 ± 100 µm (Howland and Masci 1973), with a luminal diameter of 250–300 mm diameter (Jones and Spells 1963; ten Kate and Kuiper 1970). The radius of curvature of the anterior vertical canal also appears to be greater in *P. borchgrevinki*; the allometric models predict a toroidal diameter of only 4–6 mm.

Models of semicircular canal function generally agree that the mechanical sensitivity of the canal is proportional to the fourth power of lumen radius (r), to the 1st power of endolymph density (ρ) and toroidal radius (R), and inversely proportional to endolymph viscosity (η) ($G \propto r^4 \rho R/\eta$); ten Kate et al. 1970; Oman 1981). Compared with their values in temperate fishes of equivalent size, all of these factors are increased in *P. borchgrevinki*: r is increased by a factor of 1.3–1.5; R is increased two- to threefold and η is increased almost two-fold. The relatively small increase in ρ will be ignored for the time being. Using the lesser figures for canal dimensions in *P. borchgrevinki*, the net change in sensitivity would by given by:

$$\Delta G \approx (1.3)^4\,(2)/(2) = 3;$$

thus the mechanical sensitivity of the *P. borchgrevinki* semicircular canals would appear to be more than compensated for subzero temperatures.

Temperature-related differences in synaptic function may also contribute to high sensitivity in sense organs. *P. borchgrevinki* end-plate currents show only a partial temperature compensation, in which both amplitude and duration of synaptic currents are larger than predicted by extrapolation from temperate fish (Macdonald and Montgomery 1986). Similar changes in central synaptic currents would be expected to further augment sensory gain.

The other main sense organs of the labyrinth are the otolith organs. The heavy calcareous otoliths confer a sensitivity to linear accelerations such as gravity, those produced during swimming, or movement of the fish in an acoustic field. However, there has been no physiological study of otolith function in notothenioid fishes from which to assess gravitation reception or hearing capabilities. It is unlikely that their hearing capabilities are especially good, since notothenioids do not possess swim bladders; in other fish groups such as the siluroids, the swim bladder is coupled to the labyrinth and is used to enhance the sense of hearing by responding to the pressure component of the acoustic field.

The otoliths of *T. bernacchii* are reported to be strongly piezoelectric, which could theoretically subserve a sensory function such as transduction of vibration or hydrostatic pressure (Morris and Kittleman 1967). However, there is no functional evidence for such a role.

4.4 The Lateral Line System

The lateral line system is a series of pores and canals which ramify over the surface of the head and along the lateral sides of the trunk. Between each of the pores is a vibration-sensitive patch of hair cells called a neuromast; free neuromasts may also occur independently of canals and pores, particularly on the scale-free skin of the head. The gross morphology of the anterior lateral line system in *Pagothenia* is shown in Fig. 2. Physiological studies have been undertaken in this species which show that the anterior lateral line has properties suited to the detection of planktonic prey. Primary afferent lateral line neurons respond to frequencies of vibration which overlap with the major frequency component in the swimming movements of crustacean swimming prey. Simultaneous nerve recordings and movement records of swimming prey (Fig. 8) confirm that vibration produced by crustacean swimming is indeed a potent natural stimulus of the lateral line system (Montgomery and Macdonald 1987; Montgomery et al. 1988). Recent results indicate strongly that movement of a signal source (zooplankter) past a series of pores further increases the response (Montgomery and Janssen, unpubl. observ.). The suggestion that a moving signal source is a more potent stimulator of lateral line receptors is novel, and necessitates changes in models and experimental techniques which are beyond the scope of this chapter.

The morphology of the anterior lateral line system is an important systematic characteristic at the generic level and has been studied in detail for some species (Andriashev and Jakubowski 1971; Andersen 1984). One of the interesting, but as yet unpublished observations is that the notothenioids seem to exhibit the entire range of lateral line morphologies, including free-standing neuromasts, rigid canal systems, and membranous canal systems (Montgomery, Coombs and Janssen, unpubl. observ.). The extreme development of the lateral line system incorporating membranous canals in the bathydraconid *Bathydraco nudiceps* is shown in Fig. 9. In

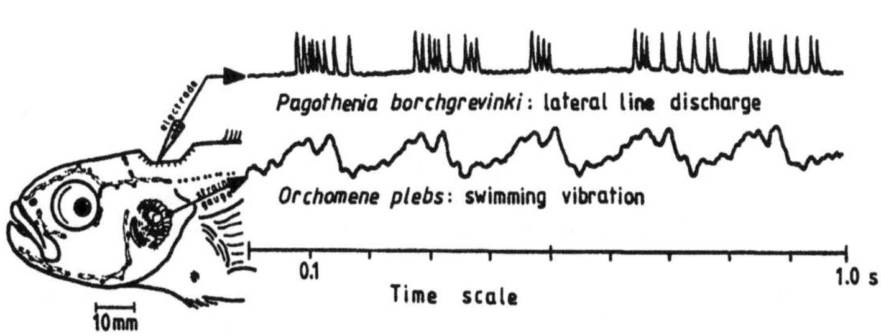

Fig. 8. Lateral line and neural response to stimulation. Nerve spikes were recorded extracellularly from the base of the anterior lateral line nerve where it enters the brain. The anterior lateral line was stimulated by a live amphipod glued to a strain gauge, which monitored the amphipod's swimming vibrations. A strong neural response was recorded in phase with the power stroke of the swimming crustacean, indicating that the anterior lateral line is capable of detecting individual swimming plankters

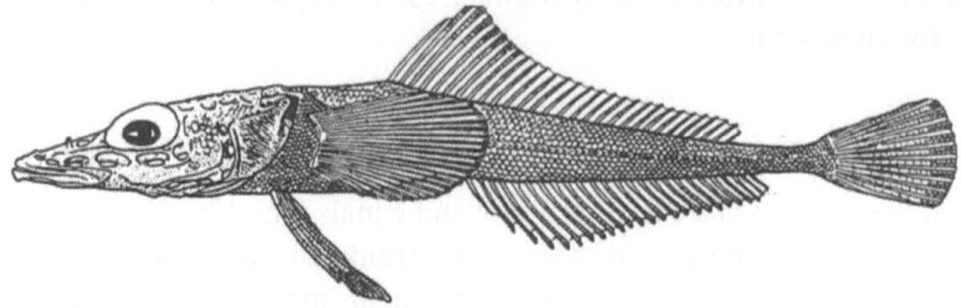

Fig. 9. *Bathydraco nudiceps* (family Bathydraconidae). This species illustrates one extreme of the development of the anterior lateral line, showing conspicuous membranous pits

the trunk region, the posterior lateral line may be divided into as many as three separate lines (dorsal, medial, ventral), which is fairly unusual amongst fish. This range of morphology within a monophyletic radiation provides an excellent opportunity to study the physiological and behavioural correlates of different lateral line structures. The diversity of lateral line specializations amongst the notothenioids also suggests that this sensory modality is of particular importance to this group of fish. A variety of factors, including ease of dissection and hardiness of individuals, have led to an intensive study of the notothenioid lateral line, to the extent that it will soon be better understood than that of many other fish, and may lead to new models of lateral line function. We believe that a study of lateral line function may provide the key to understanding how notothenioids survive the long periods when visual systems are inoperable, and that lateral line specializations are a contributing factor to the success of notothenioid fish in the antarctic ecosystem.

Acknowledgments. JAM and JCM wish to thank the manager and staff of Antarctic Division, DSIR, New Zealand for logistic support, and the now defunct University Grants Committee for funding. We are particularly grateful to the past chairman of the Ross Dependency Research Committee, Dr. Trevor Hatherton, for his interest and encouragement. We also thank the Italian National Programme for Antarctic Research for their far-sighted support of International collaboration and scientific dialog in the field of antarctic research.

References

Allum JHJ, Graf W (1977) Time constants of vestibular nuclei neurons in the goldfish: a model with ocular proprioception. Biol Cybernetics 28:95–99

Andersen NC (1984) Genera and subfamilies of the family Nototheniidae (Pisces, Perciformes) from the Antarctic and Subantarctic. Steenstrupia 10:1–34

Andriashev AP (1987) A general review of the antarctic bottom fish fauna. In: Kullander SO, Fernholm B (eds Proc Cong Europ Ichthyol, Stockholm 1985. Swed Mus Nat Hist, Stockholm pp 357–372

Andriashev AP, Jakubowski M (1971) Morphological ground for generic separation of the antarctic broad-head fishes *Trematomus borchgrevinki* Boulenger and *T. brachysoma* (Pappenheim), and a new status of the genus *Pagothenia* Nichols et Lamote (Nototheniidae) (in Russian). Zoologicheskii Zhurnal 50: 1041–1055

DeWitt HH (1971) Coastal and deep-water benthic fishes of the Antarctic. Antarctic Map Folio Ser, Folio 15, pp 1–10, 31 maps. Am Geogr Soc, New York

Eastman JT (1988) Ocular morphology in antarctic nototheniid fishes. J Morphol 196:283–306

Foster BA, Cargill JM, Montgomery JC (1987) Planktivory in *Pagothenia borchgrevinki* (Pisces: Nototheniidae) in McMurdo Sound, Antarctica. Polar Biol 8:49–54

Hartmann R, Klinke R (1980) Discharge properties of afferent fibres of the goldfish semicircular canal with high frequency stimulation. Pflugers Arch 388:111–121
Howland HC, Masci J (1973) The phylogenetic allometry of the semicircular canals of small fishes. Z Morph Tiere 75:283–296
Iwami T (1986) A note on the nasal structures of fishes of the suborder Notothenioidei (Pisces, Perciformes). Mem Natl Inst Polar Res, Spec Issue 44:151–152
Jones GM, Spells KE (1963) A theoretical and comparative study of the functional dependence of the semicircular canal upon its physical dimensions. Proc R Soc Lond B 157:403–419
ten Kate JH, Kuiper JW (1970) The viscosity of the pike's endolymph. J Exp Biol 53: 495–500
ten Kate JH, van Barneveld HH, Kuiper JW (1970) The dimensions and sensitivities of semicircular canals. J Exp Biol 53:501–514
Kestin J, Sokolov M, Wakeham WA (1978) Viscosity of liquid water in the range –8 °C to 150 °C. J Phys Chem Ref Data 7:941–948
Macdonald JA, Montgomery JC (1982) Thermal limits of neuromuscular function in an antarctic fish. J Comp Physiol 147:237–250
Macdonald JA, Montgomery JC (1986) Rate-compensated synaptic events in antarctic fish: consequences of homeoviscous cold-adoptation. Experientia 42:806–808
Macdonald JA, Montgomery JC, Wells RMG (1987) Comparative physiology of antarctic fishes. Adv Mar Biol 24:321–388
McClintock JB, Janssen J (1990) Pteropod abduction as a chemical defense in a pelagic antarctic amphipod. Nature (Lond.) 346:462–464
Meyer-Rochow VB (1981) Fish tongues - surface fine structures and ecological considerations. Zool J Linnean Soc 71:413–426
Meyer-Rochow VB, Klyne MA (1982) Retinal organization of the eyes of three nototheniid fishes from the Ross Sea (Antarctica). Gegenbaurs Morphol Jahrb 128:762–777
Meyer-Rochow VB, Vigh-Teichmann I, Vigh B (1986) Electron microscope observations on the structure of the pineal organ of the antarctic notothenioid teleost fish *Pagothenia borchgrevinki*. Int Symp Functional Morphology of Neuroendocrine Systems: Evolutionary and Environmental Aspects; August 1986, Giessen, (Abstract)
Montgomery JC (1984) Low temperature increases gain in the oculomotor system. J Neurobiol 15:295–298
Montgomery JC (1988) Sensory physiology. In: Shuttleworth TJ (ed) Physiology of elasmobranch fishes. Springer, Berlin Heidelberg New York Tokyo
Montgomery JC, Macdonald JA (1984) Performance of motor system in antarctic fish. J. Comp Physiol 154:241–248
Montgomery JC, Macdonald JA (1985) Oculomotor function at low temperature: antarctic versus temperate fish. J Exp Biol 117:181–191
Montgomery JC, Macdonald JA (1987) Sensory tuning of lateral line receptors in antarctic fish to the movements of planktonic prey. Science 235:195–196
Montgomery JC, McVean AR (1987) Brain function in antarctic fish: activity of central vestibular neurons in relation to head rotation and eye movement. J Comp Physiol 160:289–293
Montgomery JC, Macdonald JA, Housley GD (1988) Lateral line function in an antarctic fish related to the signals produced by planktonic prey. J Comp Physiol 163:827–833
Montgomery JC, Macdonald JA, Housley GD (1990) Brain function in antarctic fish: frequency response analysis of central vestibular units. J Comp Physiol A 166:407–410
Montgomery JC, Pankhurst NW, Foster BA (1989) Limitations on visual feeding in the planktivorous antarctic fish, *Pagothenia borchgrevinki*. Experientia 45:395–397
Morita Y, Meyer-Rochow VB, Tamotsu S, Uchida K (1990) Photoreception of antarctic fish- immunocytochemical analysis in pineal organ and electroretinogram. European Pineal Study Group, Sep 1990 Meeting, Univ Surrey, UK (Abstract)
Morris RW, Kittleman LR (1967) Piezoelectric property of otoliths. Science 158:368–370
Munk O (1966) Ocular anatomy of some deep-sea teleosts. Dana Rep 70:1–71
Munk O (1984) Duplex retina in the mesopelagic teleost *Radiicephalus elonagatus* Osorio, 1917. Vidensk Meddr Dansk Naturh Foren 145:183–199
Oman CM (1981) The influence of duct and utricular morphology on semicircular canal response. In: Gualtierotti T (ed) The vestibular system: function and morphology. Springer, Berlin Heidelberg New York Tokyo, pp 251–274

Pankhurst (1989) The relationship of ocular morphology to feeding modes and activity periods in shallow marine teleost from New Zealand. Environ Biol Fish 24:201–211

Pankhurst NW, Montgomery JC (1989) Visual function in four antarctic nototheniid fishes. J Exp Biol 142:311–324

Pankhurst NW, Montgomery JC (1990) Ontogeny of vision in the antarctic fish *Pagothenia borchgrevinki* (Nototheniidae). Polar Biol 10:419–422

Phan MT (1986) Histology of retina in fishes from Bransfield Strait. Acad Brasil Ciencias 58 Suppl 1:35–42

Pockett S, Macdonald JA (1986) Temperature dependence of neurotransmitter release in the antarctic fish *Pagothenia borchgrevinki* Experientia 42:414–415

Riley JP, Skirrow G (eds) (1975) Chemical oceanography, 2nd edn, vol 2. Academic Press, London

Turner JD, Schrag JD, DeVries AL (1985) Ocular freezing avoidance in antarctic fish. J Exp Biol 118:121–131

Viscosity of Body Fluids From Antarctic Notothenioid Fish

J.A. Macdonald and R.M.G. Wells

1 Introduction

Antarctica and its surrounding seas have been isolated in a progressively colder polar position since the early Tertiary (Kennett 1977). Over the antarctic continental shelf, seawater temperatures are generally below 0 °C, and near ice-covered coastal inlets and in deep basins on the shelf, water temperatures may hover year-round on the freezing point of seawater at –1.87 °C (Littlepage 1965; Andriashev 1970).

Of the 203 species of fish now known to live in waters south of the Antarctic Convergence, 53% (110 spp.) are members of a single perciform suborder, the Notothenioidei. Nearly 97% of the notothenioid species are endemic to the antarctic region (Andriashev 1987). None of the present fish species appears to have any link with earlier fossil fish faunas known from Antarctica (Eastman and Grande 1989), and the notothenioids are generally believed to have originated and diversified in frigid antarctic seas, resulting in a variety of biochemical and physiological specializations (Macdonald et al. 1987).

An all-pervading consequence of low temperature is an increase in fluid viscosity; accordingly, we propose to review in this paper the relevance of viscosity to the physiology of antarctic fish.

2 Seawater Viscosity

Pure water is 1.92 times more viscous at –2 than at 20 °C (Kestin *et al.* 1978); the viscosity of sea water at 35% salinity is about 7–8% higher than that of pure water, and roughly parallels pure water over the above temperature range, increasing 1.93-fold to give a viscosity of about 2.07 cps (1 centipoise = 1 mPa.s) at 2 °C (Horne *et al.* 1965; Riley and Skirrow 1975, Table 25). This increase in viscosity of the surrounding medium can be expected to increase the energetic cost of locomotion.

At low velocities, resistance to movement in swimming fish is dominated by frictional resistance, which is determined largely by viscosity. Antarctic fish tend to be less than 500 mm total length and are fairly slow swimmers, resulting in small Reynolds numbers: *Pagothenia borchgrevinki* (200–250 mm) reaches maximum velocities of just over 1 m s^{-1} (Montgomery and Macdonald 1984), giving a Reynolds number of about 130 000. At a similar Reynolds number, 37% of the total

Department of Zoology, University of Auckland Auckland, New Zealand drag developed in swimming tuna was frictional resistance (Magnuson 1978), which depends on viscosity. The elevated viscosity of antarctic sea water almost certainly increases the work required for swimming in antarctic fish: viscous resistance should be particularly significant for swimming in smaller larval fish, and for gill ventilation in all species.

3 Blood Viscosity

Viscosity directly affects vascular resistance and therefore the amount of work the heart must perform to pump blood (Farrell 1984). In cold-acclimated trout, an increased ventricle mass appears to compensate for increased blood viscosity (Graham and Farrell 1989). In addition to temperature, the cellular fraction of blood, or haematocrit, is an important determinant of blood viscosity in fish (Chien et al. 1971; Graham and Fletcher 1985). In several arctic fish, lower haematocrits are correlated with lower blood viscosities (Graham and Fletcher 1985), which presumably reduces the work required to maintain circulation. There are indications that antarctic fish have also achieved reductions in viscosity of body fluids which may compensate for some of the effects of low temperature. For example, it has long been known that the blood of many antarctic fish contains fewer erythrocytes and less haemoglobin than does blood of fish from warmer latitudes (Kooyman 1963; Everson and Ralph 1968: Hureau et al. 1977; Wells et al. 1980), and one notothenioid family, the Channichthyidae (icefish), is completely devoid of haemoglobin and normal erythrocytes (Ruud 1954), although some of the few remaining cells may in fact be aberrant haemoglobin-free erythrocytes (Hureau et al. 1977). Reduction of red cell mass will obviously decrease the oxygen-carrying capacity of blood, and it may be that the larger ventricular mass of antarctic fish hearts (B. Tota, pers. comm.) permits circulation of blood with reduced viscosity at high shear rates. Several investigators have reported briefly on the viscosity of channichthyid blood (Hemmingsen and Douglas 1972; Twelves 1972), and ichthyologists generally believe that it is less than that of other fish (Andriashev 1987). Detailed comparisons have been made recently between blood viscosities of channichthyids, antarctic nototheniids, and temperate New Zealand teleosts (Macdonald et al. 1989; Rickenbach et al. 1989; Wells et al. 1990) and is reviewed below.

3.1 Haematology

Haematological parameters of some antarctic and two New Zealand fish are given in Table 1. These data support the established view that antarctic fish are cytopenic in comparison to temperate fish. The two channichthyid species represent an extreme, in which cells make up about 1% of blood volume, while the antarctic nototheniids tend to have haematocrits between 15 and 18%, compared with 20–27% for the New Zealand species, blue cod (*Parapercis colias*, family Mugiloidodae), and black cod (*Notothenia angustata*, family Nototheniidae).

However, such measurements can be confounded by the effects of stress, which are particularly apparent in the data from the antarctic nototheniids. In acutely sampled *Pagothenia borchgrevinki* stressed by enforced swimming for 10 min,

Table 1. Haematology of antarctic and New Zealand fish

Species	Hct (%)	Cell count (× 10^6/ml)	Hb (g/l)	MCHC (g/l)
Antarctic species *Chionodraco hamatus* (580 ± 28g) n=6	1.23 ± 0.82	0.027 ± 0.026	0	0
Cryodraco antarcticus n=2	1.25	0.039	0	0
Trematomus bernacchii (385 ± 47g) n=5 Cannulated	7.6 ± 0.7	0.361 ± 0.0002	18.3 ± 3.6	242.0 ± 29.9
Trematomus bernacchii Acute sampling, n=10	15.2 ± 0.84	0.702 ± 0.0002	31.7 ± 1.9	208.5 ± 3.9
Pagothenia borchgrevinki Acute sampling, n=5	17.7 ± 0.60	0.830 ± 0.103	32.4 ± 2.1	183 ± 12
Pagothenia borchgrevinki Acute; stressed, n=5	31.0 ± 1.90	1.190 ± 0.173	47.0 ± 1.9	152 ± 6
Trematomus centronotus Acute sampling, n=2	16.9 ± 1.5	0.753 ± 0.186	N.A	N.A.
Dissostichus mawsoni Acute sampling, n=2	24.1 ± 0.52	1.11	N.A.	N.A.
New Zealand species *Parapercis colias* Acute sampling, n=5	21.2 ± 1.9	2.480 ± 0.031	76.9 ± 10.3	363 ± 31
Notothenia angustata Acute sampling, n=1	26.5	0.910	97.5	368

haematocrit was elevated by 75%. Acutely sampled fish are also moderately stressed by the trauma of capture, handling and insertion of the hypodermic needle, whereas cannulated fish may be considered to be unstressed. Blood from *Trematomus bernacchii*, sampled through cannulae chronically implanted in the caudal vein, had relatively low haematocrits.

For example, a healthy cannulated *T, bernacchii* had a haemotocrit of 7.7% when the fish was lying undisturbed with a heart rate of 9.8 beats min^{-1}. After gentle prodding and 30-min recovery, haemotocrit had risen to 9.4%, and heart rate of 14.5 beats min^{-1}. Acute sampling of this fish produced a haematocrit of 10.5% (elevation of 36%), while 10 min. of enforced continuous swimming further elevated the haemotocrit to 13.4% (74% increase) and heart rate to 20.9 beats min^{-1}. We would conclude that much of the data on fish haematocrits is in error. Thus a haematocrit of 15% obtained by acute sampling probably represents a haematocrit of 11–12 % in resting fish.

3.2 Viscosity

The viscosity of whole blood, plasma, and blood reconstituted from plasma and packed erythrocytes to a wide range of haematocrits was determined in a temperature-controlled Wells-Brookfield cone-plate viscometer (Brookfield Engineering Laboratories, Stoughton, Massachusetts) capable of determining the absolute viscosity of small samples (500 μl) under conditions of defined shear rate.

The effect of temperature on the viscosity of blood from five species of teleosts in comparison with the viscosity of pure water is shown in Fig.1. At the defined shear rate of 225 s^{-1}, and haematocrits between 15 and 20%, the two antarctic nototheniids (*Pagothenia borchgrevinki, Trematomus bernacchii*) and the New Zealand mugiloidid *Parapercis colias* all have quite similar viscosities between 6 and 7 cps at 0 °C. For these three species, viscosity appears to decline at about the same rate with increasing temperature, with a mean Q_{10} of 0.71. At their ambient temperature of –1.8 °C, blood viscosity for the antarctic nototheniids averaged between 6.5 and 7.5 cps, compared with about 3 cps for *P. colias* at a temperature of 20 °C in its northern New Zealand range.

Combined mean viscosities for the two channichthyid species, *Chionodraco hamatus* (= *C.kathleenae*) and *Cryodraco antarcticus* are intermediate between the viscosity of corpusculate blood from the other three species, and that of pure water. At –1.8 °C the mean viscosity for channichthyid blood was 3.6 ± 0.2, decreasing with a Q_{10} of 0.74 at higher temperatures.

One particularly notable feature in Fig. 1 is that all of the fluids tested are affected similarly by temperature over the experimental temperature range. At these haematocrits, the steeper slopes seen for corpusculate blood are the consequence of higher absolute viscosities, but Q_{10} values are all very similar to that of 0.74 for water. At these low haematocrits, the temperature dependence of blood viscosity seems to be dominated by that of water.

The influence of haematocrit on blood viscosity can be examined by reconstituting blood to a range of haematocrits from pooled separated cells and plasma (Fig. 2). This technique also cancels out possible effects of individual differences, such as plasma composition or blood cell elasticity. Clearly, haematocrit has a major effect on viscosity; 20% haematocrit has about the same effect as a 20 °C drop in temperature.

Plasma viscosity of the antarctic nototheniids is slightly higher than that of the New Zealand blue cod, and the effect of higher plasma viscosity persists as a light elevation in the reconstituted haematocrits. Plasma and most body fluids of antarctic notothenioids contain an additional 3% w/v of glycopeptide antifreeze (DeVries 1983), and this may account for the marginally higher nototheniid plasma viscosities. Plasma viscosities are considerably higher for arctic species (Graham and Fletcher 1985) than for the antarctic notothenioids reported here. The difference in underlying plasma viscosities must contribute to the difference in overall blood viscosities, if only by elevating the baseline. Surprisingly, plasma viscosities of arctic fish were much more sensitive to shear rate, even though total plasma proteins did not differ significantly between the two groups (40-50 g l^{-1}, measured by the biuret method).

Of the antarctic fish, only the channichthyids, with haemotocrits approaching zero, show a full compensation for the increase in viscosity found at subzero

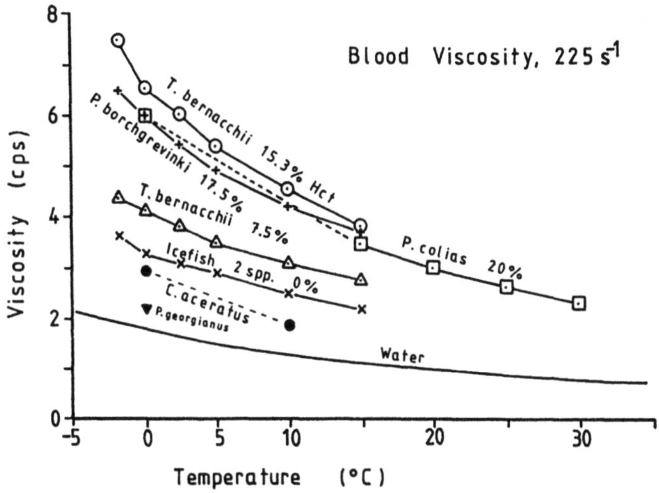

Fig. 1. Blood viscosity of antarctic notothenioids plotted against temperature. Mean viscosities are plotted for two red-blooded nototheniid species (*Trematomus bernacchii, Pagothenia borchgrevinki*), two species of channichthyid (icefish; combined mean values for *Chionodraco hamatus, Cryodraco antarcticus*), and one New Zealand mugiloidid (*Parapercis colias*). Mean Hct values are indicated for each species. Two sets of data are plotted for *T.bernacchii*, at the mean Hct for acute sampling (15.3%), and at lower Hct (7.5%) obtained from cannulated fish; the estimated true in vivo Hct is thought to be intermediate between these two values. Also shown are data on two channichthyid species, *Chaenocephalus aceratus* and *Pseudochaenichthys georgianus* (Hemmingsen and Douglas 1972)

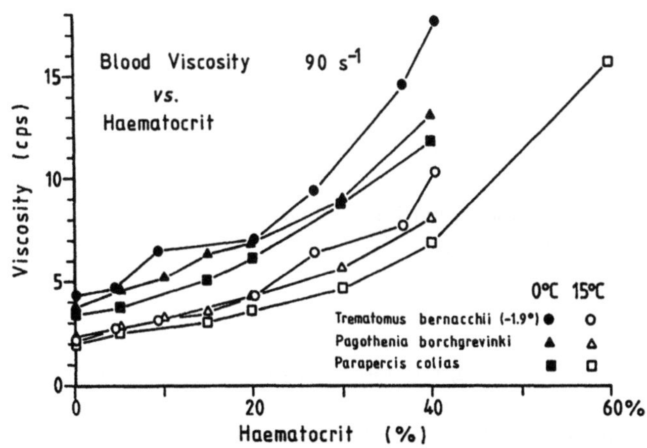

Fig. 2. Blood viscosity as a function of haematocrit. Mean viscosities for two antarctic nototheniids (*Trematomus bernacchii, Pagothenia borchgrevinki*) and a New Zealand mugiloidid (*Parapercis colias*) are plotted at 0 and 15 °C for Hct from 0% (plasma) to 60%. For *T. bernacchii*, low temperature measurements were made at -1.9 °C. The wide range of Hct was obtained by mixing plasma with packed red cells. Shear rate = 90 s^{-1}

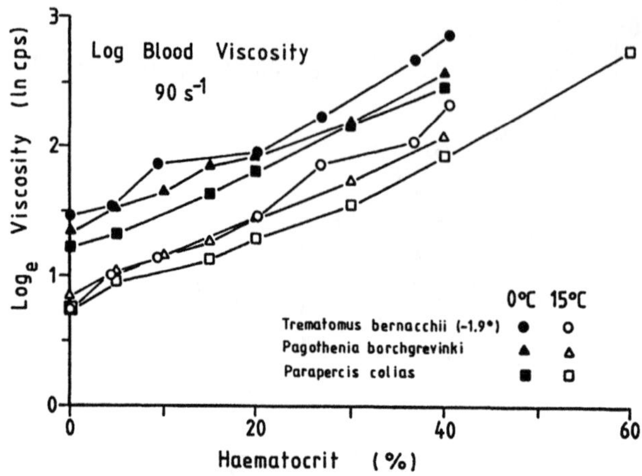

Fig. 3. Semilogarithmic plot of blood viscosity *vs* haemotocrit. Natural logarithms of mean blood viscosities for two antarctic nototheniids (*Trematomus bernacchii, Pagothenia borchgrevinki*) and a New Zealand teleost (*Parapercis colias*) are plotted at 0 and 15 °C against Hct from 0% (plasma) to 60%. For *T. bernacchii*, low temperature measurements were made at –1.9 °C. Shear rate = 90 s^{-1}

temperatures. Complete elimination of the red blood cells gives channichthyid whole blood at –1.8 °C the same viscosity as blue cod blood at 15 °C. Twelves (1972) compared flow rates of channichthyid and nototheniid blood through glass capillaries, and reported that icefish blood had half the viscosity of red nototheniid blood, a conclusion which agrees well with that reported here.

Semilogarithmic plots of viscosity *vs* haematocrit (Fig. 3) are approximately linear, suggesting an exponential relationship between haematocrit and viscosity, and that viscous interactions between individual cells are significant.

One feature which sets blood apart from ideal Newtonian fluids is its strong dependence on shear rate, as exemplified in Fig. 4 by blood from New Zealand blue cod *Parapercis colias*. At shear rates above 90 s^{-1}, whole blood seems to behave very like a Newtonian fluid, and viscosity remains constant at a level determined by temperature and haematocrit, but below 90 s^{-1}, blood viscosity increases dramatically. Blue cod plasma behaves more like a Newtonian fluid, increasing only marginally in viscosity at low shear rates.

A similar plot of viscosity *vs* shear rate at –1.8 °C is shown for antarctic nototheniids in Fig. 5. Whole blood from *Pagothenia borchgrevinki* and *Trematomus bernacchii* is slightly more viscous, but otherwise behaves much like that of *Parapercis*.

Wells et al. (1990) have found a negative correlation between mean cell haemoglobin concentration (MCHC) and blood viscosity. Low MCHC may indicate cell swelling, and is correlated with elevated haematocrit and plasma protein, which are features of stressed fish. MCHC may decrease 20 to 30% in stressed fish, and the implied increase in cell volume would also contribute to an increase in haematocrit. In antarctic fish, there are no data to clarify the relative influences of cell size and cell number on blood viscosity, but in tropical teleosts there is a

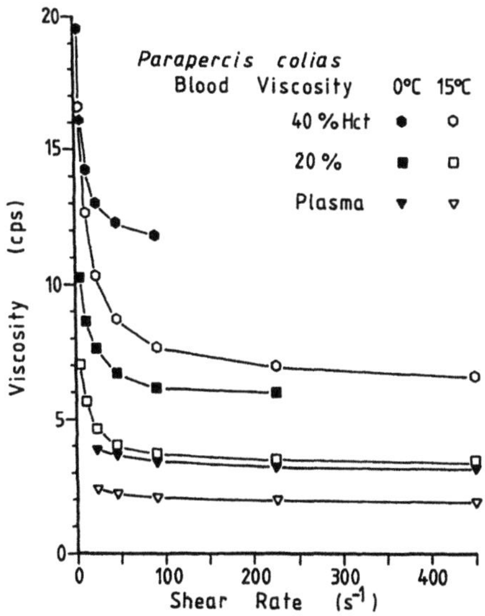

Fig. 4. Effect of shear rate on viscosity of blood from New Zealand teleost *Parapercis colias*. Viscosities at Hct of 0% (plasma), 20% (near normal for this species) and 40% are plotted for temperatures of 0 and 15 °C

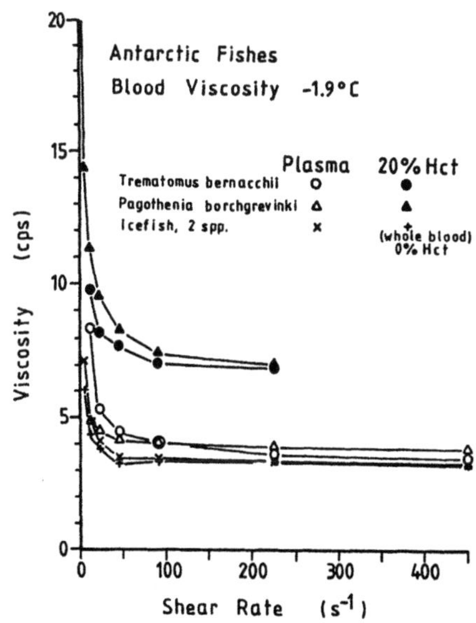

Fig. 5. Effect of shear rate on viscosity of blood from two species of antarctic nototheniids (*Trematomus bernacchii, Pagothenia borchgrevinki*) and combined data for two channichthyid species (*Chionodraco hamatus, Cryodraco antarcticus*). All determinations made at –1.9 °C. Viscosities are plotted for Hct of 0% (plasma, and channichthyid whole blood) and 20%. For the channichthyids (icefish), there is no significant difference between whole blood and plasma

suggestion that large cells increase the dependence of viscosity on shear rate, resulting in higher viscosity at low shear rates (Wells and Baldwin 1990).

Plasma from the two nototheniids, and from the channichthyid *Chionodraco hamatus*, behaves similarly to that of blue cod, although they do show greater increases in viscosity at the lowest shear rates. Absolute values of plasma viscosity are very similar between the New Zealand blue cod and all of the antarctic notothenioids, although the icefish (channichthyid) plasma contains less protein than that of *P. borchgrevinki*, and is less viscous by about 0.5 cps. Mean combined viscosities for whole blood of two channichthyid species (*Chionodraco hamatus, Cryodraco antarcticus*) follow the values for plasma very closely.

Most of the variation in blood viscosity can be explained by temperature, haematocrit and shear rate, but for any given combination of these three factors these remains some variability between species and between individuals. Some of this residual variability can be explained as a consequence of plasma composition. Dialysable salts in the body fluids of antarctic nototheniids contribute a freezing point depression of about –1 °C (O'Grady and DeVries 1982), roughly 60% that of seawater, and would also be expected to contribute proportionately to viscosity. Thus at –1.8 °C, the plasma salts alone should increase viscosity from the 1.92 cps expected for pure water to a value of about 2 cps.

Endolymph extracted from the semicircular canals and ampullae of the vestibular system is essentially protein-free, and does in fact have a viscosity remarkably close to that predicted above (Fig. 6). The significance of this finding is considered in the next section.

Most of the remaining differences in plasma viscosity can be attributed to plasma proteins, which occur commonly in concentrations between 20 and 40 g l^{-1} (Table 2). Figure 6 shows the relationship between viscosity and naturally occurring protein concentrations of a variety of cell-free fluids from antarctic fish, including endolymph, periophthalmic fluid, intracranial fluid and plasma.

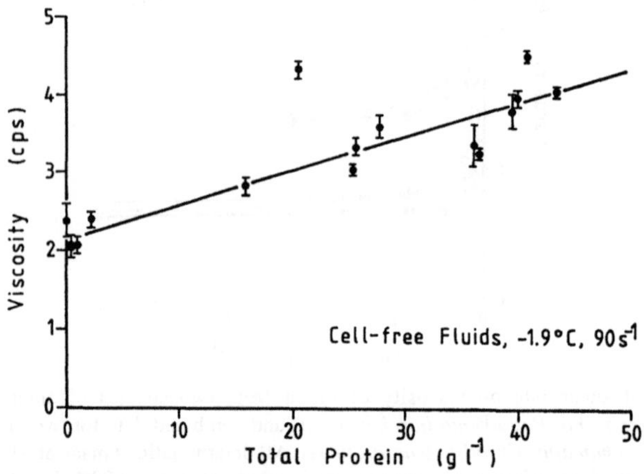

Fig. 6. Effect of protein content on viscosity of cell-free fluids from antarctic fish. Fluids include endolymph, cranial fluid, extraocular fluid and plasma. Mean viscosity ± SE is plotted against total protein, measured by the biuret method; all measurements were made at –1.9 °C and shear rate of 90 s^{-1}

Table 2. Chemical analyses of selected body fluids

	T Prot	Alb	Chol	TG	Na^+	K^+	Ca^+	HDL CH
Chionodracco hamatus								
Plasma	41.2	7.8	5.73	6.31	244	2.8	.51	2.01
n=6	± 3.7	± 1.5	± 1.39	± 1.09	± 2	± 0.1	± 0.51	± 0.43
Cryodraco antarcticus								
Plasma	42	6	11.63	8.2	239	2.8	4.23	3.14
n=1								
Dissostichus mawsoni								
Plasma	33.5	7.0	5.13	0.96	247	3.2	2.45	2.63
n=2	± 8.5	± 1.0	± 2.02	± 0.17	± 7	± 0.2	± 0.88	—
Orbital fluid	26.8	8.0	8.83	3.17	216	2.3	1.97	5.45
n=2	± 1.2	—	—	—	± 11	0.1	—	—
Cranial fluid	28.6	8.5	9.27	3.16	208	2.5	1.83	5.76
n=3	± 0.7	± 0.5	± 0.26	± 0.21	± 1	± 0.0	± 0.0	± 0.24
Endolymph	0.0	0.3	0.04	0.26	211	98.2	0.96	0.0
n=5	± 0.0	0.3 ±	± 0.0	± 0.04	± 12	± 16.2	± 0.11	± 0.0
Trematomus bernacchii								
Plasma	40.0	5.0	5.25	1.61	263	3.5	3.68	2.27
n=6	± 3.7	± 1.5	± 0.56	± 0.55	± 2	0.3 ±	± 0.35	± 0.13
Pagothenia borchgrevinki								
Plasma	38.0	3.0	7.24	3.68	250	2.9	3.55	2.31
n=3	± 1.2	± 0.0	± 0.94	± 0.97	± 16	0.3 ±	± 0.09	± 1.69
Cranial fluid	38.9	7.8	6.89	3.56	228	4.0	2.90	1.72
n=6	± 1.4	± 0.3	± 0.49	± 0.54	± 15	± 0.4	± 0.22	± 0.73
Endolymph	1.3	0.3	0.17	0.24	213	143.6	0.88	0.0
n=5	± 0.3	± 0.3	± 0.03	± 0.03	± 6	± 2.7	± 0.36	± 0.0
Trematomus centronotus								
Plasma	45	5	8.59	4.26	266	9.1	4.13	5.38
n=1								
Notothenia angustata								
Plasma	48	10	8.51	4.93	150	0.6	2.02	—
n=1								
Rhigophila dearborni								
Plasma	16	1	1.89	0.7	221	5.4	3.57	0.85
n=1								

[a]T Prot = total protein, Alb = albumin, Chol - cholesterol,TG - triglycerides, HDL CH = high density lipid; units in g l^{-1}.

4 Endolymph Viscosity

Viscosity is of theoretical importance in several other body fluids aside from those of the circulatory system. For example, the sensitivity of the semicircular canals of the vestibular system is inversely proportional to the viscosity of their contained endolymph (Jones and Spells 1963; ten Kate et al. 1970; Howland and Masci 1973; Oman 1981). In the absence of compensatory changes, the low ambient antarctic water temperature would be expected to approximately double endolymph viscosity, resulting in reduced vestibular sensitivity. However, there are indications that the vestibular system is actually *more* sensitive in antarctic fish than in temperate fish (Montgomery and McVean 1987; Montgomery et al. 1990), suggesting that temperature-compensation has occurred somewhere within the system. An investigation of endolymph viscosity was made to determine its contribution to the overall sensitivity of the vestibular system.

Endolymph was extracted from semicircular canals and ampullae of freshly-killed specimens of *Pagothenia borchgrevinki* and *Dissostichus mawsoni*, using a 30-gauge needle attached to fine polyethylene tubing. The most commonly used sampling site was the *crus commune*, where the anterior and posterior vertical canals fuse and plunge down to the utriculus. In *D.mawsoni* it was possible to get single samples of sufficient volume for viscometry (600 μl) from each semicircular canal, whereas *P. borchgrevinki* yielded only 12–15 μl per canal system, necessitating pooled samples from about 25 individuals for each determination. Most samples of endolymph were protein-free (Table 2), but a few showed slight elevations in protein up to 2.3 g l^{-1}. This is most likely explained by contamination from cranial fluid.

ten Kate and Kuiper (1970) have investigated endolymph viscosity in the pike (*Esox lucius*), and developed a model of the effect of temperature. Their published model suffers from several serious typographical errors which render it unusable; we have corrected the model used for Fig. 7 as follows:

$$\eta = 2.31/(1 + 0.036T + 0.000185T^2),$$

where η = viscosity, T = temperature in °C.

In its corrected form, the ten Kate model predicts viscosities very similar to those derived by Macdonald and Montgomery (this Vol.) by lateral displacement of the Kestin et al. (1978) model of the viscosity of pure water.

In Fig. 7, mean viscosities of endolymph from *P. borchgrevinki* and *D. mawsoni* are plotted against temperature. At a high shear rate of 450 s^{-1}, viscosity follows closely that of sea water, measuring 2.01 ± 0.06 cps at -2 °C, compared with a value of 2.49 cps predicted by the corrected ten Kate model. At this shear rate, it would seem that the viscosity of antarctic fish endolymph is temperature-compensated. However, endolymph viscosity increases slightly at low shear rates, particularly at low temperatures (Fig. 8), and biologically relevant shear rates for endolymph, though unknown are likely to be quite low. As viscosities measured at less than 225 s^{-1} overlap predicted values extrapolated from the pike determination, it appears that endolymph from antarctic fish cannot be claimed to be temperature compensated.

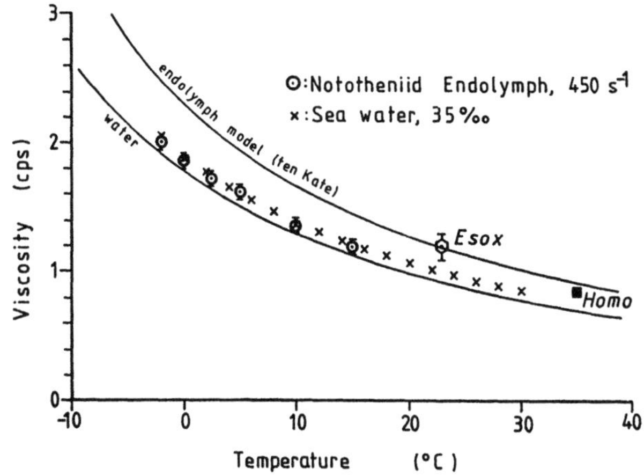

Fig. 7. Viscosity of nototheniid endolymph. Mean viscosities of endolymph from *Pagothenia borchgrevinki* and *Dissostichus mawsoni* measured at a shear rate of 450 s^{-1} are compared with the viscosity of pure water (Kestin et al. 1978), 35% seawater (Riley and Skirrow 1975), and a model of teleost endolymph (after ten Kate and Kuiper 1970). Also shown are viscosities of endolymph from *Esox* (23 °C) and *Homo* (37 °C; ten Kate and Kuiper 1970). *Error bars* represent standard errors

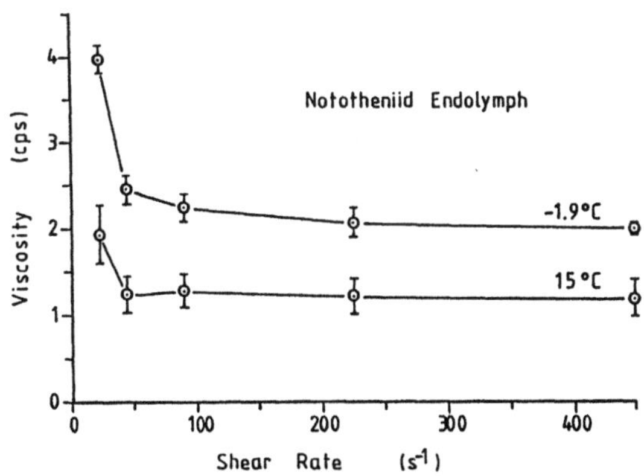

Fig. 8. Shear-dependence of nototheniid endolymph viscosity. Combined mean viscosities (± SE) of endolymph from *Pagothenia borchgrevinki* and *Dissostichus mawsoni*, measured at –1.9 and 15 °C are plotted against shear rate. Endolymph viscosity is nearly independent of shear except at very low shear rates; however, low shear rates are probably the most biologically meaningful

The viscosity determinations of ten Kate and Kuiper (1970) were not made at a defined shear rate, and were probably made at a much lower shear rate than 450 s^{-1}, which could account for the disparity at that rate.

5 Other Fluids

In tetrapods, the viscosity of synovial fluid is important for proper function of the joints, but teleosts lack comparable skeletal articulations with large fluid-lubricated spaces. The principal ball-and-socket in fish is the eye and its orbit, which is lubricated by orbital fluid.

Composition and rheology of orbital fluid from *D. mawsoni* are very similar to those of intracranial fluid (Table 2). These fluids resemble plasma rheologically, although they differ in composition, containing slightly less protein and cations, but more lipids.

6 Membrane Viscosity

There is considerable evidence that one component of cold adaptation involves the regulation of cell membrane fluidity through changes in the composition of membrane lipids. Poikilotherms inhabiting cold environments have long been known to have higher proportions of unsaturated fatty acids in nervous tissues as well as in stored lipids (Lewis 1962: Hilditch and Williams 1964), and antarctic nototheniids conform to this trend (Bottino *et al.* 1967; Morris and Schneider 1969; Patton 1975; Meyer-Rochow and Pyle 1980). At low temperatures, unsaturated lipids tend to be more fluid than saturated lipids of the same molecular size, and when incorporated into cell membranes serve to maintain molecular mobility in the membrane, a process which has been termed "homeoviscous adaptation" (Sinensky 1974). Correlations between acclimation temperature, the proportion of unsaturated lipids and membrane fluidity have been reported for preparations of synaptic membrane (synaptosomes) from the brain of goldfish (*Carassius auratus*; Cossins 1977), and similar correlations are known from other northern species inhabiting a wide range of natural temperatures, from the Bering Sea to hot springs in California's Death Valley (Cossins and Prosser 1978). So far, similar direct studies of membrane fluidity have not been made on antarctic fish, but there is strong circumstantial evidence for increased membrane fluidity playing a significant role in cold adaptation of antarctic nototheniids (Macdonald 1983; Macdonald and Montgomery 1986).

Miniature end plate currents (MECPs) at the neuromuscular synaptic junction are thought to follow closely the molecular events involved in opening and closing of transmembrane ion channels (Gage 1976), and have been used as physiological probes to assess changes in the viscosity of the membrane in which the channels are embedded. Increased viscosity of the lipid matrix will impede allosteric changes in the proteins making up the channels, or the gates to the channels. As one might expect, the decay of MECPs is temperature-sensitive: in MEPCs of a temperate fish (*Trachurus novaezelandiae*) at 15 °C, the mean time constant (τ) for return to

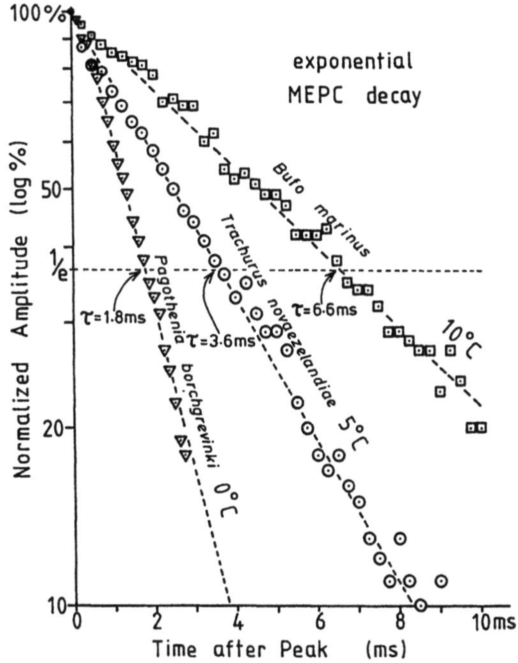

Fig. 9. Decay of miniature end-plate currents (MECPs) in tropical, temperate and antarctic animals. Semilogarithmic plots approximate straight lines in all cases, indicating that decay is an exponential process. Temperature compensation in the antarctic fish is indicated by faster decay, which is consistent with decreased viscosity of the synaptic membrane. The recording temperatures indicated are near the minimum temperatures at which MEPCs could be recorded from each species (Macdonald 1983)

baseline is about 1.0 ms, while at 5 °C τ equals 3.2 ms (Macdonald and Balnave 1984). However in *Pagothenia borchgrevinki*, τ is only 1.6 ms at 5 °C, and about 2 ms at the ambient seawater temperature of -1.9 °C (Macdonald and Montgomery 1986). Thus, MEPC recovery at low temperatures in the antarctic fish is faster than one would predict from the behaviour of MEPCs in temperate fish, and is consistent with the occurrence of less viscous membranes (Fig. 9).

7 Discussion

Clearly, increases in the viscosity of biological fluids induced by low temperature can have ranging implications for antarctic fish. In some cases, the tendency for viscosity to increase at low temperatures has been offset by reductions in other factors contributing to viscosity, resulting in a partial reduction in the high viscosity otherwise expected at subzero temperatures (blood haematocrit, membrane viscosity). In other cases, there appear to have been no significant compensations (endolymph), or even, as in the case of plasma, an increase in viscosity. Viscosity is only one of many temperature-dependent factors with which antarctic fish must cope, and any potential advantages of reducing viscosity must be weighed against potentially conflicting factors. In the case of plasma, the utility of glycopeptide antifreezes may preclude further reduction of plasma proteins, while for endolymph, it may simply be impossible to further reduce proteins from a level which is already minimal.

Acknowledgements. JAM and RMGW wish to thank the manager and staff of Antarctic Division, DSIR, New Zealand for logistic support, and the now defunct University Grants Committee for funding. We are particularly grateful to the past chairman of the Ross Dependency Research Committee, Dr. Trevor Hatherton, for his interest and encouragement. We also thank the Italian National Programme for Antarctic Research for the opportunity to conduct collaborative research at the Italian Terra Nova Bay station, and for providing generous support.

References

Andriashev AP (1970) Cryopelagic fishes and their significance in polar ecosystems. In: Holdgate MW (ed) Antarctic ecology, vol 1. Academic Press London, pp 297–204

Andriashev AP (1987) A general review of the antarctic bottom fish fauna. In: Kullander SO, Fernholm B (eds) Proc V Cong Europ Ichthyol, Stockholm 1985. Swed Mus Nat Hist, Stockholm, pp 357–372

Bottino NR, Jeffrey LM, Reiser R (1967) The lipids of antarctic fish. Antarctic J US 2:194–195

Chien S, Usami S, Dellenmack DJ, Bryant CA (1971) Comparative hemorheology-hematological implications of species differences in blood viscosity. Biorheology 8:35–37

Cossins AR (1977) Adaptation of biological membranes to temperature. The effect of temperature acclimation of goldfish upon the viscosity of synaptosomal membranes. Biochim Biophys Acta 470:395–411

Cossins AR, Prosser CL (1978) Evolutionary adaptation of membranes to temperature. Proc Natl Acd Sci USA 75:2040–2043

DeVries AL (1983) Antifreeze peptides and glycoproteins in cold-water fishes. Annu Rev Physiol 45:245–260

Eastman JT, Grande L (1989) Evolution of the antarctic fish fauna with emphasis on the recent notothenioids. In: Crame JA (ed) Origins and evolution of the antarctic biota. Geol Soc Lond, pp 241–252

Everson I, Ralph R (1968) Blood analyses of some antarctic fish. Bull Br Antarct Surv 15:59–62

Farrell AP (1984) A review of cardiac performance in the teleost heart: intrinsic and humoral regulation. Can J Zool 62:523–536

Gage PW (1976) Generation of end-plate potentials. Physiol Rev 56:177–247

Graham MS, Farrell AP (1989) The effect of temperature acclimation and adrenaline on the performance of a perfused trout heart. Physiol Zool 62:38–61

Graham MS, Fletcher GL (1985) On the low viscosity blood of a cold-water marine sculpin: a comparison with winter flounder. J Comp Physiol B155:455–459

Hemmingsen EA, Douglas EL (1972) Respiratory and circulatory responses in a hemoglobin-free fish, *Chaenocephalus aceratus*, to changes in temperature and oxygen tension. Comp Biochem Physiol 43A:1031–1043

Hilditch TP, Williams PN (1964) The chemical constitution of natural fats, 4th edn. Chapman and Hall, London

Horne RA, Courant RA, Johnson DS, Margosian FF (1965) The activation energy of viscous flow of pure water and sea water in the temperature region of maximum density. J Phys Chem 69:3988–3991

Howland HC, Masci J (1973) The phylogenetic allometry of the semicircular canals of small fishes. Z Morph Tiere 75:283–296

Hureau J-C, Petit D, Fine JM, Marneaux M (1977) New cytological, biochemical and physiological data on the colorless blood of the Channichthyidae (Pisces, Teleosteans, Perciformes). In: Llano GA (ed) Adaptation within antarctic ecosystems. Smithsonian Inst Washington, pp 459–477

Jones GM, Spells KE (1963) A theoretical and comparative study of the functional dependence of the semicircular canal upon its physical dimensions. Proc R Soc Lond B 157:403–419
Kennett JP (1977) Cenozoic evolution of antarctic glaciation, the circum-Antarctic Ocean and their impact on global palaeoceanography. J Geophys Res 82:3843–3876
Kestin J, Sokolov M, Wakeham WA (1978) Viscosity of liquid water in the range 8 °C to 150 °C. J Phys Chem Ref Data 7:941–948
Kooyman GL (1963) Erythrocyte analysis of some antarctic fishes. Copeia 1963:457–458
Lewis RW (1962) Temperature and pressure effects on the fatty acids of some marine ectotherms. Comp Biochem Physiol 6:75–89
Littlepage JL (1965) Oceanographic investigations in McMurdo Sound, Antarctica. In:Llano GA (ed) Biology of the antarctic seas, vol 2, pp 1–37. Antarctic Res Ser 5, Am Geophys Union, Washington, DC
Macdonald JA (1983) Neuromuscular cold-adaptation in an antarctic fish. NZ Antarctic Record 5(2):1–10
Macdonald JA, Balnave RJ (1984) Miniature end plate currents from teleost extraocular muscle. J Comp Physiol A 155:649–659
Macdonald JA, Montgomery JC (1986) Rate-compensated synaptic events in antarctic fish: consequences of homeoviscous cold-adaption. Experientia 42:806–808
Macdonald JA, Montgomery JC, Wells RMG (1987) Comparative physiology of antarctic fishes. Adv Mar Biol 24:321–388
Macdonald JA, Wells RMG, di Prisco G (1989) Thin-blooded antarctic fishes: absence of erythrocytes is a major factor determining low blood viscosity in ice-fishes. Proc Physiol Soc NZ 9:25
Magnuson JJ (1978) Locomotion by scombrid fishes: hydromechanics, morphology, and behavior. In: Hoar WS, Randall DJ (eds) Fish physiology, vol 7. Academic Press New York, pp 239–313
Meyer-Rochow VB, Pyle CA (1980) Fatty acid analysis of lens and retina of two antarctic fish and of the head and body of the antarctic amphipod *Orchomene plebs*. Comp Biochem Physiol 65B:395–398
Montgomery JC (1984) Low temperature increases gain in the oculomotor system. J Neurobiol 15:295–298
Montgomery JC, Macdonald JA (1984) Performance of motor systems in antarctic fishes. J Comp Physiol 154:241–248
Montgomery JC, Macdonald JA (1985) Oculomotor function at low temperature: antarctic versus temperate fish. J Exp Biol 117:181–191
Montgomery JC, McVean Ar (1987) Brain function in antarctic fish: activity of central vestibular neurons in relation to head rotation and eye movement. J Comp Physiol 160:289–293
Montgomery JC, Macdonald JA, Housley GD (1990) Brain function in antarctic fish: frequency response analysis of central vestibular units. J Comp Physiol A 166:407–410
Morris RW, Schneider MJ (1969) Brain fatty acids of an antarctic fish, *Trematomus bernacchii*. Comp Biochem Physiol 28:1461–1465
O'Grady SM, DeVries AL (1982) Osmotic and ionic regulation in polar fishes. J Exp Mar Biol Ecol 57:219–228
Oman CM (1981) The influence of duct and utricular morphology on semicircular canal response. In: Gualtierotti T (ed). The vestibular system: function and morphology. Springer Berlin Heidelberg, New York, Tokyo, pp 251–274
Patton JS (1975) The effect of pressure and temperature on phospholipid and triglyceride fatty acids of fish white muscle: a comparison of deepwater and surface marine species. Comp Biochem Physiol 52B:105–110
Rickenbach AF, Macdonald JA, Wells RMG (1989) Blood viscosity in antarctic *vs* New Zealand fishes. Proc Physiol Soc NZ 9:20
Riley JP, Skirrow G (eds) (1975) Chemical oceanography, 2nd edn. vol 2. Academic Press, London
Ruud JT (1954) Vertebrates without erythrocytes and blood pigment. Nature Lond 173:848–850
Sinensky M (1974) Homeoviscous adaption- a homeostatic process that regulates the viscosity of membrane lipids in *Escherichia coli*. Proc Natl Acad Sci USA 71:522–525
ten Kate JH, Kuiper JW (1970) The viscosity of the pike's endolymph. J Exp Biol 53:495–500
ten Kate JH, van Barneveld HH, Kuiper JW (1970) The dimensions and sensitivities of semicircular canals. J Exp Biol 53:501–514

Twelves EL (1972) Blood volumes of two antarctic fishes. Bull Br Antarct Surv 31:85–92

Wells RMG, Ashby MD, Duncan SJ, Macdonald JA (1980) Comparative study of the erythrocytes and haemoglobin in nototheniid fishes from Antarctica. J Fish Biol 17:517–527

Wells RMG, Baldwin J (1990) Oxygen transport potential in tropical reef fish with special reference to blood viscosity and haematocrit. J Exp Mar Biol Ecol 141:131–143

Wells RMG, Macdonald JA, di Prisco G (1990) Thin-blooded antarctic fishes: a rheological comparison of the haemoglobin-free icefishes *Chionodraco kathleenae* and *Cryodraco antarcticus* with a red-blooded nototheniid, *Pagothenia bernacchii*. J Fish Biol 36:595–609

Low Temperature Limits Burst Swimming Performance in Antarctic Fish

I.A. Johnston[1,2], T.P. Johnson,[1,2] and J.C. Battram[1]

1 Introduction

The temperature of the Southern Ocean varies from -1.86 °C at high latitudes to around 5 °C towards the Antarctic Convergence. Seasonal variations in temperature are less than 0.2 °C close to the permanent ice shelf (Littlepage 1965), and only 2.5 °C in the northern Antarctic (Everson 1970). There is evidence that these stenothermal conditions have existed relatively unchanged for several million years (Kennett 1977). Antarctic fish have therefore become highly specialized to cold conditions, and their upper lethal temperatures are often only 5-6 °C (Somero and DeVries 1967). Shallow-water species synthesize a variety of glycopeptide or peptide antifreezes to prevent the growth of ice crystals in the blood down to -2.7 °C (DeVries 1988). The nature of other adaptations which confer cold tolerance are relatively poorly understood (Clarke 1983; Johnston 1990). For example, although brain tubulins from antarctic fish assemble in vitro to form microtubules at -2.3 °C, they have broadly similar isoelectric points and amino-acid compositions to their mammalian counterparts which are cold-labile (Detrich and Overton 1988). The rates of molecular diffusion and enzyme reactions slow markedly at low temperatures. Thus, in the absence of compensating mechanisms, physiological processes would be expected to proceed more slowly in antarctic than in temperate or tropical fish. The rate of embryonic development would appear to confirm this prediction; this is exemplified by the time from fertilization to hatching in *Harpagifer antarcticus*, which is around 100 days at 0 °C, compared with 36–48 h in warm temperate fish at 25 °C (Blaxter 1988; Johnston 1990). However, there is evidence that other processes, such as locomotion and respiration, show varying degrees of temperature compensation (Clarke 1983, 1987; Johnston 1990). During burst swimming, ATP utilization by the fast muscle fibres increases over 100-fold very rapidly. Since maximum speed is an important factor determining the success of prey capture and predator avoidance it is crucial to the survival of individuals, and presumably subject to high selective pressures. Larval stages are capable of much higher tail-beat frequencies and length-specific speeds than adult fish because of scaling effects (Blaxter 1986). Thus the escape behaviour of larvae is an ideal place to look for the limits to cold adaptation in polar species.

[1]Gatty Marine Laboratory, Department of Biology & Preclinical Medicine, University of St. Andrews, St. Andrews, Fife, Scotland KY16 8LB, UK
[2]Pacific Gamefish Research Foundation, 74-425 Kealakehe Parkway, 15 Kailua Kona, Hawaii 96740, USA

2 Materials and Methods

2.1 Fish

Harpagifer antarcticus (Hureau) were obtained from the South Orkney Islands, Antarctica and transported to St. Andrews where they were maintained in a cold aquarium at 0-0.5 °C (12 h light:12 h dark). They were fed three times a week on squid and live shrimps (*Crangon crangon*). Gravid fish were transferred to darkened nursery tanks with a gravel substrate 3-4 weeks prior to spawning. The eggs hatched in 101-105 days. Larvae were fed brine shrimp nauplii and natural plankton which included various copepod species. The larvae began to feed prior to the exhaustion of the yolk sac, which occurred after 12-16 days.

Fourteen-day larvae of the dolphin fish (*Coryphaena hippurus*), reared at 24-26 °C, were obtained from the Waikiki Aquarium, Honolulu, Hawaii.

2.2 Measurement of Burst Swimming Speed

Measurements of burst swimming speed were made in a water-jacketed perspex tank 50 x 50 x 2 cm deep which had a 1-cm-square grid glued to the bottom of the tank. Antifreeze solution was circulated through the water jacket in order to maintain a constant temperature. The water in the tank was circulated by using a small electric pump. Individual larvae were introduced into the tank and observed using a National Panasonic WVP-F10E video camera with a strobe-effect shutter (1/1000th s exposure). Light from a 20–W bulb was reflected off a semi-silvered mirror set at an angle of 45° in front of the camera lens. Sharp silhouettes were obtained by using a "Scotchlite" reflex reflector background beneath the tank. Larvae were observed until they settled in the field of view of the camera (5 × 5 cm). Burst swimming sequences were stimulated by touching the larvae close to the caudal fin with a 1-mm-diameter glass pipette. Burst swimming sequences were recorded at a rate of 50 frames/s. Those sequences begining with a startle or C-response, i.e. involving a sharp flexion of the body following stimulation, were analyzed frame-by-frame. Several burst swimming sequences were recorded over a 30–40-min period and the fastest selected for detailed analyses. The recordings were played back using a 63.5-cm-wide screen monitor and the outline of the larvae and the calibration grid traced onto a transparent overlay. The first two frames were excluded from the analysis due to the considerable lateral movement of the head following initiation of the startle response. The "true" mean speed was calculated by aggregating the displacements of the snout frame by frame over the first ten frames following the C-start (200 ms) (Yin and Blaxter 1987). The maximum speed was estimated from the largest straight-line displacement of the snout between any pair of consecutive frames between frame 3 and frame 10. At the end of the experiments, the total length of each larvae was recorded.

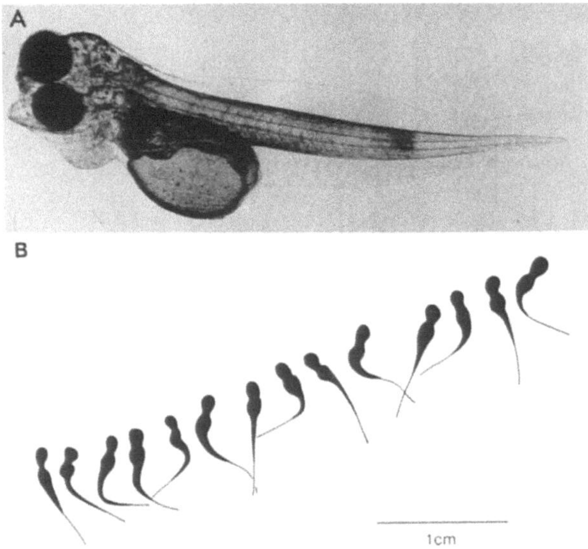

Fig. 1. **A** *Harpagifer antarcticus* larva at hatching. **B** Frame-by-frame tracing of an escape swimming sequence in response to tactile stimulation. Each successive frame is 20 ms apart

2.3 Results

An example of a C-start response in a newly hatched *Harpagifer antarcticus* larvae is shown in Fig. 1. All of the larvae were responsive to the probe at hatching. Escape swimming was characterized by large-amplitude body movements with a wavelength fractionally greater than one body length (Fig. 1). Frame-by-frame changes in speed during an escape response for 1-day-old yolk-sac larvae are shown in Fig. 2. Maximum speeds were attained after 40–60 ms and maintained for up to 240 ms (Fig. 2). Absolute true mean speed and maximum speed increased with length to 49 and 76 mm s^{-1} in 9.2-mm end-of-yolk sac stage larvae. However, specific escape speeds calculated as body lengths s^{-1} were similar in 1-day-old and 21–27-day-old *Harpagifer* larvae (Table 1). The escape speed of 21–27-day-old larvae also reached maximum values after 40–60 ms, but fell to around 40 mm s^{-1} at 140 ms and were maintained at this level for up to 360 ms (not illustrated). Mean escape speeds (200 ms) and maximum speeds of 8-mm dolphin fish (*Coryphaena hippurus*) larvae in body lengths s^{-1} were 2.4 times and 1.8 times higher than for equivalently sized *Harpagifer* larvae ($P < 0.01$); Table 1). The recording rate was not fast enough to accurately measure the tail-beat frequency of dolphin fish larvae at 25 °C, though in most cases it was close to 50 Hz over the first three to four strides, compared with around 15 Hz for *Harpagifer* larvae at 0 °C (Table 1).

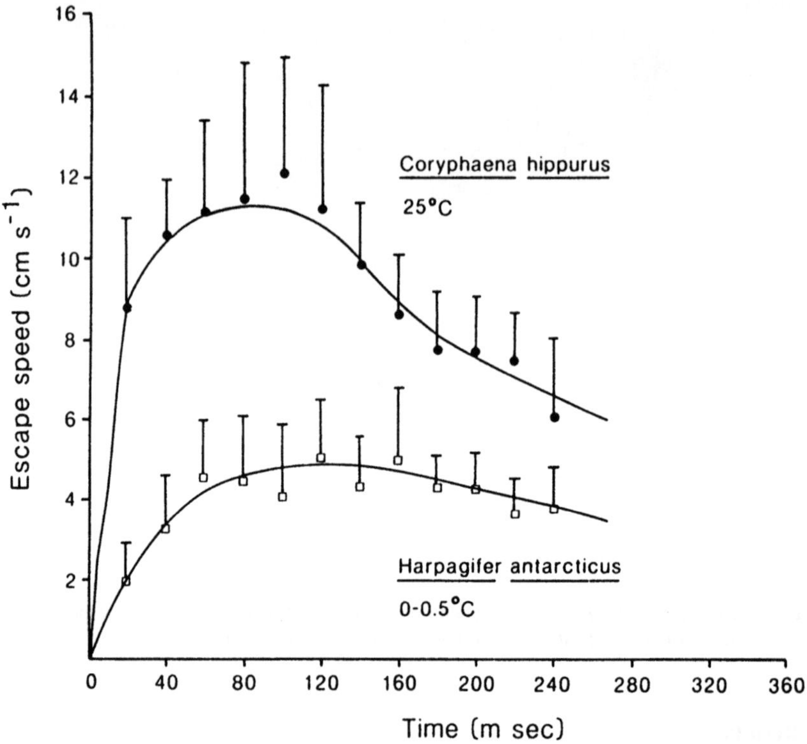

Fig. 2. Escape speeds of 1-day-old *Harpagifer antarcticus* larvae at 0 °C (n=9), and 14-day-old *Coryphaena hippurus* (14-day) larvae at 25 °C (n=9). Larvae were approximately 8 mm total length

Table 1. Burst swimming performance of fish larvae[a]

Species	Stage	Total length (mm)	True mean speed (body lengths s^{-1})	Maximum speed (body lengths s^{-1})	Maximum tail-beat frequency (Hz)
Harpagifer antarcticus	Yolk sac 1 day	7.8 ± 0.6 (n = 9)	5.2 ± 1.0	8.2 ± 1.7	15.0 ± 2.1
Harpagifer antarcticus	End yolk sac 21-27 days	9.2 ± 0.5 (n = 7)	5.3 ± 0.3	8.3 ± 1.0	16.0 ± 1.7
Coryphaena hippurus	End yolk sac 14 days	8.0 ± 0.7 (n =9)	12.6 ± 2.3	14.4 ± 2.4	30-50

[a]Values represent mean ± SD.

3 Discussion

3.1 Swimming Behaviour

The larvae of *Harpagifer antarcticus* are relatively large and well advanced at hatching (see Fig. 1). The organization of the myotomal muscles in *Harpagifer* is typical of other fish larvae (Batty 1984; El-Fiky et al. 1987). Sustained swimming activity is supported by a thin layer of aerobic muscle fibres (2–4 μ diameter) situated immediately beneath the skin (Fig. 3). These fibres probably receive their oxygen supply directly across the skin, since functional gills are not present at this stage. In other species, the loss of these outer aerobic muscle fibres is correlated with the development of secondary gill lamallae (El-Fiky and Wieser 1988). Escape swimming behaviour is powered by the inner mass of anaerobic fibres (6-12 μm diameter). These fibres contain densely packed myofibrils and have relatively few mitochondria (Fig. 3). Although these two larval fibre types are analogous to the slow and fast muscles of adult stages, we have recently shown that they contain different isoforms of the contractile proteins (Crockford and Johnston, unpubl. results).

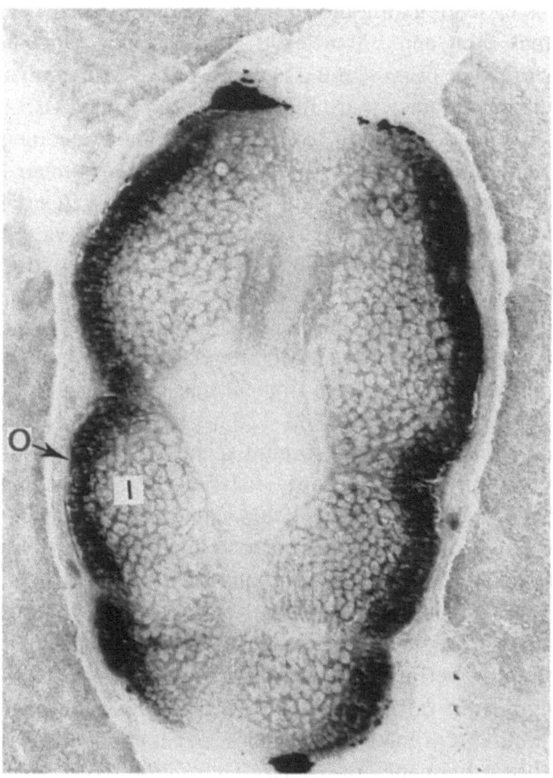

Fig. 3. Transverse frozen section through the trunk of a 1-day-old *H. antarcticus* larva stained for succinic dehydrogenase activity. Note the well-developed superficial layer of darkly stained aerobic muscle fibers (*O*) and the larger inner mass (*I*) of lightly stained anaerobic fibres

A variety of methods have been used to initiate startle responses in fish larvae, including electric shocks (Webb and Corolla 1981), natural predators (Webb 1981; Bailey and Batty 1984), acoustic stimuli (Blaxter and Batty 1985), and tactile stimuli (Yin and Blaxter 1987). Yin and Blaxter (1987) triggered C-starts either using a 0.2-mm-diameter probe or by attempting to suck the larvae into a 1.5-mm-diameter pipette. For cod, plaice, and herring, but not flounder, significantly higher maximum speeds and response rates were obtained using the pipette. Preliminary experiments with *Harpagifer* showed that touching and sucking with the pipette produced similar maximum speeds, suggesting it is the size of the probe which is important. The maximum swimming speed recorded between consecutive frames is dependent on the filming speed and is lower using conventional video (Bailey and Batty 1984; Blaxter and Batty 1985; Yin and Blaxter 1987) than when using cine film at 150–200 frames s^{-1} (Fuiman 1986). At 17 °C, northern anchovy larvae (8 mm length) accelerate over 8–16 ms to reach their prey, travelling a distance of about 0.4 mm (Hunter 1972). Thus maximum speed is likely to be the major performance component of larval behaviour determining the outcome of encounters with prey, whereas both speed and endurance will affect interactions with predators and nets (Webb and Corolla 1981). In the present study we have calculated the mean escape speed over 200 ms by aggregating the straight-line displacement of the snout over the first ten frames (Yin and Blaxter 1987). In order to evaluate interspecific differences in performance, it is essential to compare larvae of equivalent size, since the speed and tail-beat frequency are highly scale-dependent (Blaxter 1986). For example, speeds as high as 64 body lengths s^{-1} have been recorded at 28 °C, for newly hatched zebra danios which are only 2–3 mm long (Fuiman 1986). Data on swimming speeds for approximately 8-mm larvae analyzed in a similar manner are presented in Fig. 4. *Harpagifer antarcticus* larvae have mean escape speeds which are about half that for plaice and herring at 9–11 °C, and around 40% of that for anchovy and dolphin fish at 17–25 °C. Perhaps more significantly, the maximum tail-beat frequency in dolphin fish and plaice is around 30–50 Hz, compared with only 15 Hz in *Harpagifer* (see Table 1). Slightly larger post-larvae (1.07 cm) of the flying fish (*Hirundichthys affinis*) from tropical seas have been reported to achieve 35.9 body lengths s^{-1} for short periods (80–120 ms) at 20–25 °C (Davenport 1990). Unfortunately, comparisons of the maximum speeds achieved by adult stages are complicated by variations in body shape, different methods of analysis, and the limited and variable nature of the available data. Archer and Johnston (1989) carried out a detailed kinematic study of burst swimming in static water in juvenile (7–8 cm) *Notothenia neglecta*. They found that values for length-specific maximum swimming speed (8.7 body lengths s^{-1}) and tail-beat frequency (11.1 Hz) were lower than reported for temperate species of similar size (Archer and Johnston 1989). A modest burst swimming performance has also been reported for the cryopelagic antarctic fish *Pagothenia borchgrevinki* (Montgomery and Macdonald 1984), which at –1.5 °C and 23 cm total length had a maximum speed of only 4.9 body lengths s^{-1} at a tail-beat frequency of 5.7 Hz. Thus, although limited, the available data are consistent with the view that low temperature constrains burst swimming performance in polar fish, particularly for the early life stages.

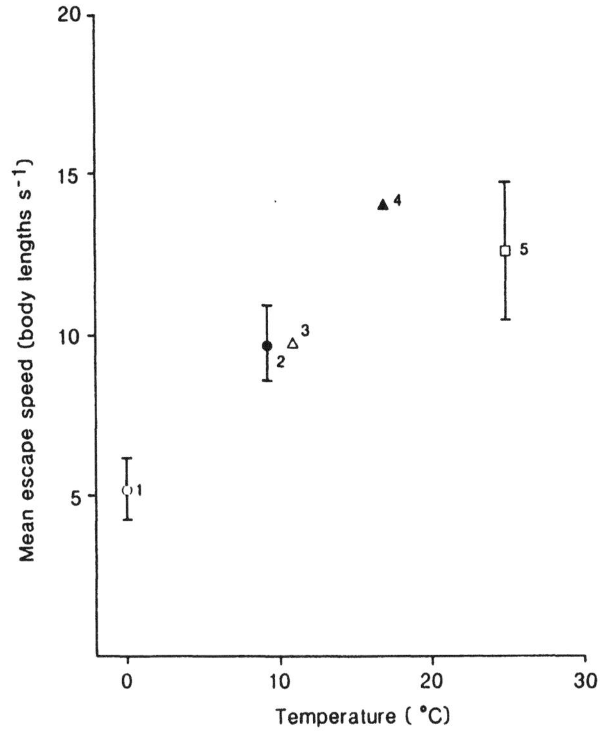

Fig. 4. Mean escape speeds of antarctic, temperate and tropical fish larvae, all 8 mm total length. *1 Harpagifer antarcticus* (this study). *2* Baltic herring (*Clupea harengus*; Yin and Blaxter 1987). *3*. Plaice (*Pleuronectes platessa*; Bailey and Batty 1984). *4*. Northern anchovy (*Engraulis mordax*; Webb and Corolla 1981. *5*. *Coryphaena hippurus* (this study)

3.2 Muscle Contractile Properties

During swimming, muscle fibres alternately shorten and lengthen and are phasically stimulated during each tail-beat cycle. Under these conditions, power output is a complex function of the amplitude and frequency of the locomotory movements and the timing of nervous stimulation in relation to changes in muscle length (Johnston and Altringham 1988; Van Leeuwen et al. 1990). Although the power output of fish muscle has been measured under conditions simulating swimming (Altringham and Johnston 1990a), this approach has not yet been applied to the study of temperature adaptation. In teleosts, the temperature range over which isolated muscle fibres remain viable is strongly correlated with the habitat temperature (Fig. 5). Thus, myotomal fibres from the tropical species (*Thalossoma duperreyi*) develop little force below 10 °C, whereas conversely, fibres from the Antarctic species (*Pseudochaenichthys georgianus*) are inexcitable at this temperature (Fig. 5). Altringham and Johnston (1986) used skinned muscle fibres to determine the energy cost of maintaining tension under force-generating conditions. They found that whilst the force-temperature relation differed markedly between species, ATPase activity was remarkably similar. Thus, at 0 °C, the economy of contraction decreased markedly in the order antarctic > North Sea > tropical species. The

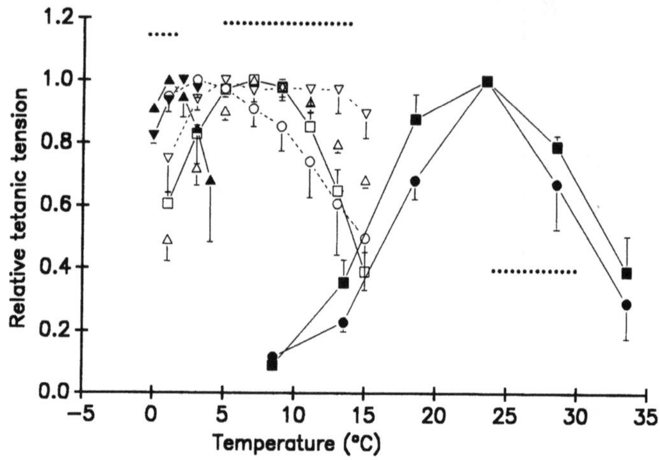

Fig. 5. The relationship between relative tetanic tension (normalized to the maximum at the normal habitat temperature) and experimental temperature for live, fast, myotomal and fin muscle preparations from antarctic, temperate and tropical marine teleosts. The *symbols* refer to the following species: Antarctic ▼—▼ *Trematomus lepidorhinus*, ▲—▲ *Notothenia neglecta*; Temperate △—△ Saithe (*Pollachius virens*), ○—○ Dab (*Limanda limanda*), □—□ Dragonet (*Callionymus lyra*), ▽—▽ Pogge (*Agonis cataphractus*); Tropical ■—■ Hawaiian Sergeant (*Abidefduf abdominalis*), ●—● Saddle Wrasse (*Thalassoma duperryi*). Data represent the mean ± SE. The normal environmental temperature range (Johnson and Jonhston 1991)

economy of contraction of fibres from *Notothenia neglecta*, at normal body temperatures, is around twice as high as that of representative temperate species, and four times that of tropical species (Altringham and Johnston 1986). These results may be explained by (1) an uncoupling of ATPase activity and force generation at low temperatures in warm adapted species; or (2) an increase in the force produced per cross bridge cycle in the cold water species. The latter could be achieved through an increase in bound time (and hence the number of attached cross-bridges), and/or more force per cycle with no change in bound time. Experiments with skinned rabbit psoas muscle fibres show that at 5 °C, about two-thirds of the cross bridges are in the "force generating" state which suggests that both mechanisms are probably needed to explain increases in force with temperature within species (Brenner 1986). These adaptations in force generation are associated with changes in myosin structure (Johnston and Walesby 1977). Myosins from cold-water fish belong to an unstable type which rapidly aggregate on isolation, and are readily denatured by heat and urea (Connell 1961). However, the amino acid composition and subunit composition of these and mammalian myosins are similar (Connell 1958a, b; Rowlerson et al. 1985). Johnston et al. (1975) reported that the -SH groups of myosin from the antarctic species, *Notothenia rossii*, reacted with 5,5'-dithio bis-2-nitrobenzoic acid 15 times faster than myosin from the Indian Ocean species *Amphiprion sebea*. This suggests that -SH groups in the cold-water species are more accessible, possibly reflecting a relatively open tertiary structure.

Studies with skinned fibres have shown, that, in contrast to force production, unloaded contraction velocity (Vmax) shows no evidence for temperature compensation (Johnston and Brill 1984). Thus Vmax is around 1 muscle lengths s^{-1} for the antarctic species *Trematomus hansoni* at 0 °C and > 9 muscle lengths s^{-1} in the tropical scrombroid (*Euthynmus affinis*) at 28 °C (Johnston and Brill 1984; Johnston and Harrison 1985). Recently, similar results have also been reported for live muscle fibres (Johnson and Johnston 1991). The force-velocity (P-V) relationship becomes less curved as experimental and/or body temperature decreases, and this could provide a mechanism for partially compensating for the reduction in power output as both force and velocity decline (Johnston and Altringham 1985; Langfeld *et al.* 1989). For example, the change in curvature for fast muscles fibres from *Myoxocephalus scorpius* between 8 and 1 °C has been calculated to increase relative maximum power output by around 15% (Langfeld et al. 1989). This is because at a given load, a less curved force-velocity relationship yields a higher velocity, and thus a greater power output. However, such effects could only account for a modest temperature compensation of contraction velocity under load in vivo.

Twitch duration is greatly prolonged at low temperatures in tropical compared with cold-water species; this is largely due to differences in relaxation rate (Fig. 6). The mechanism underlying cold-adaptation of relaxation includes faster rates of calcium sequestration by the sarcoplasmic reticulum (SR). For example, at 0 °C, SR isolated from the fast muscle of *Notothenia rossii* accumulates Ca^{2+} at six times the rate of SR from the fast muscle of several tropical fish (McArdle and Johnston 1980). However, the degree of temperature compensation shown by relaxation is only partial, and on average, twitch duration is around four times shorter at the characteristic body temperature of Indo-West Pacific (23–26 °C) than antarctic (-1 to 1 °C) fish (Fig. 6). These results would suggest that maximum tail-beat frequency is limited by low temperature in polar species.

Altringham and Johnston (1990b) measured the power output of fast muscle fibres from cod (*Gadus morhua* L.) under conditions simulating swimming. They found that relative to isometric conditions, force, and the rates of rise and fall of force, are increased, and the duration of the contractile event is decreased. Therefore positive oscillatory work can be performed at somewhat higher frequencies than might be predicted from the time course of isometric contractions alone (Altringham and Johnston 1990b). Indeed, studies with isolated lizard (Marsh and Bennett 1985) and salamander (Else and Bennett 1987) muscles have shown a poor correlation between the time course of isometric twitches and limb cycling times, particularly at low temperatures. For example, during burst running in the salamander *Ambystoma tigrinum nebulosum*, the extension phase of the limb requires 203 ms at 10 °C, whereas the isometric twitch duration was in excess of 400 ms (Else and Bennett 1987). However, a much closer agreement between locomotory and muscle properties was found at preferred body temperatures (> 20 °C ; Else and Bennett 1987). Similarly, the twitch duration of fast myotomal muscle in adult *Notothenia neglecta* is 114 ms at 1 °C (Johnson and Johnston 1990), and is comparable with the tail-beat cycle of 152 ms (Archer and Johnston 1989).

Modelling studies have shown that for anterior myotomes, positive work is produced throughout each tail-beat cycle, whereas net negative work is done in posterior myotomes (Hess and Videler 1984; Van Leeuwen et al. 1990). These differences result from systematic changes in the phase relation between length and force cycles (Johnston and Altringham 1988). Further complications result from the variable and complex orientation of fibres within myotomes, patterns of fibre

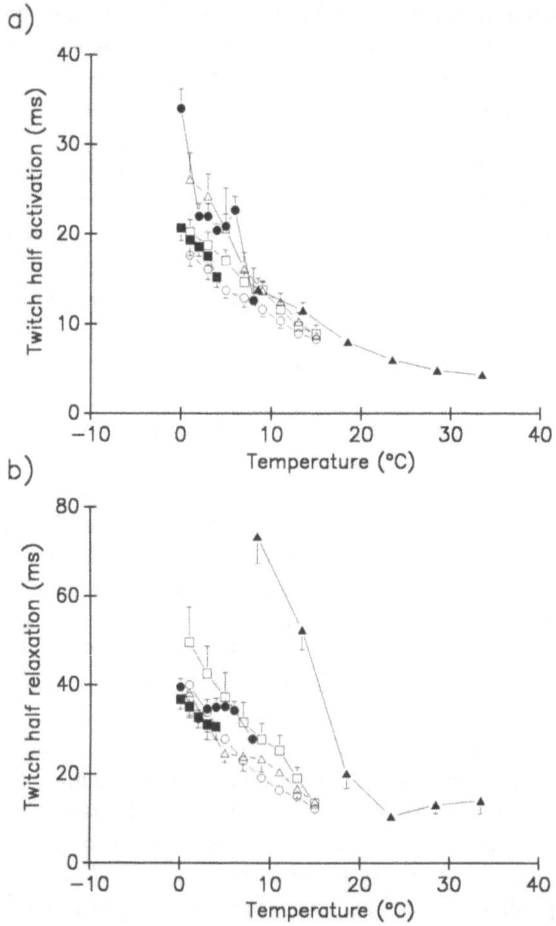

Fig. 6. The relationship between experimental temperature and (***a***) twitch half-activation time (*ms*) and (***b***) twitch half-relaxation time (*ms*), for live, fast myotomal muscle preparations from antarctic, temperate and tropical teleost marine species. The *symbols* refer to the following species: Antarctic ● icefish (*Pseudochaenichtys georgianus*), ■ *Notothenia neglecta*; temperate ○ saithe (*Pollachius virens*, △ dab (*Limanda limanda*), □ dragonet (*Callionymus lyra*); tropical ▲ saddle wrasse (*Thalassoma duperreyi*). Data represent mean ± SE (Johnson and Johnston 1991)

recruitment and the passive properties of series and parallel elastic components. Although there is no simple relationship between whole animal performance and the properties of isolated muscle fibres, such studies have provided important insights into the mechanisms underlying cold adaptation of locomotion.

Acknowledgments. This work was supported by an Antarctic Special Topics Grant (GST/02/447) from the Natural Environment Research Council. We are grateful to Dr. David Grobecker for providing facilities at the PGRF laboratories and to Dr. Syd Kraul and the Waikiki Aquarium for supplying the dolphin fish larvae. Our visit to Hawaii was supported by travel grants from the Royal Society (IAJ) and the Russell Trust (TPJ). TPJ was in receipt of a NERC research studentship. Finally, we are grateful to Professor M.S. Laverack for photographing the *Harpagifer antarcticus* larvae.

References

Altringham JD, Johnston IA (1986) Evolutionary adaptation to temperature in fish muscle cross-bridge mechanisms: tension and the ATP turnover. J Comp Physiol B 156:819–821
Altringham JD, Johnston IA (1990a) Modelling muscle power output in a swimming fish. J Exp Biol 148:395–402
Altringham JD, Johnston IA (1990b) Scaling effects on muscle function: power output of isolated fish muscle fibres performing oscillatory work. J Exp Biol 151:453-467
Archer SD, Johnston IA (1989) Kinematics of labriform and subcarangiform swimming in the antarctic fish *Notothenia neglecta*. J Exp Biol 143:195–210
Bailey KM, Batty RS (1984) Laboratory study of predation by *Aurelia aurita* on larvae of cod, flounder, plaice and herring: development and vulnerability to capture. Mar Biol 83:287–291
Batty RS (1984) Development of swimming movements and musculature of larval herring (*Clupea harengus*). J. Exp Biol 110:217–229
Blaxter JHS (1986) Development of sense organs and behaviour of teleost larvae with special reference to feeding and predator avoidance. Trans Am Fish Soc 115:98–114
Blaxter JHS (1988) Pattern and variety in development. In: WS Hoar, Randall DJ (eds) *Fish physiology*, vol 13, Part A, Eggs and larvae. Academic Press; London, pp 1–58
Blaxter JHS, Batty RS (1985) The development of startle responses in herring larvae. J Mar Biol Assoc UK 65:737–750
Brenner B (1986) Kinetics of the crossbridge cycle derived from measurements of force, rate of force development and isometric ATPase. J Muscle Res Cell Motl 7:75–76
Clarke A (1983) Life in cold water: the physiological ecology of polar marine ectotherms. Oceanogr Mar Biol 21:341–453
Clarke A (1987) The adaptation of aquatic animals to low temperatures. In: Grout BMW, and Morris GJ (eds) The effects of low temperatures on biological system. Arnold, London, pp 315–348
Connell JJ (1958a) Studies on the protein of fish skeletal muscle 4. Ultracentrifugal analysis of codling extracts. Biochem J 69:5–12
Connell JJ (1958b) Studies on the protein of fish skeletal muscle 5. Molecular weight and shape of cod fibrillar proteins. Biochem J 70:81–91
Connell JJ (1961) The relative stabilities of skeletal muscle myosins of some animals. Biochem J 80:503–538
Davenport J (1990) Observations on the locomotion of post-larval and juvenile flying fish. J Mar Biol Assoc UK 70:311–320
Detrich WH III, Overton SA (1988) Antarctic fish tubulins: heterogeneity, structure, amino acid compositions and charge. Comp Biochem Physiol 90B:593–600
DeVries AL (1988) The role of antifreeze glycopeptides and peptides in the freezing avoidance of antarctic fishes. Comp Biochem Physiol 90B:611–622
El-Fiky N, Hinterleitner S, Wieser W (1987) Differentiation of swimming muscles and gills, and development of anaerobic power in the larvae of cyprinid fish (Pisces, Teleostei). Zoomorphology 107:126–132
El-Fiky N, Wieser W (1988) Life styles and patterns of development of gills and muscles in larval cyprinids (Cyprinidae; Teleostei). J Fish Biol 33:135–145
Else PL, Bennett AF (1987) The thermal dependence of locomotor performance and muscle contractile function in the salamander *Ambystoma tigrinum nebulosum*. J Exp Biol 128:219–233
Everson I (1970) The population dynamics and energy budget of *Notothenia neglecta* Nybelin at Signy Island, South Orkney islands. Bull Br Antarct Surv 23:25–50
Fuiman LA (1986) Burst-swimming performance of larval zebra danios and the effects of diel temperature fluctuations. Trans Am Fish Soc 115:143–148
Hess F, Videler JJ (1984) Fast continuous swimming of saithe (*Pollachius virens*): a dynamic analysis of bending moments and muscle power. J Exp Biol 109:229–251
Hunter JR, (1972) Swimming and feeding behaviour of anchovy, *Engraulis mordax*, larvae. US Natl Mar Fish Bull 70:821–838
Johnson TP, Johnston IA (1991) Temperature adaptation and the contractile properties of live muscle fibres from the teleost fish. J Comp Physiol B 161:27–36
Johnston IA (1990) Cold adaptation in marine organisms. Phil Trans R Soc Lond B 326:655–667

Johnston IA, Altringham JD (1985) Evolutionary adaptation of muscle power output to environmental temperature: force velocity characteristics of skinned fibres isolated from antarctic, temperate and tropical marine fish. Pflugers Arch 405:136–140

Johnston IA, Altringham JD (1988) Muscle function in locomotion. Nature (Lond) 335:767–768

Johnston IA, Brill R (1984) Thermal dependence of contractile properties of single skinned fibres isolated from antarctic and various Pacific marine fishes. J Comp Physiol B 155:63–70

Johnston IA, Harrison P (1985) Contractile and metabolic characteristics of muscle fibres from antarctic fish. J Exp Biol 116:223–226

Johnston IA, Walesby NJ (1977) Molecular mechanisms of temperature adaptation in fish myofibrillar adenosine triphosphatases. J Comp Physiol B 119:195–206

Johnston IA, Walesby NJ, Davison W, Goldspink G (1975) Temperature adaptation in myosin of antarctic fish. Nature (Lond) 254:74–75

Kennett JP (1977) Cenozoic evolution of antarctic glaciation, the circum-Antarctic Ocean, and their impact on global paleoceanography. J Geophys Res 82:3843–3860

Langfeld KS, Altringham JD, Johnston IA (1989) Temperature and the force-velocity relationship of live muscle fibres from the teleost *Myoxocephalus scorpius*. J Exp Biol 144:437–448

Littlepage JL (1965) Oceanographic investigations in McMurdo sound, Antarctica. In: Llan GA (ed) Biology of the Antarctic seas, vol. 2. Am Geophys Union; Washington, pp 1–37

Marsh RL, Bennett AF (1985) Thermal dependence of isotonic contractile properties of skeletal muscle and sprint performance of the lizard *Dipsosaurus dorsalis*. J Comp Physiol B 155:541–551

McArdle HJ, Johnston IA (1980) Evolutionary temperature adaptation in fish muscle sarcoplasmic reticulum. J Comp Physiol B 135:157–164

Montgomery JC, Macdonald JA (1984) Performance of a motor systems in antarctic fishes. J Comp Physiol A 154:241–248

Rowlerson A, Sapolo PA, Mascarello F, Carpene E, Veggetti A (1985) Comparative study of myosins present in the lateral muscle of some fish: species variations in myosin isoforms and their distribution in red, pink and white muscle. J Musc Res Cell Motility 6:601–640

Somero G, DeVries AL (1967) Temperature tolerance of some antarctic fishes. Science 156:257–258

Van Leeuwan JL, Lankheet MJM, Akster HA, Osse JWM (1990) Function of red axial muscles of carp (*Cyprinus carpio*): recruitment and normalized power output during swimming in different modes. J Zool Lond 220:123–145

Webb PW (1981) Responses of northern anchovy, *Engraulis mordax*, larvae to predation by a biting planktivore, *Amphiprion percula*. US Natl Mar Serv Fish Bull 79:727–735

Webb PW, Corolla RT (1981) Burst swimming performance of northern anchovy, *Engraulis mordax*, larvae. Fish Bull 79:143–150

Yin MC, Blaxter JHS (1987) Escape speeds of marine fish larvae during early development and starvation. Mar Biol 96:459–468

Respiratory and Cardiovascular Adaptations in Hemoglobin-Free Fish: Resolved and Unresolved Problems

E.A. Hemmingsen

1 Introduction

Due to the low solubility which oxygen has in aqueous solutions, the presence of an oxygen-carrying pigment in the blood has been regarded as a necessity in all but the simplest types of animals. Yet, all species of the antarctic and subantarctic fish family *Channichthyidae* lack such a pigment. The family is comprised of 16 known species among 10 genera, with adult sizes ranging from about 20 to about 60 cm. The distribution is circumpolar, with some species extending north to such areas as Patagonia, the Falkland Islands, and the South Georgia, Bouvet and Kerguelen Islands.

These so-called icefish have attracted considerable attention since their lack of hemoglobin was first pointed out in the scientific literature by Ruud (1954,1965), and a good deal of information has been obtained on their physiological and morphological characteristics. Yet, many of the physiological consequences of the absence of hemoglobin are unclear, and certainly, any functional advantages of abandoning the hemoglobin have not become apparent. Even though their existence no doubt is closely tied to the particular food-chain system that is unique to the antarctic waters, low temperature is the most tangible factor that appears to be of critical importance in their survival. The most obvious advantages to the low temperature are the reduction in metabolic requirements for oxygen and the increase in the amount of oxygen that can be dissolved in the water and the plasma. For example, the solubility of oxygen in seawater at 0 °C is 0.83 vol % per atm of air, while that at 20 °C is 0.53 vol % (Weiss 1970); a similar relative solubility difference would exist in the plasma. However, this gain in solubility is offset by the approximate doubling of the water viscosity for the same temperature interval. The increased viscosity adds substantially to the circulatory and ventilatory work that must be performed.

Most of the physiological studies have been performed on two of the largest species, *Chaenocephalus aceratus* and *Pseudochaenichthys georgianus*, which are commonly found in the waters off the Antarctic Peninsula. While *C. aceratus* is a benthic species and is probably relatively inactive, *P. georgianus* is pelagic and therefore must be at least modestly active. Some degree of activity also is indicated

Physiological Research Laboratory, Scripps Institution of Oceanography, University of California, San Diego, La Jolla, California 92093, USA by stomach contents as well as by migration patterns (e.g.,Olsen 1955; Permitin 1970; Twelves 1972). Both species have proved to be quite rugged and well suited for physiological experiments in the laboratory.

This chapter will provide an overview of the results obtained in the various physiological studies on the channichthyids, specifically with regard to adaptations which appear to have occurred in response to the absence of hemoglobin. It will be necessary to compare certain functions with those in other cold-water teleosts which have hemoglobin. Although truly comparative data often do not exist, some of the differences are of such magnitude that it is not critical to have precise quantitative data.

2 Oxygen Consumption

A key question is whether or not the oxygen requirements of the channichthyids are different from those of other teleosts at similar temperatures. In general, the resting, or standard, metabolic rate has proved to vary a great deal among fish, and among different specimens of the same species. This is partly because some unavoidable, but undefined minor swimming activity takes place, and partly because fish in general seem to be very sensitive to even minor handling and noncontact disturbances. Another factor is the time elapsed between the capture of the fish and the rate determinations; this alone readily can change the rate by a factor of two over the period of a day (Holeton 1974). Furthermore, as fish usually are quite sensitive to even modest short-term changes in temperature by not necessarily following a normal Q_{10} for the temperature change, it is important that oxygen consumption be determined at a temperature close to that of their normal habitat. For these and other reasons, using the existing data for discerning minor trends may pose some problems.

The rate of oxygen uptake has been determined for several channichthyid species (Ralph and Everson 1968; Hemmingsen et al. 1969; Hemmingsen and Douglas 1970, 1972, 1977; Holeton 1970; Hureau et al. 1977). Examples of the results obtained are shown in Fig. 1. The mean rates of oxygen uptake from various studies on channichthyids and other antarctic fish are given in Table 1. Even though the oxygen consumption rates for the channichthyids clearly tend to be lower than those of the other fish, they are not greatly suppressed. The average value for the channichthyids is merely 27% lower than that for the hemoglobin-containing fish represented in this table. Another factor to consider in this comparison is that axial musculature of, for example, *C. aceratus* constitutes a smaller portion of the total weight than it does in some other fish (Holeton 1976); hence, the oxygen consumption is deceptively low on a weight basis.

The question of whether the lowered rate of oxygen consumption in the channichthyids reflects a special adaptation in tissue metabolism has not yet been resolved. Certainly, some fraction of the decreased consumption must be ascribed to behavioral factors, such as the suppression of all movements that are not essential. Such capability is characteristic at least for specimens maintained in captivity or used for laboratory experiments. However, certain biochemical metabolic processes

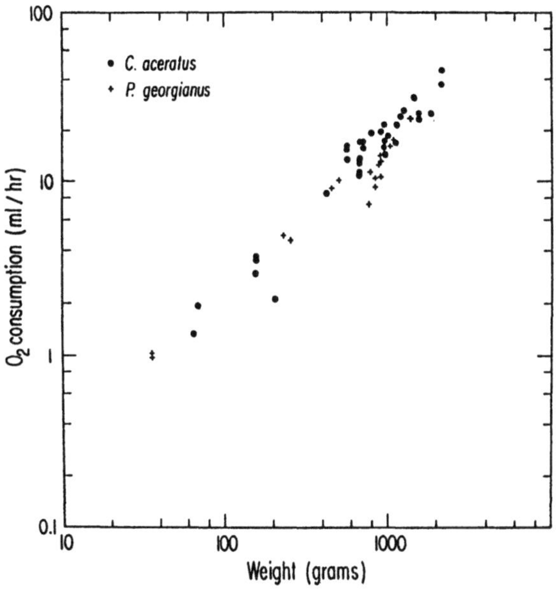

Fig. 1. Oxygen consumption in *Chaenocephalus aceratus* and *Pseudochaenichthys georgianus* as a function of weight. Temperatures: 0.5–2 °C (Hemmingsen and Douglas 1977)

Table 1. Rates of oxygen consumption in some antarctic fish[a]

Species[a]	ml O_2/kg min	Reference
Pagetopsis macropterus	16	Hemmingsen et al. (1969)
Pseudochaenichthys georgianus	22	Hemmingsen and Douglas (1970)
–	22	Hemmingsen and Douglas (1977)
Chaenocephalus aceratus	39	Ralph and Everson (1968)
–	21	Holeton (1970)
–	25	Hemmingsen and Douglas (1970)
–	23	Hemmingsen and Douglas (1977)
Pagothenia borchgrevinki	34	Wohlschlag (1964)
Trematomus bernacchii	35	Wohlschlag (1964)
T. hansoni	42	Wohlschlag (1964)
T. loennbergi	26	Wohlschlag (1964)
T. centronotus	32	Macdonald et al. (1987)
Notothenia coriiceps	66	Hemmingsen et al. (1966)
N. nudifrons	30	Hemmingsen et al. (1969)
N. gibberifrons	48	Hemmingsen et al. (1969)
–	19	Holeton (1970)
N. neglecta	61	Ralph and Everson (1968)
–	25	Holeton (1970)
N. rossi	29	Ralph and Everson (1968)
Gymnodraco acuticeps	32	Macdonald et al. (1987)
Rhigophila dearborni	14	Wohlschlag (1964)
–	7	Wells (1986)

[a] The top three species are hemoglobin-free channichthyids. For comparison purposes, all rates are calculated for 100 g weight, using a weight regression coefficient of 0.9 (Macdonald et al. 1987). Temperature range: -1.5 to +2 °C.

not directly related to activity also appear to proceed at lower levels in the channichthyids. For example, protein synthesis in most tissues of *C. aceratus* is lower than in two nototheniids tested (Haschemeyer 1983).

It is reasonable to assume that most of the oxygen uptake occurs in the gill lamellae because of their large surface area and, in particular, because of the short diffusion path for the oxygen between the blood and the water. Yet, some oxygen is taken up through the skin, which lacks scales. Measurements show that cutaneous uptake in just the tail section of *C. aceratus* at rest is up to 8% of the total uptake (Hemmingsen and Douglas 1970). The tail section tested was estimated to be less than one-fourth of the total surface area. If the uptake per unit area is the same for the rest of the surface, including the buccal cavity, the cutaneous uptake could be as much as 40 % of the total uptake at rest, but probably is less. This is not an unreasonable fraction as it is similar to that estimated for some other teleosts (Nonnotte and Kirch 1978; Wells 1986; Takeda 1989). Because the simple diffusion of gas to the underlying cells and vessels occurs without any internal energy expenditure, any amount of oxygen gained without taxing the limited transport capacity of the blood obviously is very advantageous.

3 Cardiovascular Functions

The blood of the channichthyids is colorless with a slight turbidity. Although it contains no functional erythrocytes, erythrocytelike cells have been reported to be present in half of the species (Martsinkevich 1958; Hureau 1977; Barber and Westermann 1981). The blood of *C. aceratus* has been examined for porphyrin, and none has been found (Douglas et al. 1973). Oxygen is carried exclusively in physical solution; the oxygen capacity is about 0.7 vol % (Ruud 1954), which is about one-tenth of the amount carried by hemoglobin in most other cold-water fish (e.g.,Scholander and Van Dam 1957; Grigg 1967; Hemmingsen et al. 1969; Holeton 1972).

Even with such a severely limited oxygen capacity, the rate of oxygen consumption is only modestly lower than that of the other fish. This has been made possible be several interrelated adjustments in their cardiovascular functions. One of the easily identifiable adjustments is the development of a very large blood volume. This has been found to be about 8–9 vol% for *C. aceratus* by using the Evan's Blue dilution method (Hemmingsen and Douglas 1970; Twelves 1972). A volume of similar magnitude was indicated in *P. georgianus* by the fact that exsanguination yielded similar volumes for the two species. These large blood volumes are two to four times those typically found for teleosts containing hemoglobin (e.g, Thorson 1961; Smith 1966). Because most of the blood resides outside the heart and the larger vessels of animals, the exceptionally high volumes observed reflect a well-developed capillary bed in the tissues.

The usefulness of such a high volume and low unit volume capacity system depends on a high rate of turnover, and this is made feasible by a well-developed heart. Thus, the ventricle weight constitutes a substantially larger fraction of the total body weight than it does in other fish (Twelves 1972; Johnston et al. 1983), and the ventricular wall has ultrastructural characteristics (Tota et al. 1988) that are

Table 2. Mean cardiac output in two channichthyid species and some other teleosts at rest, determined by implanted flow probes or by the Fick principle. In the latter cases, no correction were made for cutaneous oxygen uptake or for oxygen utilization in the gills

Species	Approx. temp. (°C)	Cardiac output (ml kg $^{-1}$ min^{-1})	Method	Reference
Pseudochaenichthys georgianus	+1	67	Probe	Hemmingsen and Douglas (1977)
–	+1	94	Fick	Hemmingsen and Douglas (1977)
Chaenocephalus aceratus	+1	61	Fick	Holeton (1970)
–	+1	66	Probe	Hemmingsen and Douglas (1977)
–	+1	104	Fick	Hemmingsen and Douglas (1977)
Gadus morhua	+10	21	Probe	Jones et al. (1974)
–	+10	29	Probe	Pettersson and Nilsson (1980)
Cyprinus carpio	+10	18	Fick	Garey (1970)
Salmo gairdneri	+10	18	Fick	Kiceniuk and Jones (1977)
Ophiodon elongatus	+10	11	Probe	Farrell (1982)

compatible with and indicative of high levels of activity. Although the cardiac muscles of most animals normally contain substantial concentrations of myoglobin, it is absent in those of *Champsocephalus gunnari* (Hamoir and Gerardin-Otthiers 1980), but present in trace amounts in those of *C. aceratus, P. georgianus,* and *Chaenodraco wilsoni* (Douglas et al. 1985).

It has been possible to determine directly the cardiac output in *C. aceratus* and *P. georgianus* by means of electromagnetic flow probes implanted on the ventral aorta of unrestrained specimens, and indirectly by means of the Fick principle (Holeton 1970, 1972; Hemmingsen and Douglas 1972, 1977). With the Fick principle, the cardiac output is obtained by dividing the rate of oxygen uptake by the difference in oxygen content between the dorsal and ventral aortic blood. The assumption is that the oxygen is taken up by the gills, with no cutaneous uptake, and that no oxygen is consumed in the gills. Because both processes occur, this method gives too low output values. The cardiac output values obtained so far on channichthyids are given in Table 2, together with examples of values for hemoglobin-containing teleosts. The cardiac output of both channichthyids is several-fold higher than that in the other fishes and, in fact, is similar to that in larger mammals. The high output is maintained with a relatively low heart rate, and hence, with stroke volumes that are extremely high for fish.

In order to maintain the very high blood flow rates observed, it is crucial that flow resistance in the circulatory system be minimized; otherwise, the heart would be subjected to prohibitive work demands. Blood pressure measurements confirm that, indeed, the channichthyids have low systemic flow resistances (Table 3). The aortic blood pressures are substantially lower than those commonly observed in other teleosts (e.g., Johansen and Martin 1965; Robertson et al. 1966; Stevens and

Table 3. Mean venous and arterial oxygen tensions, systolic/diastolic blood pressures and heart rates in *Pseudochaenichthys georgianus* and *Chaenocephalus aceratus* at rest

Species	Aortic PO_2 Ventral (torr)	Dorsal	Aortic pressure Ventral (torr)	Dorsal	Heart rate (beats/min)	Reference
P. georgianus	60	115	16/12	14/11	13	Hemmingsen and Douglas (1977)
C. aceratus	34	101	22/15	17/11	16	Hemmingsen and Douglas (1977)
	25	120	20/12	15/10	14	Holeton (1970)

Randall 1967; Farrell 1982), even though the cardiac output has been greatly increased. The two main factors that appear to be responsible for this are the absence of erythrocytes and the increased vascularization of the tissues.

An increase in cross-sectional area of the vascular bed probably is more important than is the decrease in relative viscosity due to the absence of erythrocytes, because the resistance for laminar flow in a tube, as given by Poiseuille's equation, varies inversely with the radius to the fourth power but varies only in proportion to the viscosity. Therefore, relatively small increases in the diameter of the capillaries would be very effective in decreasing the flow resistance, particularly if combined with an increase in capillary density. It appears from studies of white trunk muscles of *C. aceratus* and other fish (Fitch et al. 1984) that enlargement of the bore of the capillaries, rather than an increase in their density, is the predominant adaptation that has occurred in the deeper tissues of the channichthyids. Other morphological studies have shown that the vascularization in subcutaneous body tissues and fins is exceptionally high, and also, that the capillaries are larger (Walvig 1960; Jakubowski and Byczkowska-Smyk 1970; Jakubowski 1982). The developments of the vascular system appears to extend to the larger vessels as well. The main arteries and veins are more prominent and are easier to cannulate and work with than those encountered in other teleosts of similar size.

The gills of the channichthyids do not appear to be particularly well developed to enhance the gas exchange capacity or efficiency, as the total surface area of the lamellae and the length of the diffusion path between blood and water are quite similar to those of many fish containing hemoglobin (Steen and Berg 1966; Jakubowski and Byczkowska-Smyk 1970; Vogel and Kock 1981). One question which has not been resolved by either physiological or morphological studies is the degree to which the secondary lamellae generally are more perfused relative to their maximum capacity that they are in other fish, where such regulation can occur (Steen and Kruysse 1964; Booth 1978). Conceivably, major differences may exist between the two groups with regard to shunting between afferent and efferent blood. The number of anastomoses between afferent and efferent vessels in the gill filaments of *C. aceratus* is lower than between those of hemoglobin-containing teleosts (Vogel and Kock 1981). This indicate that the lamellae generally are more perfused in the channichthyids. Because other features of the gills are not especially well developed, the diffusive gas exchange at the surface of the lamellae may not be a limiting step in the overall gas exchange and transport process in these animals.

The pressure drops from the ventral to the dorsal aorta are uncommonly low in the channichthyids (Table 3); this indicates enlargement of the vessels in the gills. Morphological studies have revealed that the lacunae (Steen and Berg 1966) as well as the marginal channels of the lamellae and the main branchial vessels (Vogel and Kock 1981) of *C. aceratus* are unusually large.

4 Metabolic Scope

In fish with hemoglobin, only a small portion of the hemoglobin-bound oxygen usually is removed from the blood on its passage through the vascular system under conditions of low activity. Indeed, it has been shown that when the oxygen binding to hemoglobin is blocked, or when most of the erythrocytes are removed, inactive fish will survive for extended periods without any clear ill effects (e.g., Nicloux 1923; Anthony 1961; Steen and Berg 1966; Cameron and Wohlschlag 1969).

Because the respiratory reserve for activity provided by the hemoglobin is absent in the channichthyids, the question arises as to their ability to meet the oxygen needs above their resting level when activity is required. Ideally, this can be answered only by the direct measurement during activity, but some information can be obtained from respiratory and circulatory performance under hypoxia. Thus, rather than increasing the oxygen supply, the supply is decreased while maintaining a reasonably constant demand.

The experimental approach has been to use a closed respirometer, and to allow the fish to decrease the oxygen content of the water, while monitoring the oxygen content of the water and the blood, and in some cases, blood pressure and blood flow (Hemmingsen and Douglas 1970, 1972, 1977; Holeton 1970). Table 3 shows the blood oxygen tensions in channichthyids at rest in aerated water. Normally, the arterial blood, sampled from the dorsal aorta, is not fully oxygenated; its oxygen tension is about two-thirds that in the water. The venous blood, sampled from the ventral aorta, contains a substantial amount of oxygen. The fact that less than one-half of the oxygen-carrying capacity of the blood is utilized under resting conditions is indicative of a substantial, unused reserve.

With progressing hypoxia, the oxygen content of the water decreases linearly with time; usually, the oxygen uptake remains constant until the oxygen tension drops below 50 torr (Fig. 2). This constancy shows that the uptake is a regulated process. Without regulation, the uptake would decrease steadily with the decreasing gradient. This alone indicate that some reserve uptake capability exists. The arterial oxygen tensions further substantiate the existence of a respiratory reserve; they decrease at a slower rate than the water oxygen tension, particularly in *C. aceratus*. The ability to maintain a constant oxygen uptake with a diminishing gradient between water and blood means that the efficiency of the gills, that is, the amount of oxygen taken up per unit oxygen gradient, increases severalfold over during hypoxia. The efficiency increase probably is accomplished primarily by opening up new flow channels in the lamellae, and may be aided by reducing the amount of blood shunted via anastomoses between efferent and afferent branchial vessels. The additional blood perfusion was confirmed by the approximately 50% increase in

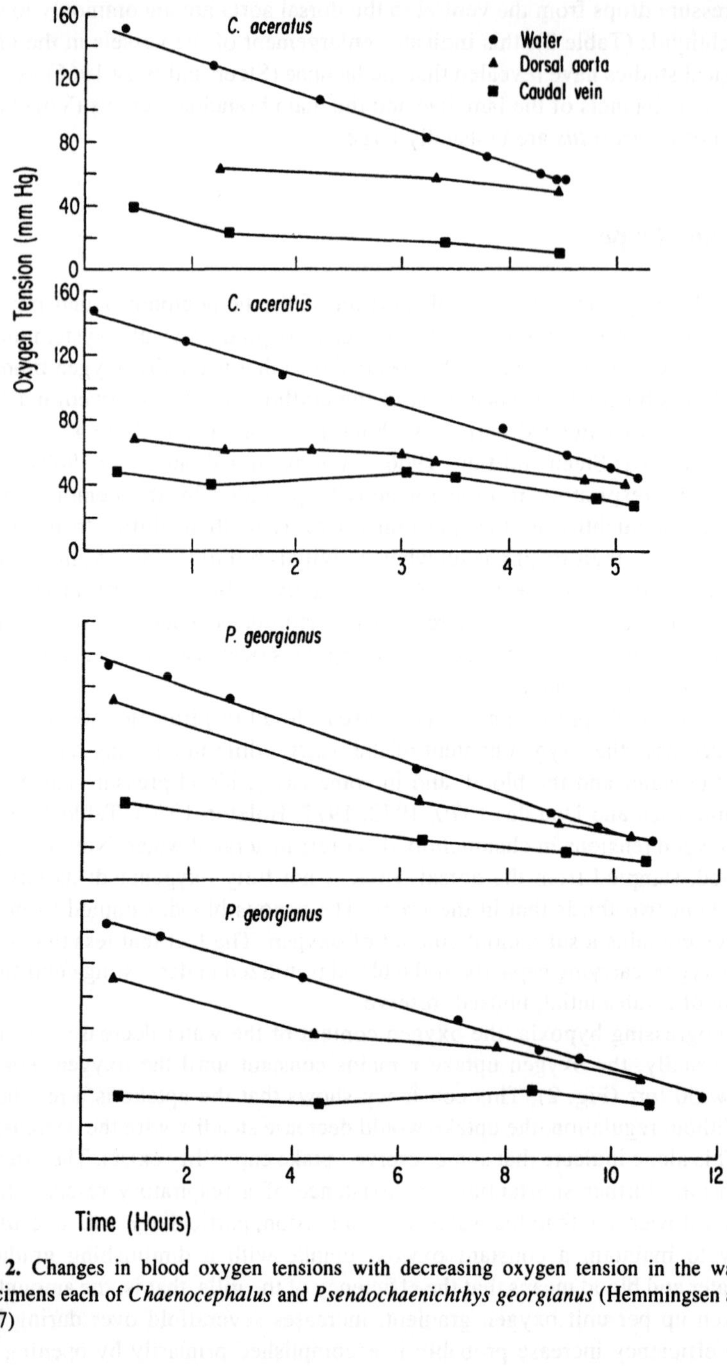

Fig 2. Changes in blood oxygen tensions with decreasing oxygen tension in the water for two specimens each of *Chaenocephalus* and *Psendochaenichthys georgianus* (Hemmingsen and Douglas 1977)

cardiac output during hypoxia (Hemmingsen et al. 1972). A clear increase occurred concurrently in the ventilatory water flow, as revealed by the amplitude of the opercular breathing movements.

During prolonged exposure of *C. aceratus* to hypoxia, blood lactic acid levels increased from a few to about 20–25 mg per 100 ml (Hemmingsen and Douglas 1972), indicating that some anaerobic metabolism takes place. These modest increases in lactic acid effectively rule out that anaerobic metabolism contributes in a major way to the overall metabolism of the channichthyids. Most of the rise in lactic acid in *C. aceratus* may be caused by the unusually high cardiac or ventilatory work that must be performed during the imposed hypoxia.

It is clear from these various results that the channichthyids, as exemplified by *C. aceratus* and *P. georgianus*, have a considerable reserve capacity for uptake and transport of oxygen which allows them to perform a certain degree of activity without extensive anaerobic metabolism.

5 Energetic Consequences of Hemoglobin Absence

The energy cost of maintaining circulation is a substantial part of the metabolism in any animal, even when it is at rest, as are such processes as ventilation, maintenance of osmotic equilibria and protein synthesis. Unique to the channichthyids is the extra cost of the dramatically increased rate of circulation. The underlying, overall cardiac work can be estimated from the cardiac output. The work equals the blood flow times the net pressure generated by the heart (i.e., ventral aortic minus caudal venous pressure). Assuming a muscle work efficiency of 20%, and energy conversion based on fat metabolism, the cardiac work for *C. aceratus* and *P. georgianus* would be at least 22% of the total energy production, or far higher than that which is obtained for other fish (Hemmingsen and Douglas 1977). Because the assumed muscle efficiency of 20% probably is too high, the cardiac work readily could consume between one-fourth and one-third of the metabolic energy produced.

The magnitude of other energy-requiring processes are not known for the channichthyids and can be estimated less readily. Some values exist for other teleosts. For example, ventilatory work in fish may be as high as 15–30% of the resting metabolism (Schumann and Piiper 1966; Edwards 1971). If a similar value is applicable to the channichthyids, then the total energy cost for gas exchange and gas transport within the animals would be close to half of the resting metabolism. The other half must be used for all of the remaining processes, including osmotic regulation and protein synthesis.

Osmotic regulation appears to impose a substantial energy load. It has been estimated that 20–30% of the oxygen consumption of fish is associated with osmotic regulation (Rao 1968; Febry and Lutz 1987). In trout, this fraction remains unchanged with increased rates of oxygen consumption, probably because of the increased regulatory load induced by the increased water irrigation and blood perfusion of the gills. This raises the question of whether the very high perfusion of the gills and cutaneous tissues in the channichthyids which expose a very large area to osmotic gradients, over short physical distances, may require an even larger portion of the metabolism for osmotic work than is the case is trout.

The last major fraction of the metabolism involves protein synthesis, which in mammals is typically about 20% of the metabolism; presumably it is similar in ectotherms. Little information is available with respect to the cost of protein synthesis in fish, and in particular, in the channichthyids. However, relative values for fish are becoming available (e.g., Smith and Haschemeyer 1980; Haschemeyer 1983; Dunn 1988), and it is hoped that this will lead to quantitative estimates of the total protein metabolism in channichthyids and other fish. Certainly, some savings will result from the fact that no synthesis of hemoglobin or of erythrocytes is required.

6 Concluding Remarks

In summary, life without hemoglobin has been made feasible in the channichthyids by three differently directed developments.

1. The metabolic requirements have been suppressed to the maximum extent compatible with a competitive existence. This has been accomplished by (1) selecting a low temperature habitat; (2) eliminating all but the absolutely essential muscle activities; and (3) reducing protein synthesis and other biochemical processes at least as a consequence of reduced activity, and possibly, slow growth.

2. Features that provides for better gas exchange have been maximized, most notably the vascularization of cutaneous tissues and fins. The added capabilities, combined with the lowered metabolic needs for oxygen, have aided to reduce substantially the demands placed on the circulatory system.

3. Most prominently, the circulatory volume and the rate of turnover have been greatly increased. The many different elements embodied in this development make it difficult to suggest an evolutionary sequence of events. Whereas the loss of hemoglobin could have occurred spontaneously, none of the circulatory adjustments need not to have been. At least initially, the channichthyids probably only depended on modest expansion of existing features, based on the physiological pliability that appears to exist in all animals. Furthermore, the evolutionary paths diverge somewhat among the various species. For example, whereas the cardiac output in *C. aceratus* is regulated by adjusting the stroke volume and not the frequency, that of *P. georgianus* is regulated by frequency rather than stroke volume. But irrespective of these variations in the adaptations, increasing the turnover rate is not a clear-cut simple solution due to the energetic consequences involved. Certainly, it is not a solution adaptable to other animals that do not exists within the unique habitat of the antarctic waters.

References

Anthony EH (1961) Survival of goldfish in presence of carbon monoxide. J Exp Biol 38:109–125

Barber DL, Westermann JEM (1981) The blood cells of the antarctic icefish *Chaenocephalus aceratus* Lonnberg: light and electron microscopic observations. J Fish Biol 19:11–28

Booth JH (1978) The distribution of blood flow in the gills of fish: application of a new technique to rainbow trout (*Salmo gairdneri*). J Exp Biol 73:119–129

Cameron JN, Wohlschlag DE (1969) Respiratory response to experimentally induced anemia in the pinfish (*Lagodon rhomboides*). J Exp Biol 50:307–317

Douglas EL, Chapman DJ, Hemmingsen EA (1973) Absence of porphyrin respiratory pigments in the blood in the antarctic icefish *Chaenocephalus aceratus*. Cryobiology 10:260–261

Douglas EL Peterson KS, Gysi JR, Chapman DJ (1985) Myoglobin in the heart tissue of fishes lacking hemoglobin. Comp Biochem Physiol 81A:855–888

Dunn JF (1988) Muscle metabolism in antarctic fish. Comp Biochem Physiol 90B:539-545

Edwards RRC (1971) An assessment of the energy cost of gill ventilation in the plaice (*Pleuronectes platessa L.)*. Comp Biochem Physiol 40A:391–398

Farrell AP (1982) Cardiovascular changes in the unanaesthetized lingcod (*Ophiodon elongatus*) during the short-term, progressive hypoxia and spontaneous activity. Can J Zool 60:933–941

Febry R, Lutz P (1987) Energy partitioning in fish: the activity-related cost of osmoregulation in a euryhaline ciclid. J Exp Biol 128:63–85

Fitch NA, Johnston IA, Wood RE (1984) Skeletal muscle capillary supply in a fish that lacks respiratory pigments. Respir Physiol 57:201–211

Garey W (1970) Cardiac output of the carp (*Cyprinus carpio*). Comp Biochem Physiol 33:181–189

Grigg GC (1967) Some respiratory properties of the blood of four species of antarctic fishes. Comp Biochem Physiol 23:139–148

Hamoir G, Gerardin-Otthiers N (1980) Differentiation of the sarcoplasmic proteins of white, yellowish and cardiac muscles of an antarctic hemoglobin-free fish, *Champsocephalus gunnari*. Comp Biochem Physiol 65B:199–206

Haschemeyer AEV (1983) A comparative study of protein synthesis in nototheniids and icefish at Palmer Station, Antarctica. Comp Biochem Physiol 76B:541–544

Hemmingsen EA, Douglas EL (1970) Respiratory characteristics of the hemoglobin-free fish *Chaenocephalus aceratus*. Comp Biochem Physiol 33:733–744

Hemmingsen EA, Douglas EL (1972) Respiratory and circulatory responses in a hemoglobin-free fish, *Chaenocephalus aceratus*, to changes in temperature and oxygen tension. Comp Biochem Physiol 43A:1031–1043

Hemmingsen EA, Douglas EL (1977) Respiratory and circulatory adaptations to the absence of hemoglobin in chaenichthyid fishes. In: Llano GA (ed) Adaptations within antarctic ecosystems. Smithsonian Inst, Washington, DC, 479–487 pp

Hemmingsen EA, Douglas EL, Grigg GC (1969) Oxygen consumption in an antarctic hemoglobin-free fish, *Pagetopsis macropterus*, and in three species of Notothenia. Comp Biochem Physiol 29:467–470

Hemmingsen EA, Douglas EL, Johansen K, Millard RW (1972) Aortic blood flow and cardiac output in the hemoglobin-free fish, *Chaenocephalus aceratus*. Comp Biochem Physiol 43A:1045–1051

Holeton GF (1970) Oxygen uptake and circulation by a hemoglobinless antarctic fish (*Chaenocephalus aceratus* Lonnberg) compared with three red-blooded antarctic fish. Comp Biochem Physiol 34:457–471

Holeton GF (1972) Gas exchange in fish with and without hemoglobin. Resp Physiol 14:142–150

Holeton GF (1974) Metabolic cold adaptation of polar fish: fact or artefact? Physiol Zool 47:137–152

Holeton GF (1976) Respiratory morphometrics of white and red blooded antarctic fish. Comp Biochem Physiol 54A:215–220

Hureau JC (1966) Biologie de *Chaenichthys rhinoceratus* Richardson et problemes du sang incolore des *Chaenichthyidae*, poissons des mers australes. Bull Soc Zool Fr 91:735–751

Hureau JC, Petit D, Fine JM, Marneux M (1977) New cytological, biochemical, and physiological data on the colorless blood of the *Channichthyidae* (Pisces, Teleosteans, Perciformes). In: Llano GA (ed) Adaptations within antarctic ecosystems. Smithsonian Inst, Washington, DC, 459–477 pp

Jakubowski M (1982) Dimensions of respiratory surfaces of the gills and skin in the antarctic white-blooded fish *Chaenocephalus aceratus*, Lönnberg (Chaenichthyidae). Z Mikiosk - Anat Forch 96:145–156

Jakubowski M, Byczkowska-Smyk (1970) Respiratory surfaces of the white-blooded fish *Chaenichthys rugosus*, Regan (Perciformes). Polish Arch Hydrobiol 17:273–281

Johansen K, Martin AW (1965) Comparative aspects of carsiovascular function in vertebrates. In: Hamilton WF (ed) Circulation, vol 3, Handbook of physiology. Am Physiol Soc, Washington, DC, 2583–2613 pp

Johnston IA, Fitch N, Zummo G, Wood RE, Harrison P, Tota B (1983) Morphometric and ultrastructural features of the ventricular myocardium of the haemoglobin-less icefish *Chaenocephalus aceratus*. Comp Biochem Physiol 76A:475–480

Jones DR, Langille BL, Randall DJ, Shelton G (1974) Blood flow in dorsal and ventral aortas of the cod, *Gadus morhua*. Am J Physiol 226:90–95

Kiceniuk JW, Jones DR (1977) The oxygen transport system in trout (*Salmo gairdneri*) during sustained exercise. J Exp Biol 69:247–260

Macdonald JA, Montgomery JC, Wells RMG (1987) Comparative physiology of antarctic fishes. In: Blaxter JHS, Southward AJ (eds) Advances in marine biology, vol 24. Academic Press, London, 321–388 pp

Martsinkevich LD (1958) Cellular makeup of the blood of white-blooded fish (*Chaenichthyidae*) from the Antarctic. Inform Byul Sov Antarkt Eksp 3:67–68

Morris DJ, North AW (1984) Oxygen consumption of five species of fish from the South Georgia. J Exp Mar Biol Ecol 78:75–86

Nicloux M (1923) Action de l'oxyde de carbone sur les poissons et capacite respiratoire du sang de ces animaux. C R Seances Soc Biol 89:1328-1331

Nonnotte G, Kirsch R (1978) Cutaneous respiration in seven sea-water teleosts. Resp Physiol 35:111-118

Olsen S (1955) A contribution to the systematics and biology of chaenichthyid fishes from South Georgia. Nytt Mag Zool 3:79-93

Permitin YY (1970) The consumption of krill by antartic fishes. In: Holdgate M (ed) Antarctic ecology, vol 1. Academic Press, London, 177 pp

Pettersson K, Nilsson s (1980) Drug induced changes in cardio-vascular parameters in the Atlantic cod, *Gadus morhua*. J Comp Physiol 137:131-138

Ralph R, Everson I (1968) The respiratory metabolism of some antarctic fish. Comp Biochem Physiol 27: 299-307

Randall DJ, Daxboeck C (1982) Cardiovascular changes in the rainbow trout (*Salmo gairdneri* Richardson) during exercise. Can J Zool 60 :1135-1140

Rao GMM (1968) Oxygen consumption of rainbow trout (*Salmo gairdneri*) in relation to activity and salinity. Can J Zool 46:781-786

Robertson OH, Krupp MA, Thompson N, Thomas SF, Hane S (1966) Blood pressure and heart weight in immature and spawning Pacific salmon. Am J Physiol 210:957-964

Ruud JT (1954) Vertebrates without erythrocytes and blood pigment. Nature (Lond) 173:848-850

Ruud JT (1965) The icefish. Sci Am 213:108-114

Scholander PF, Van Dam L (1957) The concentration of hemoglobin in some cold water arctic fishes. J Cell Comp Physiol 49:1-4

Schumann d, Piiper J (1966) Der Sauerstoffbedarf der Atmung bei Fischen nach Messungen an der narkotisierten Schleie (*Tinca tinca*). Arch Ges Physiol 288:15-26

Smith LS (1966) Blood volumes of three salmonids. Fish Res Bd Canada 23: 1439-1446

Smith MAK, Haschemeyer AEV (1980) Protein metabolism and cold adaptation in antarctic fishes. Physiol Zool 53:373-382

Steen JB, Berg T (1966) The gills of two species of haemoglobin-free fishes compared to those of other teleosts–with a note on severe anaemia in an ell. comp Biochem Physiol 18:517-526

Steen JB, Kruysse A (1964) The respiratory function of teleostean gills. Comp Biochem Physiol 12:221-241

Stevens ED, Randall DJ (1967) Changes in blood pressure, heart rate and breathing rate during moderate swimming activity in rainbow trout. J Exp Biol 46:307-315

Takeda T (1989) Cutaneous and gill O_2 uptake in the carp *Cyprinus carpio*, as a function of ambient PO_2. Comp Biochem Physiol 94A:205-208

Thorson TB (1961) The partitioning of body water in *Osteichthyes*: phylogenetic and ecological implications in aquatic vertebrates. Biol Bull 120:238-254

Tota B, Farina F, Zummo G (1988) Ultrastructural aspects of functional interest in the ventricular myocardial wall of the antarctic icefish *Chaenocephalus aceratus*. Comp Biochem Physiol 90B:561-566

Twelves EL (1972) Blood volumes of two antarctic fishes. Bull Br Antarct Surv 31:85-92

Vogel W, Kock K-H (1981) Morphology of gill vessels in icefish. Arch Fisch Wiss 31:139-150

Walvig F (1960) The integument of the icefish *Chaenocephalus aceratus* (Lonnberg). Nytt Mag Zool 9:31-36

Weiss RF (1970) The solubility of nitrogen, oxygen and argon in water and seawater. Deep-sea Research 12:721-735

Wells RMG (1986) Cutaneous oxygen uptake in the antarctic icequab, *Rhigophila dearborni* (Pisces: Zoarcidae. Polar Biol 5:175-179

Wells RMG (1987) Respiration of antarctic fish from McMurdo Sound. Comp Biochem Physiol 88A:417-424

Wohlschlag DE (1964) Respiratory metabolism and growth of some antarctic fishes. In: Carrick R, Holdgate MW, Prevost J (eds) Biologie antarctique. Proc 1st Symp on Antarctic Biology, Paris. Hermann, Paris, 489–502 pp

Structural and Mechanical Characteristics of the Heart of the Icefish *Chionodraco hamatus* (Lönnberg)

B. Tota[1,3], C. Agnisola[2,3], M. Schioppa[1,3], R. Acierno[3,4], P. Harrison[5], and G. Zummo[6]

1 Introduction

The absence of haemoglobin from the blood of icefish, antarctic teleosts of the family Chaenichthydae, has stimulated important physiological mechanisms which serve to compensate for the absence of the respiratory pigment. In view of the central role played by the circulation in meeting the oxygen demands of the tissues, it is not surprising to find that the cardiocirculatory system has been extensively remodelled in icefish. These adaptations include: (1) increases in blood volume, from two to four times larger than that in many teleosts (Hemmingsen and Douglas 1970; Holeton 1970); (2) increases in blood flow associated with lowered oxygen demands (Hemmingsen and Douglas 1970; Holeton 1970); (3) corresponding increase in cardiac output by means of very large stroke volume (6-15 times greater than in other teleosts) at low heart rate and at relatively low ventral aortic pressure (Hemmingsen and Douglas 1972; Hemmingsen et al. 1972); (4) gill blood vessels (Vogel and Koch 1981) and muscle capillaries (Fitch et al. 1984) which have larger diameters compared to other teleosts; (5) increased vascularity in the skin and in the fins (Walvig 1960; Jakubowsky 1982); (6) impressive cardiac enlargement resulting in a relative heart weight similar to that of small mammals (Johnston et al. 1983; Feller et al. 1985; Tota et al. 1990). The interrelationships between these compensatory adaptations are illustrated in Fig 1.

To date, physiological data concerning these cardiocirculatory adjustments have been obtained from unrestrained specimens of two species of icefish: the benthic *Chaenocephalus aceratus* and the pelagic *Pseudochaenichthys georgianus* with intact humoral and nervous control (Hemmingsen and Douglas 1972; Hemmingsen et al. 1972). The question arises therefore, whether the homeostatic mechanisms apparent in the whole animal can be related directly to characteristics at the organ level, i.e., in the heart itself. Such a reductionist approach may serve to reveal hidden mechanisms of cardiovascular adaptation in the cardiac design of the

[1] Dipartimento di Biologia Cellulare, Università della Calabria, I-87030 Arcavacata di Rende (CS), Italy
[2] Dipartimento di Fisiologia Generale ed Ambientale, Università di Napoli, I-80134 Napoli, Italy
[3] Stazione Zoologica "A. Dohrn" di Napoli, I-80121 Napoli, Italy
[4] Istituto di Fisiologia, Università di Lecce, Lecce, Italy
[5] School of Biological Sciences, University of Birmingham, Edgbaston, Birmingham, England, UK
[6] Dipartimento di Discipline Chirurgiche ed Anatomiche, Università di Palermo, Palermo, Italy

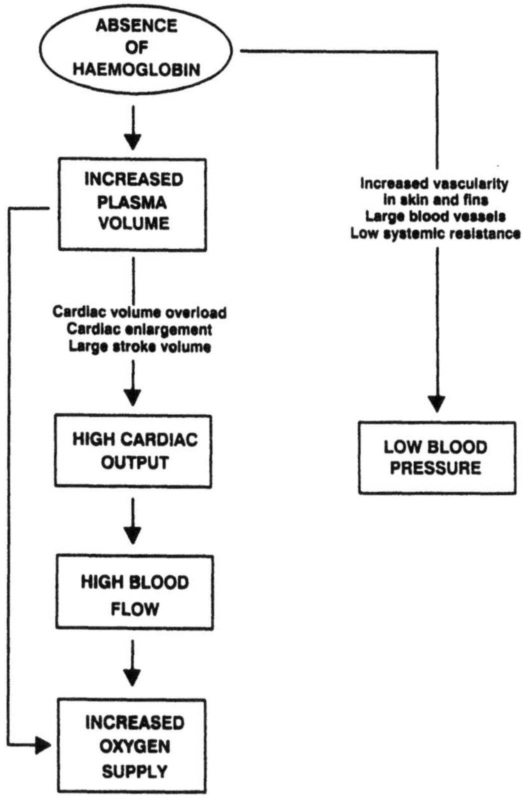

Fig. 1. Scheme of the circulatory adaptations in icefish

icefish which are important for both basic and comparative cardiology. Until now, no attempt has been made to answer this question. It is the purpose to this paper to focus on the characteristics of the mechanical performance of the heart of the icefish *Chionodraco hamatus*, revealed by an isolated and perfused working heart preparation, and to report data on the structural arrangement of the heart in this species and in *Chaenocephalus aceratus*, previously studied in our laboratory. The aim is to draw attention to the icefish heart as a muscular pump and to compare it to what is known in other teleosts. The specimens of *Chionodraco hamatus*, the most common icefish in Terranova Bay, Antarctica, were collected during the 5th Italian Antarctic Expedition (December 1989-February 1990). All the physiological experiments were carried out in the laboratories of the Italian Antarctic Base, Terranova Bay (74° 42' S, 164° 06' E).

2 Scaling Aspects and Ventricular Myoarchitecture

The "heart index" and "ventricle index" indicate the percentage of body weight that is accounted for by the heart and its ventricular chamber, respectively. In all

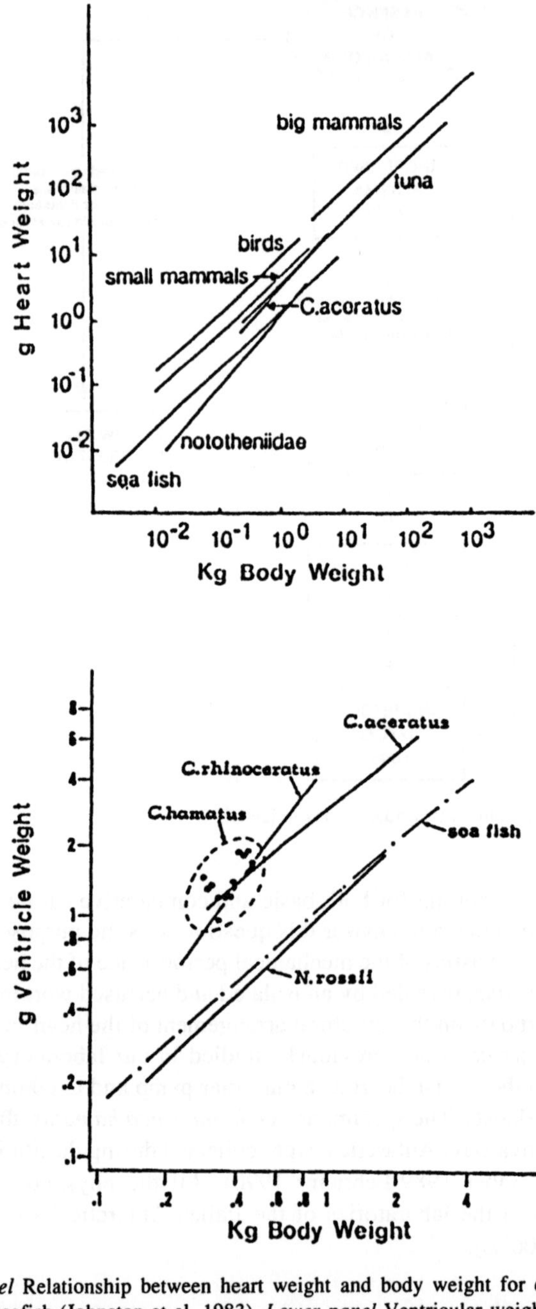

Fig. 2. *Upper panel* Relationship between heart weight and body weight for different groups of animals including icefish (Johnston et al. 1983). *Lower panel* Ventricular weight and body weight values for *C. hamatus* plotted together with data for other species taken from the literature: *Chaenocephalus aceratus* and *Notothenia rossii* (Johnston *et al.* 1983), *Channichthys rhinoceratus* (Feller *et al.* 1985) and a composite temperate sea–fish line (Poupa *et al.* 1981). It can be noted that the values from *C. hamatus* fit the curves of the other icefish, *C. aceratus* and *C. rhinoceratus*

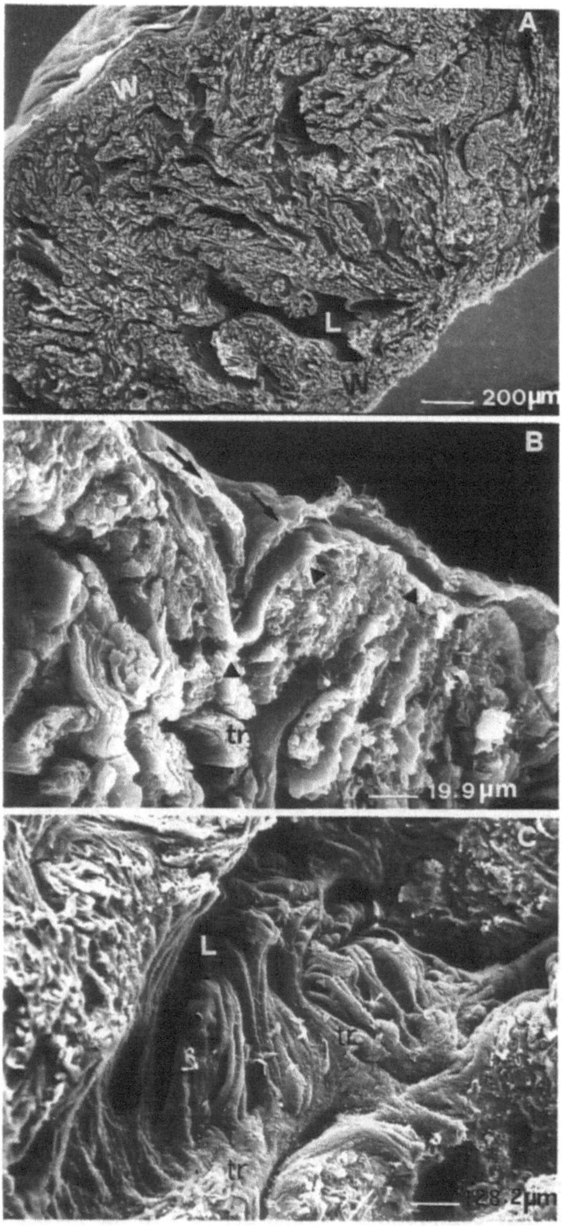

Fig. 3. SEM pictures of the cardiac ventricular wall of *C. hamatus*. **A** Spongy type of ventricular myoarchitecture. It is apparent the trabeculated organization with a wide lacunary system *(L); W* ventricular wall. **B** The epimyocardial region showing the origin of the muscular trabeculum and the relationship with the pericardial lining. Arrows Parietal pericardium; *arrowheads* visceral pericardium; *tr* trabeculae. **C** Ventricular lumen: note the trabaculae (*tr*) and the intertrabecular lacunary spaces (*L*)

vertebrates, heart weight increases with body weight: in ectotherms the cardiac mass has an allometric exponent of 0.743, whereas in endotherms it approximately scales with body weight to the power of 1 (Poupa and Lindstrom 1983; Seymour 1987). In most antarctic teleosts (Notothenioids) the relative heart weight is similar to that of most teleosts, including those from temperate waters, while in the icefish the heart weight is similar to that of small mammals. The only teleost that exhibits such remarkable increase of cardiac mass is the tuna fish (Fig. 2). The comparison between the general myoarchitecture of the ventricular wall in the icefish and in tuna can shed light on the different structural strategies by which cardiac growth is attained (Figs. 3, 4). The ventricle of tuna is pyramidal in shape and it is endowed with the highest amount of compacta found in teleosts (Tota 1978; Santer and Greer Walker 1980). Thus it represents in the spectrum of cardiac designs in fish one extreme, the opposite being the icefish heart.

The ultrastructure of the icefish heart ventricle provides useful information on how this cardiac increase is reflected at the individual fiber.

3 Ultrastructural Aspects of Ventricular Myocytes

There is considerable variation in the myofibrillar and mitochondrial packing within the ventricular myocytes of icefish. In *Channichthyis rhinoceratus* (Feller et al. 1985) the volume densities of these organelles are similar to those of *Notothenia neglecta* and *Notothenia rossii*, antarctic nototheniids with myoglobin-rich ventricles (Table 1). In contrast, in *Chaenocephalus aceratus* (Harrison et al. 1991) and Chionodraco hamatus (the present authors, unpubl. results) the mitochondrial volume densities [$V_{v(mt,f)}$= about 0.43] are the highest reported in any teleosts (Table 1). Indeed, with the exception of the Etruscan shrew [$V_{v(mt,f)}$ = 0.45; Weibel 1985] the mitochondrial compartments in these icefish are larger than in any vertebrate. The space devoted to mitochondria appears to be at the expense of force generating elements; the mean myofibrillar volume densities in *C. aceratus* (0.31, Harrison et al. 1991) and *C. hamatus* (0.25, the present authors, unpubl. results) are low compared to other teleosts (Table 1).

The myocytes of all notothenioids, both red-blooded and haemoglobinless, are large cells with diameters between 4 and 6 μm, which are towards the upper end of the range reported in other teleosts (Table 1). This, in combination with the low environmental temperatures experienced by these animals, presents particular problems with respect to metabolite diffusion. In icefish, these are compounded by the lack of haemoglobin and low myocardial myoglobin concentrations (Douglas et al. 1985) which may compromise oxygen delivery to the myocardial cells. The remarkable proliferation of mitochondria in the myocytes of some icefish probably compensates for diffusional limitations and so maintains functional levels of ATP synthesis.

It is well established that modifications in the morphometric characteristics of oxidative skeletal muscle fibres serve a similar purpose. For example, in their study on two red-blooded antarctic teleosts (*Trematomus newnesi* and *Notothenia gibberifrons*), Londraville and Sidell (1990) suggest that the high mitochondrial

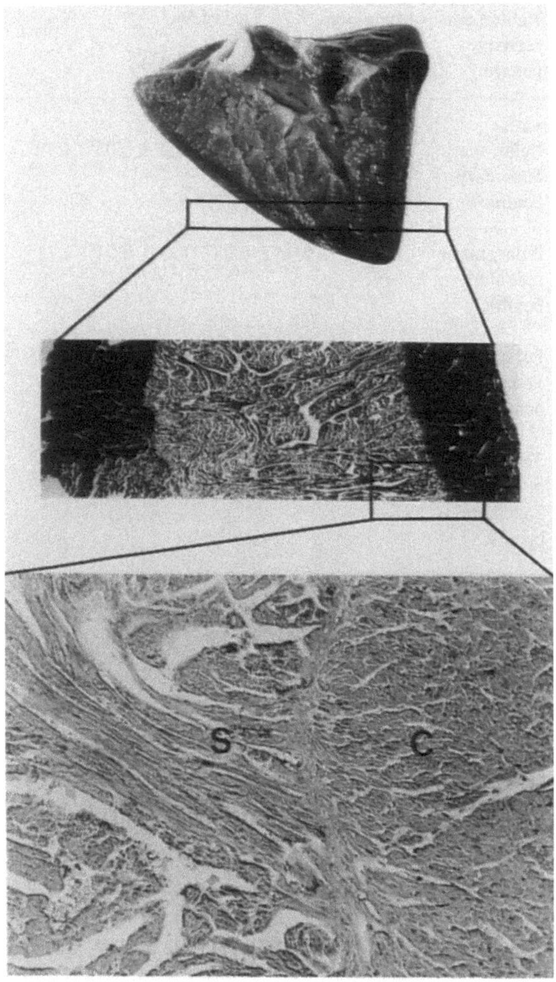

Fig. 4. Shape and structural organization of the ventricular chamber of tuna fish (*Thunnus thynnus thynnus*). Body weight of the juvenile specimen was 0.7 kg. The cross-section of the ventricle (*upper panel*) shows the extremely muscular, thick-walled chamber with a typical triangle pyramidal shape. It is evident the thickened epimyocardial layer (compacta) and the inner trabeculated myocardium (spongiosa) that make up the mixed type of heart ventricle (*middle panel*). Note that in comparison with the ventricular wall of the icefish (Fig. 3A), in tuna the spongiosa is more densely packed with consequent lesser extension of the lacunary spaces. The *lower panel* shows the histological texture of the compacta (*C*) and the spongiosa (*S*) separated by a thin layer of connective tissue (Azan staining, 60x). On the whole, the pyramidal configuration and the high muscularity of the wall are well designed to develop interfascial tension, with very high resistance to stretching, especially at the apices. This makes the ventricle architecturally designed to function as a high-pressure pump, which is consistent with the kind of performance needed in the high-pressure, high–resistance systemic circulation typical of tunas (Tota 1978)

Table 1. Quantitative ultrastructural characteristics of teleost ventricular myocytes

Species	Habitat and activity pattern	Myoglobin (nmol g^{-1} wet wt.)	$V_{v(mt,f)}$	$V_{v(my,f)}$	$V_{v(mt,f)}$ / $V_{v(my,f)}$	Myocyte diameter µm)
Myoglobin-poor hearts						
Chaenocephalus aceratus	Polar, marine Sedentary, benthic	1.02-1.51[a]	0.43 ± 0.04[b]	0.31 ± 0.03[b]	1.39[b]	5.95 ± 1.92[b]
Chionodraco hamatus	Polar, marine sedentary, benthic	nd	0.44 ± 0.03[c]	0.25 ± 0.04[c]	1.72[c]	Nd
Chaenichthyis rhinoceratus	Polar, marine Sedentary, benthic	nd[d]	0.24 ± 0.04[e]	0.44 ± 0.07[e]	0.54[e]	5.70 ± 2.30[e]
Lumpsucker (*Cyclopterus lumpus*)	Temperate, marine Sedentary, benthic	nd[e]	0.27 ± 0.01[b]	0.42 ± 0.02[b]	0.64[b]	2.17 ± 1.00[b]
Myoglobin-rich hearts						
Notothenia neglecta	Polar, marine Largely benthic	19.6 ± 0.1[b]	0.25 ± 0.01[b]	0.43 ± 0.02[b]	0.58[b]	5.29 ± 1.51[b]
Notothenia rossii	Polar, marine Bentho/ pelagic	—	0.23 ± 0.04[e]	0.45 ± 0.02[e]	0.51[e]	4.30 ± 1.30[e]
Nezumia aequalis	Mesopelagic marine Bentho/ pelagic	—	0.20 ± 0.07[g]	0.51 ± 0.11[g]	0.39[g]	3.68 ± 2.92[g]
Cottus (*Myoxocephalus scorpius*)	Temperate, marine Sedentary, benthic	20.7 ± 0.4[b]	0.17 ± 0.06[g]	0.51 ± 0.15[g]	0.33[g]	3.59 ± 2.59[g]
Thunnus alalunga	Marine Active pelagic Tropical/ temperate	— —	0.24 ± 0.02[h] 0.32 ± 0.02	0.52 ± 0.01[h] 0.55 ± 0.02	0.50[h] 0.61	4-6[h] Compacta 2.5-5.0 Spongiosa

Table 1. (*continued*)

Species	Habitat and activity pattern	Myoglobin (nmol g^{-1} wet wt.)	$V_{v(mt,f)}$	$V_{v(my,f)}$	$\frac{V_{v(mt,f)}}{V_{v(my,f)}}$	Myocyte diameter (μm)
Thunnus thynnus	Marine Active pelagic Tropical/ temperate	581-1175[i] 581-1175	0.39[l] 0.32	0.47[l] 0.49	0.62[l] 0.65	— Compacta — Spongiosa
Oscar cichlid (*Astronotus ocellatus*)	Tropical Fresh-water Pelagic	—	0.15 ± 0.01[b]	0.15 ± 0.01[b]		2.01 ± 1.12[b]

Abbreviations: Vv(mt,f)=mitochondrial volume density. Vv(my,f)=Myofibrillar volume density. Nd-not detected.

[a] Calculated from Douglas et al. (1985) assuming the molecular weight of myoglobin = 1660 and the ratio of ventricular dry weight:wet weight is 1:5.2.
[b] Harrison et al. (1991).
[c] These authors (unpubl. results).
[d] Feller (1987).
[e] Feller et al. (1985).
[f] Driedzic and Stewart (1982).
[g] Morris and Johnston (unpubl. results).
[h] Breisch et al. (1983).
[i] Giovane et al. (1980).
[l] Tota (1978).

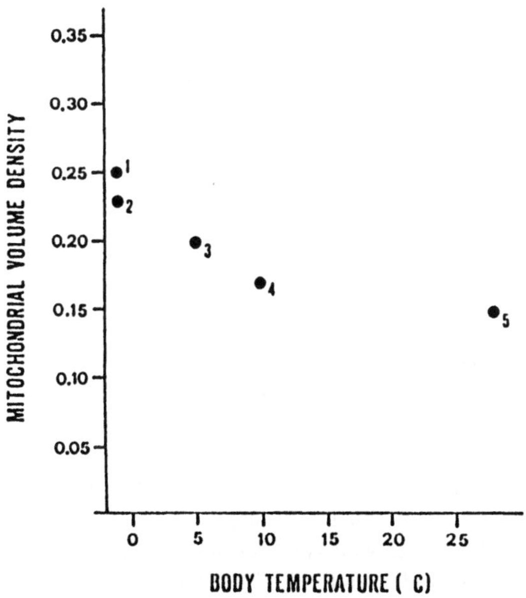

Fig. 5. Relationship between body temperature and mitochondrial volume density in myocytes from the spongy ventricular myocardia of sedentary species of teleosts with myoglobin-rich hearts. *1 Notothenia neglecta* (Harrison, unpubl. results); *2 Notothenia rossii* (Feller et al. 1985); *3 Nezumia aequalis* (Morris and Johnston, unpubl. results); *4 Myoxocephalus scorpius* (Morris and Johnston, unpubl. results); *5 Astronotus ocellatus* (Harrison, unpubl. results)

densities found in the oxidative fibres may serve to decrease mean intracellular diffusion distance. Moreover, in these fibre types in eurythermal teleosts there is an increase in mitochondrial volume density during acclimation to low temperatures (Johnston and Maitland 1980; Tyler and Sidell 1984), which is thought to offset the rate depressing effects of low temperature on the activities of oxidative enzymes and on the diffusion of metabolites across the mitochondrial membranes (Sidell 1983; Londraville and Sidell 1990). The data in Fig. 5, which considers only sedentary species with myoglobin-rich ventricles, illustrate an evolutionary analogue of this situation in ventricle myocytes, since there is a trend towards increasing values of $V_{v(mt,f)}$ with decreasing adaptation temperature.

In a detailed comparison of the characteristics of the myocytes in the antarctic nototheniid *N. neglecta* and in the icefish *C. aceratus*, Johnston and Harrison (1987) showed that there was no significant difference between the activities of cytochrome oxidase in the two species. They concluded therefore that the higher value of $V_{v(mt,f)}$ in the icefish (0.43) compared to the red-blooded species (0.25) was not related to increases in aerobic capacity, but served to minimize oxygen diffusion distances between the lacunae and the outer mitochondrial membrane, thereby compensating for the low intracellular myoglobin content. If this is so, there remains an open question: how can the myocardial function be maintained in *C. rhinoceratus*, which has mitochondrial volume densities similar to those of nototheniids (Table 1).

4 Bulbus Arteriosus:Structural Aspects and ANP Binding Study

In common with other teleosts, the wall of the bulbus arteriosus *C. hamatus* consists of three layers, the innermost tunica intima, the middle tunica media and the outermost adventitia. Morphological analysis on these layers has been obtained by conventional histology and electron microscopy (see SEM view, Fig. 6). Tunica intima is characterized by a folding arrangement which is responsible for the longitudinal ridges seen in the luminal surface. The endothelial lining consists of rounded (in surface view) elongated cells bulging into the lumen. The subendothelial connective tissue is extremely rich in elastic fibres and is responsible for the relevant thickness of this layer. Tunica media is made up of elastic laminae that are separated from each other by a weakly basophilic matrix. The connective andowement consists mainly of elastic fibres circumferentially oriented. The adventitia is made up of mesothelial cells that abut on the pericardial cavity and, beneath them, of numerous bundles of collagen fibres.

An important of the bulbar wall is represented by its richness in binding sites for the atrial natriuretic peptide (ANP; Fig. 7). This suggests that this cardiopeptide has a role in the regulation of blood flow between the ventricular chamber and branchial vasculature. According to the windkessel theory applied to the function of the bulbus arteriosus of teleost fish (see for review Priede 1976 and Benjamin *et al.* 1983), during ventricular ejection the inflow allow the elastic bulbus wall to store

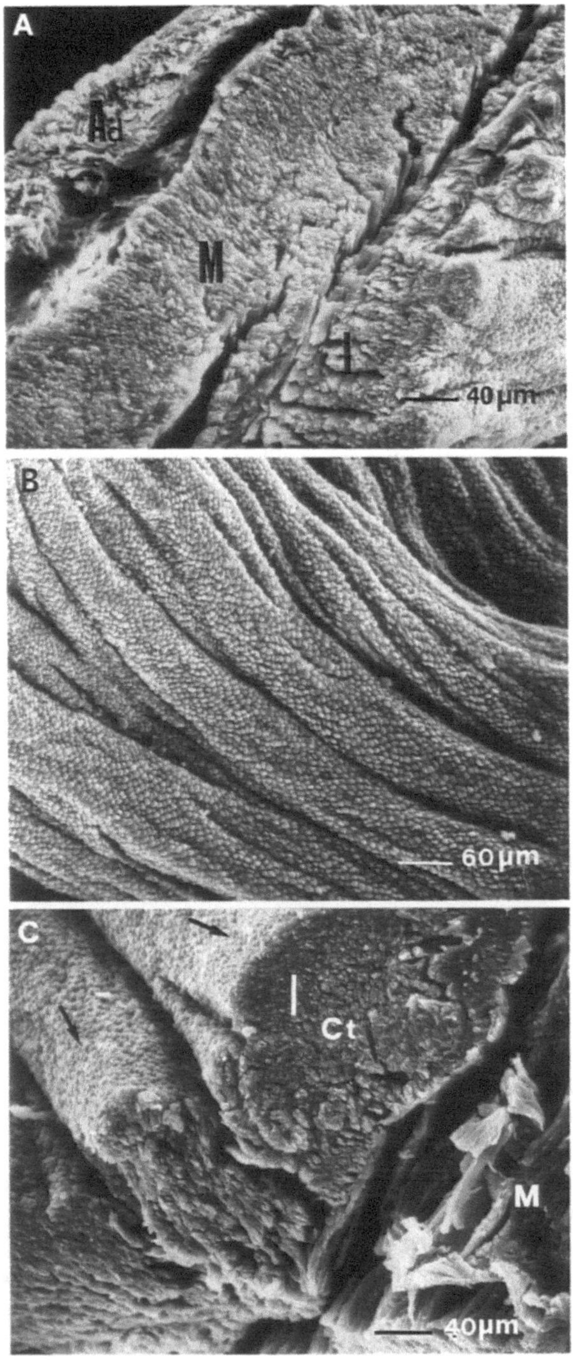

Fig. 6. The bulbus of *C. hamatus* with the structural components of the wall as revealed by SEM. **A** Cross-section of the wall showing the innermost tunica intima (*I*), the very thick tunica media (*M*) and the outermost adventitia (*Ad*). Tunica intima is characterized by numerous longitudinal ridges. **B** A view from the bulbar lumen showing the ridges with their endothelial lining. **C** The ridges are made up of an endothelial lining (*arrows*) and by a very thick connective layer **(Ct)**. The depth of the ridges is about 150 µm

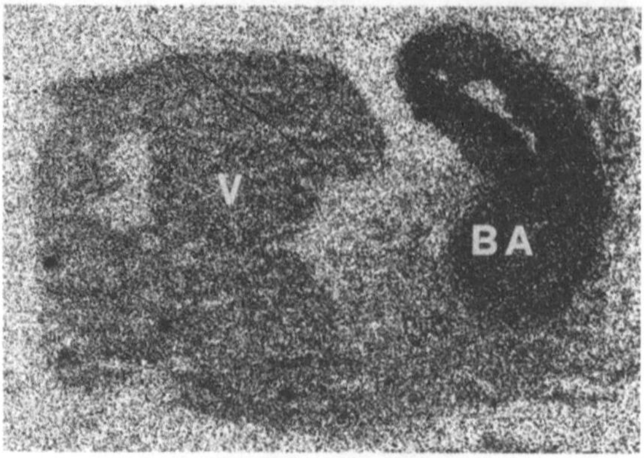

Fig. 7. Localization of binding sites for rat [^{125}I]atrial natriuretic peptide (ANP) in the icefish (*C. hematus*) heart by *in vitro* autoradiography. Cross-section showing the bulbus arteriosus (*BA*) and the ventricle (*V*). On the basis of preliminary quantitative analysis, a higher labeling (388 fmol/mg proteins) of the bulbar wall is evident in comparison with the labeling of the ventricular wall (90 fmol/mg proteins). This indicates that the bulbus of the icefish possesses ANP-binding sites in a significant amount

blood which eventually manifests as outflow to the ventral aorta in diastole (Johansen 1962). A putative endocrine and/or paracrine function of the bulbus in the icefish would be of relevance particularly in relation to the large stroke volume and the very long diastolic period characteristic of the icefish.

5 Mechanical Characteristics

To evaluate the basic mechanical properties of the heart of the icefish *Chionodraco hamatus*, an isolated and perfused heart preparation has been used according to Acierno et al. (1990). This kind of preparation allows the control and measurement of many variables of cardiac function, independently from neurohumoral control, and in conditions that permit an atrial filling with suctional mechanism similar to the *vis a fronte* mechanism operating in the intact fish. The important functional parameters which have been determined are stroke volume and stroke work as measures of systolic function and power output, used as measure of the ability of the heart to generate sufficient pressure to overcome vascular resistance and to supply the tissues with sufficient blood flow (cardiac output) to meet their metabolic requirements (Tota et al. 1991).

Figure 8 shows the results of experiments of volume and pressure loadings on the isolated heart. Volume loading was obtained by changing the preload from –0.7 to –0.4 cm H_2O. Each curve performance-preload was obtained at different values of afterload. The experiments were performed at 3 °C and were completed within 2 h. The heart rate was unaffected by both preload and afterload, with a mean value of 26.3 ± 1.9 beats min^{-1}. Figure 8 shows clearly that the icefish heart responds to the Frank-Starling mechanism, with relevant changes of stroke volume and stroke work in a narrow range of preloads. On the other hand, the icefish heart is able to

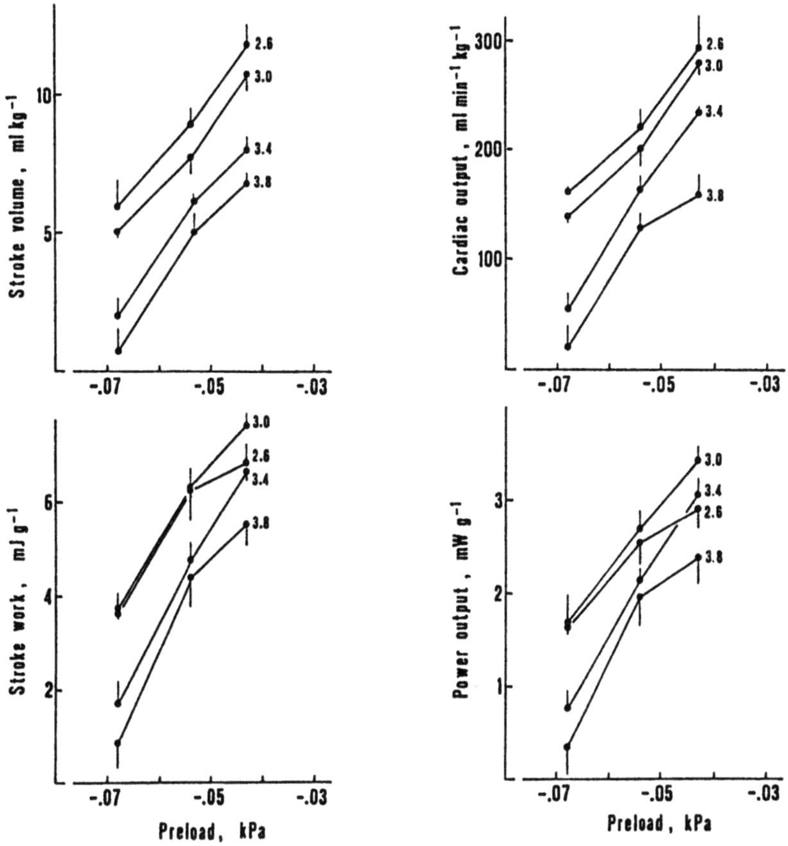

Fig. 8. Relationship between stroke volume, cardiac output, stroke work or power output and preload in the isolated and perfused heart of *C. hamatus*. *Numbers* on the *curves* represent the afterload values (kPa, 1kPa = 10.2 cm H_2O). The experiments were performed at 3 °C. The heart rate was independent from both preload and afterload. Its mean value was 26.2 ± 1.1 beats min^{-1}. See Tota et al. (1991) for details of the experimental procedure used

maintain constant stroke work only in a relatively narrow range of afterloads, the upper value of which corresponds to the physiological value of ventral aorta pressure found in *C. aceratus* (from 18 to 27 cm H_2O, Hemmingsen et al. 1972).

As the heart rate is unchanged in all loading conditions used, the behaviour of cardiac output and power output reflects that of stroke volume and stroke work. In the physiological range of afterloads, values of 150-300 ml min^{-1} kg^{-1} of cardiac output have been found, which are comparable with those found *in* vivo by Hemmingsen et al. (1972). These values are up to ten times higher than the typical values reported in temperate teleosts (Farrell 1984).

Since the performance is quickly impaired with afterloads higher than 30 cm H_2O, normal myocardial function in *C. hamatus* appear to be impossible unless the blood pressure in the ventral aorta is kept low. This dynamic behaviour represents a clear *in vitro* counterpart to the in vivo situation described by Hemmingsen et al. (1972) on unrestrained specimens of *C. aceratus*, and further epitomized by

Hemmingsen and Douglas (1977) as follows: "Without a low flow resistance, cardiac work probably would be intolerable for the Chaenichthyids".

In conclusion, the results in vitro demonstrate that the heart of *C. hamatus* is designed as a very specialized volume pump, i.e., it is able to displace at low heart rate large volumes of blood at high flow rate and low pressure.

The effects of temperature on cardiac performance in teleosts consist of positive chronotropic and inotropic responses, probably representing a compensatory mechanism by which cardiac output is intrinsically increased to meet the metabolic needs of the tissues increased by temperature (see Farrell 1984 for review). Icefish live in a very stable environment, characterized by a very stable temperature (Macdonald et al. 1987). The temperature tolerance range of these animals is relatively narrow (from -1.9 to 6.0 °C, Macdonald et al. 1987), and it is highly probable that this icefish never experiences temperatures higher than 1 °C. To evaluate the response of the icefish heart to temperature changes, we have determined the cardiac parameters of the isolated heart of *C. hamatus*, mentioned above, under constant preload (-0.3 cm H_2O) and afterload (21 cm H_2O) at an initial temperature of 0.5 °C and after the stabilization at a temperature of 5.8 °C. Figure 9 shows that both stroke volume and stroke work are unaffected by the change in temperature. On the contrary, there was an increase of heart rate (21%) which was reflected in an increase of cardiac output (18%) and power output (24%). These effects were reversed by relowering the temperature to basal values. In conclusion, in *C. hamatus*, the main cardiac intrinsic response to temperature changes is positive chronotropism, while, in contrast with other teleosts, the positive inotropism is absent.

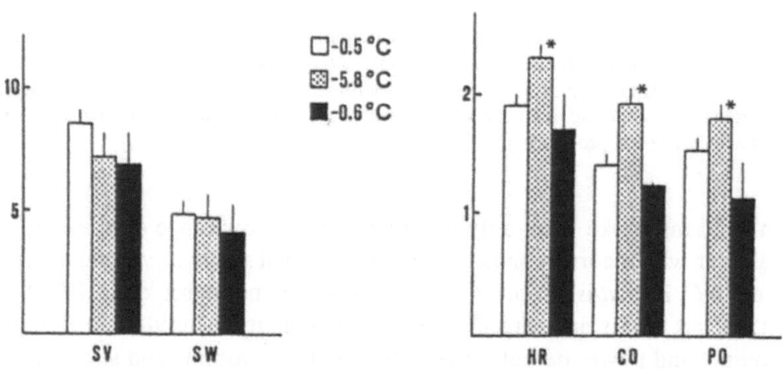

Fig. 9. Effect of temperature on the performance of the isolated heart of *C. hamatus*. The mean preload was -0.03 ± 0.01 kPa. The mean afterload was 2.3 ± 0.1 kPa. *Left SV* stroke volume (ml kg^{-1}); *SW* stroke work (mJ g^{-1}). *Right* HR heart rate (x 10 beats min^{-1}); *CO* cardiac output (x 10^2 ml min^{-1} g^{-1}); *PO* power output (mW g^{-1}). *Asterisks* Differences statistically significant ($P < 0.05$). See Tota et al. (1991) for details of the experimental procedure used

6 Morphofunctional Synopsis

An impressive amount of data, particularly from mammalian and medical literature, has shown that the heart grows in response to increasing circulatory demands by adjusting its mass to compensate for augmented load. This growth is attained either by hypertrophy or by enlargement (Braunwald 1984). In the strict sense, hypertrophy is utilized to mean increased weight, while enlargement means increased weight and decreased cavity size. Enlargement characterizes the response of the heart primarily to conditions of volume overload, while under conditions of pressure overload, the heart increases its wall thickness by hypertrophy (Braunwald 1984). It has been suggested that this increase in wall thickness is an adaptive mechanism to maintain wall stress at constant value (Linzbach 1960) and that the relatively high mean blood pressure and high resistance to the blood leaving the heart are the major factors for increasing stress on the ventricular wall (Hjalmarsson and Isaksson 1972). Since these concepts have been scarcely applied to the morphofunctional design of the fish heart, we have drawn attention to the epigenetic factors, particularly haemodynamic, that can modulate heart-shape changes in fish (Tota 1978, 1983, 1989; Tota et al. 1983). Indeed, the different histological myoarchitectures of the ventricular wall of tuna and icefish reported in Figs. 3 and 4 illustrate how the impressive cardiac growth common to these two teleosts results in divergent remodellings of the ventricular chamber, with typical pressure pump and volume pump designs in tuna and icefish, respectively. The above concepts are summarized in Fig. 10.

A closer look at the morphofunctional design of the icefish heart reveals that the ventricular enlargement attained by a spongy type of myoarchitecture is the only structural arrangement compatible with a cardiac performance characterized by large ventricular filling at relatively low pressure-high ventricular compliance and by large stroke volumes at low afterloads. This arrangement is well complemented by the structural characteristics of the bulbus arteriosus that indicate extreme distensibility and important endocrine properties, as revealed by the ANP-binding sites of the wall. The cardiomegaly of the spongy type represents a well-balanced increased of myocardial trabeculum and intertrabecular lacunary spaces, for which

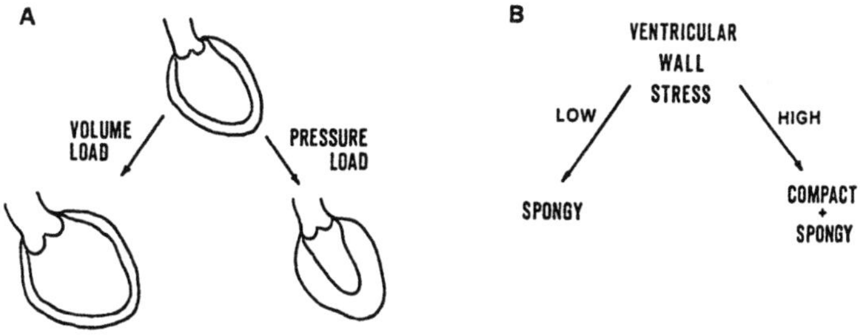

Fig. 10A. Schematic representation of the concept of cardiac enlargement as derived from mammalian and medical literature. Both dilatation (increase of stretch) and hypertrophy (increase of basic contractility) are compensatory adjustments to a rising load/strength ratio. Both processes, when excessive, predispose to heart failure. **B** Type of shape change and myoarchitecture in the teleost heart ventricle large ventricular volume loads can be adjusted with low stress on the myocardial fibres (Tota et al. 1988). The maintenance of low stress during systole prevents increased mechanical demands upon the contractile apparatus, and should thus prevent, at the ultrastructural level, the discrepancy between the low myofibrillar endowment and the very large mitochondrial compartment necessary for increasing intracellular diffusion capacity. Hence, for this kind of ventricular chamber, it is impossible to face the relatively high wall stresses required to pump blood in a circulatory tree with high resistance. Both structural and mechanical data indicate that type of cardiomegaly without true hypertrophy can only be functional in a very narrow spectrum of compensations. This in turn mirrors the very specialized eco-physiological traits of the icefish that can be successful in the very stable and unique antarctic environment.

Acknowledgment. This work was supported by the Italian National Research Program in Antarctica (PNRA).

References

Acierno R, Agnisola C, Venzi R, Tota B (1990) Performance of the isolated and perfused working heart of the teleost *Conger conger*: study of the inotropic effect of prostacyclin. J Comp Physiol B 160:365–371

Benjamin M, Norman D, Santer RM, Scarborough D (1983) Histological, histochemical and ultrastructural studies on the bulbus arteriosus of the sticklebacks, *Gasterosteus aculeatus* and *Pungitius pungitius* (Pisces:Teleostei). J Zoll Lond 200:325–346

Braunwald E (1984) Heart disease: a textbook of cardiovascular medicine. 2nd edn. WB Saunders, Philadelphia, London, Toronto

Breisch EA, White F, Jones HM, Laurs RM (1983) Ultrastructural morphometry of the myocardium of *Thunnus alalunga*. Cell Tissue Res 233:427–438

Douglas EL, Peterson KS, Gysi JR, Chapman DJ (1985) Myoglobin in the heart tissue if fishes lacking hemoglobin. Comp Biochem Physiol 81A:855–888

Driedzic WR, Stewart JM (1982) Myoglobin content and activities of enzymes of energy metabolism in red and white fish hearts. J Comp Physiol B 149: 67–73

Farrell AP (1984) A review of cardiac performance in the teleost heart: intrinsic regulation. Can J Zool 62:523–536

Feller G (1987) Les channichthyidae: poissons de l'Antarctque. Adaptations metaboliques et structurales a l'absence d'hemoglobine et de myoglobine. These pour le grade de Docteur on Sciences Biomedicales Experimentales.

Feller G, Goessens G, Gerday C,Bassleer R (1985) Heart structure and ventricular ultrastructure of hemoglobin- and myoglobin-free icefish *Channichthys rhinoceratus*. Cell Tissue Res 242:669–676

Fitch NA, Johnston IA,Wood RE (1984) Skeletal muscle capillarity supply in a fish that lacks respiratory pigments. Resp Physiol 57:201–211

Giovane A,Greco G,Maresca A,Tota B (1980) Myoglobin in the heart ventricle of tuna and other fishes. Experientia 36:219–220

Harrison P, Zummo G, Farina F, Tota B, Johnston IA (1991) Gross anatomy, myoarchitecture and ultrastructure of the heart ventricle in the haemoglobinless icefish *Chaenocephalus aceratus*. Can J Zool 69

Hemmingsen EA, Douglas EL (1970) Respiratory characteristics of the hemoglobinfree fish *Chaenocephalus aceratus*. Comp Biochem Physiol 33:733–744

Hemmingsen EA, Douglas EL (1972) Respiratory and circulatory responses in a hemoglobin-free fish, *Chaenocephalus aceratus*, to changes in temperature and oxygen tension. Comp Biochem Physiol 43A:1031–1043

Hemmingsen EA, Douglas EL (1977) Respiratory and circulatory adaptations to the absence of hemoglobin in chaenichthyid fishes. In Llano GA (ed) Adaptation within antarctic ecosystems. Smithsonian Inst, Washington, pp 479–487

Hemmingsen EA, Douglas EL, Johansen K, Millard RW (1972) Aortic blood flow and cardiac output in the hemoglobin-free fish, *Chaenocephalus aceratus*. Comp Biochem Physiol 43A:1045-1051

Hjalmarsson A, Isaksson O (1972) In vitro work load and rat heart metabolism I. Effect on protein synthesis . Acta Physiol Scand 86:126–144

Holeton GF (1970) Oxygen uptake and circulation by a hemoglobinless antarctic fish (*Chaenocephalus aceratus* Lonnberg) compared with three red-blood antarctic fish. Comp Biochem Physiol 34:457–471

Jakubowski M (1982) Dimensions of respiratory surfaces of the gills and skin in the antarctic white-blooded fish, *Chaenocephalus aceratus* Lonnberg (Chaenichthyidae). Z Mikiosk Anat Forch 96:145–156

Johansen K (1962) Cardiac output and pulsatile aortic flow in the teleost, *Gadus morhua*. Comp Biochem Physiol 7:169–174

Johnston IA, Harrison P (1987) Morphometrics and ultrastructure of myocardial tissue in notothenioid fishes. Fish Physiol Biochem 3:1–6

Johnston IA, Maitland B (1980) Temperature acclimation in crucian carp, *Carassius carassius* L., morphometric analysis of muscle fibre ultrastructure. J Fish Biol 17:113–125

Johnston IA, Fitch N, Zummo G, Wood RE, Harrison P, Tota B (1983) Morphometric and ultra-structural features of the ventricular myocardium of the haemoglobinless icefish *Chaenocephalus aceratus*. Comp Biochem Physiol 76A:475–480

Linzbach A (1960) Heart failure from the point of view of quantitative anatomy. Am J Cardiol 5:370–376

Londraville RL, Sidell BD (1990) Ultrastructure of aerobic muscle in antarctic fishes may contribute to maintenance of diffusive fluxes. J Exp Biol 150:205–220

Macdonald JA, Montgomery JC, Wells RMG (1987) Comparative physiology of antarctic fishes. Adv Mar Biol 24:321–388

Poupa O, Lindstrom L, (1983) Comparative and scaling aspects of heart and body weights with reference to blood supply of cardiac fibers. Comp Biochem Physiol 76A:413–422

Poupa O, Lindstrom L, Maresca A, Tota B (1981) Cardiac growth, myoglobin, proteins and DNA in developing tuna (*Thunnus thynnus thynnus L.*). Comp Biochem Physiol 70A:217–222

Priede IG (1976) Functional morphology of the bulbus arteriosus of rainbow trout (*Salmo gairdneri* Richardson). J Fish Biol 9:209–216

Santer RM, Greer Walker M (1980) Morphological studies on the ventricle of teleost and elasmo-branch hearts. J. Zool Lond 190:259–272

Seymour RS (1987) Scaling of cardiovascular physiology in snakes. Am Zool 27:97–109

Sidell BD (1983) Cellular acclimatization to environmental change by quantitative alterations in enzymes and organelles. In: Cossins AR, Sheterline P (eds) Cellular acclimatization to environ-mental change. Soc Exp Biol Sem Ser 17. Univ Press, Cambridge, New York, pp 103–120

Tota B (1978) Functional cardiac morphology and biochemistry in Atlantic bluefin tuna. In: Sharp GD, Dizon AE (eds) The physiological ecology of tunas. Academic Press, New York, pp 89–112

Tota B (1983) Vascular and metabolic zonation in the ventricular myocardium of mammals and fishes. Comp Biochem Physiol 76A:423–437

Tota B (1989) Myoarchitecture and vascularization of the elasmobranch heart ventricle. J Exp Zool Suppl 2:122–135

Tota B, Cimini V, Salvatore G, Zummo G (1983) Comparative study of the arterial and lacunary systems of ventricular myocardium of elasmobranch and teleost fishes. Am J Anat 167:15–32

Tota B, Farina F, Zummo G (1988) Ultrastructural aspects of functional interest in the ventricular myocardial wall of the antarctic icefish *Chaenocephalus aceratus*. Comp Biochem Physiol 90B:561–566

Tota B, Acierno R, Agnisola C (1991) Mechanical performance of the isolated and perfused heart of the haemoglobinless antarctic icefish *Chionodraco hamatus*: effects of loading conditions and temperature. Philos Trans R Soc Lond B 332:191–198

Tyler S, Sidell BD (1984) Changes in mitochondrial distribution and diffusion distances in muscle of goldfish (*Carassius auratus*) upon acclimation to warm and cold temperatures. J Exp Zool 232:1–10

Vogel WOP, Koch K-H (1981) Morphology of gill blood vessels in icefish. Archiv Fischwiss 31:139–150

Walvig F (1960) The integument of the icefish *Chaenocephalus aceratus* (Lonnberg). Nytt Mag Zool 9:31–36

Weibel ER (1985) Design and performance of muscular systems: an overview. J Exp Biol 115:405–412

Physiological Roles of High Lipid Content in Tissues of Antarctic Fish Species

B.D. Sidell

1 Introduction

Antarctic fish show many adaptations to their stenothermic cold life histories. The majority of antarctic fish produce glycoprotein antifreeze compounds which provide resistance to freezing of their hypoosmotic tissues by the subzero environment (DeVries 1983; DeVries, this Vol.). In many species, hematocrit of the blood is either substantially reduced or red cells and hemoglobin are absent (as in Channichthyidae), conditions which lower viscosity of the blood at severely cold temperature (Hemmingsen and Douglas 1977; Wells this Vol.). Lack of hemoglobin in Channichthyids is compensated for by increased heart size, blood volume and cardiac output (Hemmingsen et al. 1972; Hemmingsen, this Vol.).

The majority of antarctic notothenioid fish also show substantial corporeal accumulations of lipid which are considered to be buoyancy adaptations in this swimbladderless group (Eastman and DeVries 1982). For example, lipid has been reported as accounting for up to 37% of the dry weight of oxidative muscle tissues from *Pagothenia borchgrevinki* and *Disstostichus mawsoni* from McMurdo Sound (Lin et al. 1974). This feature has also led to a prevailing view that lipid is the primary fuel for energy metabolism of antarctic fish, an idea which has received direct experimental support only recently (Crockett and Sidell 1990; Sidell et al., in prep.). In fact, we still know surprisingly little about the characteristics of energy metabolism in antarctic fish.

This essay will begin by briefly reviewing current information regarding the question of metabolic cold adaptation of antarctic fish. The discussion will then be directed primarily toward physiological significance of the very high lipid content of these animals. I will present arguments based upon recent data from our laboratory that suggest: (1) aerobic energy metabolism of antarctic fish is predominantly reliant upon the oxidation of fatty substrates; and (2) the high fat content of tissues may also affect movement of respiratory gases between capillaries and mitochondria.

Department of Zoology and Center for Marine Studies, University of Maine, 203 Murray Hall, Orono, Maine 04469, USA

2 Antarctic Fish Display Metabolic Cold Adaptation

Are the metabolic rates of polar fish higher than those predicted for temperate zone species at equivalent cold temperature? This question has been the focus of lively debate during the last decade and has yet to be resolved conclusively. Early studies by Scholander et al. (1953) with fish from antarctic waters, and those of Wohlschlag (1960, 1964) and Ralph and Everson 1968) with antarctic species suggest that mechanisms of metabolic compensation to cold temperature have evolved in polar fish. Measurements of routine respiratory metabolism at approximately 0 °C in each of these studies resulted in oxygen utilization rates for polar fish [55-65 ml O_2 (kg . h)$^{-1}$] which are comparable to values characteristic of temperate zone species at approximately 20 °C. Holeton (1974), however, measured routine metabolism of 11 species of arctic marine fish at -1.5 °C and found rates of oxygen utilization which consistently were lower than the "cold-adapted" values previously reported. Marked reduction in oxygen uptake by these specimens occurred during a 48-h period after capture and handling. Holeton suggested that the decline was due to recovery from post-capture stress, and criticized the earlier studies by Scholander and Wohlschlag for not carefully controlling this factor. More recent measurements of oxygen utilization by antarctic marine invertebrates also have prompted Clarke (1983) to conclude that the standard metabolic rates of these animals are not elevated to "cold-adapted" levels.

The controversy surrounding metabolic cold adaptation of polar fish, however, inappropriately may have emphasized Standard or Routine metabolic processes, i.e. maintenance costs. *A priori*, a more compelling argument can be made for compensation of active metabolism to support such behaviors as locomotion of the animal. Mechanical work (and therefore energetic cost) of swimming by antarctic fish should be similar to that of species of comparable ecotype from temperate waters. Biochemical machinery of energy metabolism in locomotory musculature consequently must be of sufficient capacity to provide ATP at cold body temperature. Indeed, this is exactly what available data on enzyme activities appear to tell us.

We recently examined the activities at 1 °C of enzymes from oxidative locomotory and cardiac muscles of ecotypically similar antarctic and northern temperate-zone fish species (Crockett and Sidell 1990; Fig. 1). The enzymes selected are indicators of maximal potential for flux through specific pathways of energy metabolism (Crabtree and Newsholme 1972a, b). Maximal activities of enzymes from central pathways of aerobic energy metabolism (citrate synthase, cytochrome oxidase) were 1.5- to 5-fold higher in oxidative muscles from polar species than from temperate zone counterparts. Enzyme markers for fatty acid oxidation (carnitine palmitoyltransferase, ß-hydroxyacyl CoA dehydrogenase) were 1.3- to 27-fold higher in tissues from polar species, while those for both aerobic (hexokinase) and anaerobic (6-phosphofructokinase, pyruvate kinase, lactate dehydrogenase) metabolism of carbohydrates were generally lower in tissues of polar than of temperate zone fish. These data lead us to conclude that antarctic fish display significant metabolic cold adaptation of aerobic energy metabolism and suggest that this metabolism is reliant primarily upon oxidation of lipid rather than carbohydrate fuels.

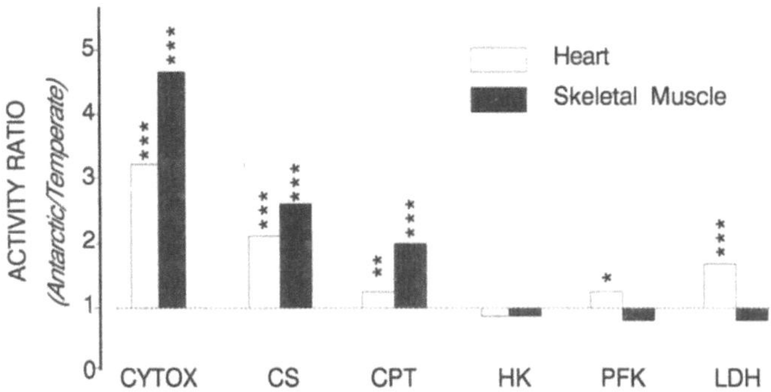

Fig. 1. Ratio of enzyme activities between homologous or analogous muscles of the antarctic fish, Notothenia gibberifrons and ecotypically similar temperate zone fish, Myoxocephalus octodecimspinosus. All assays were performed at 1 C. CYTOX Cytochrome oxidase; CS citrate synthase; CPT carnitine palmitoyltransferase; HK hexokinase; PFK 6-phosphofructokinase; LDH lactate dehydrogenase. * $P < 0.05$; ** $P < 0.01$; *** $P < 0.001$ (Sidell *et al* 1988)

3 Monoenoic Fats are a Preferred Metabolic Substrate

Enzyme activities, although diagnostic of maximal capacities, nonetheless are indirect indicators of which alternative metabolic fuels are most suitable substrates for supporting the energy metabolism of intact tissue systems. Recognizing this limitation, we recently have completed studies with isolated intact oxidative muscles from antarctic fish that corroborate our conclusion of the primacy of fatty acid oxidation in these animals.

We have measured at 1 °C the rates of $^{14}CO_2$ evolution from radiolabeled carbohydrate and lipid substrates by isolated electrically paced heart ventricle and isolated pectoral adductor muscles from the antarctic fish, *Notothenia gibberifrons* (Sidell et al., in prep.). Results of these experiments, shown here for the heart ventricle (Table 1), clearly indicate two features. First, the monounsaturated fatty acid, oleic acid (18:1), is oxidized much more readily by the tissue than the fully saturated compound, palmitic acid (16:0). Second, the rate of ATP regeneration calculated from oxidation rates of these compounds (at yields of 36 ATP/glucose oxidized, 15 ATP/lactate and 128 ATP/oleate) shows that an approximately three-

Table 1. Rates of 14C substrate oxidation and calculated ATP yields for Notothenia gibberifrons ventricular muscle at 1 °C [a]

Substrate	Oxidation rate [nmol $(g\ h)^{-1}$]	Calculated ATP yield [umol $(g\ h)^{-1}$]
5 mM ^{14}C-6-glucose	34.70 ± 13.96	1.25 ± 0.50
0.2 mM ^{14}C-U-lactate	7.29 ± 4.17	0.11 ± 0.06
0.5 mM ^{14}C-1-palmitate	8.77 ± 2.90	1.43 ± 0.37
0.5 mM ^{14}C-1-oleate	27.47 ± 4.20	3.54 ± 0.54

[a] All entries are mean ± SEM for n = 6(Sidell et al., in prep.).

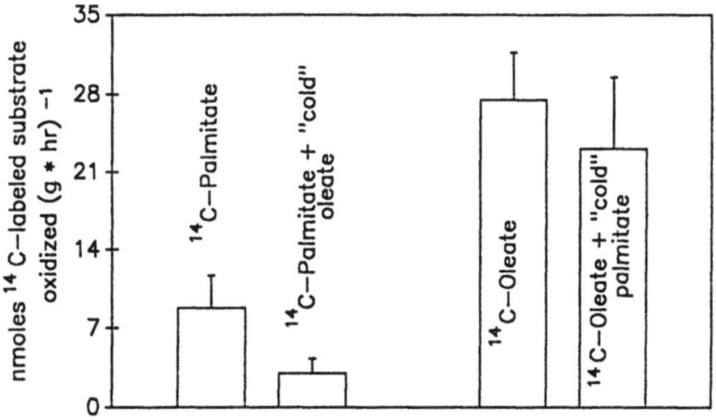

Fig. 2. Rates of oxidation of radiolabeled fatty acyl substrates by heart ventricle of N. gibberifrons in the presence and absence of unlabeled "competing" fatty acid. All experiments were performed at an incubation temperature of 1 C. n = 6 for all treatments (Sidell et al., in prep)

fold greater rate of ATP regeneration can be supported by oxidation of fatty fuel than that from glucose oxidation.

We were able further to confirm the distinctly greater suitability as a substrate for energy metabolism of monounsaturated oleic acid compared to saturated palmitic acid in a series of substrate competition experiments (Sidell et al., in prep.; Fig. 2). Simultaneous presence of unlabeled oleic acid with ^{14}C-labeled palmitic acid greatly depresses the oxidation rate of the saturated compound, compared to that observed with palmitate alone in the incubation medium. In the reciprocal experiment (unlabeled palmitate, ^{14}C-1-oleate). however, oxidation of oleic acid is not depressed significantly in the presence of competing cold palmitate.

The results described above clearly indicate that intact oxidative muscular tissue from antarctic fish displays a marked preference for monounsaturated fatty acyl compounds as substrates for energy metabolism. Lipid metabolism within any somatic cell, however, is an extremely complex, multistep pathway (Fig. 3). Selectivity of oxidative muscle for monounsaturated acyl compounds may be imposed or modified by preferential binding and/or catalysis at any one or more of these steps. Consequently, we have initiated a series of experiments to assess the roles of several key stages in cellular processes of handling fatty acyl compounds.

Intracellular breakdown of fatty acids for energy metabolism takes place primarily within the mitochondrial ß-oxidation pathway. However, the inner mitochondrial membrane is impermeable to fatty acids greater than eight carbons in length. Two metabolic steps must first occur in the cytoplasm before the fatty acid may be transported into the mitochondrion (Fig. 3). First, fatty acyl CoA synthetases of the cytosol activate and link the acyl chain with free CoASH to form an acyl CoA thioester, which also is incapable of penetrating the mitochondrial inner membrane. Next, the acyl moiety is transferred from CoA to free L-carnitine to form acyl-carnitine in a reaction catalyzed by carnitine palmitoyltransferase I (CPT I), located on the outer surface of the inner mitochondrial membrane, or perhaps associated with the mitochondrial outer membrane (Murthy and Pande 1987). L-carnitine, typically found at 1–4 mM concentration intracellularly, is required for transfer of the acyl group into the mitochondrial compartment. This is accomplished

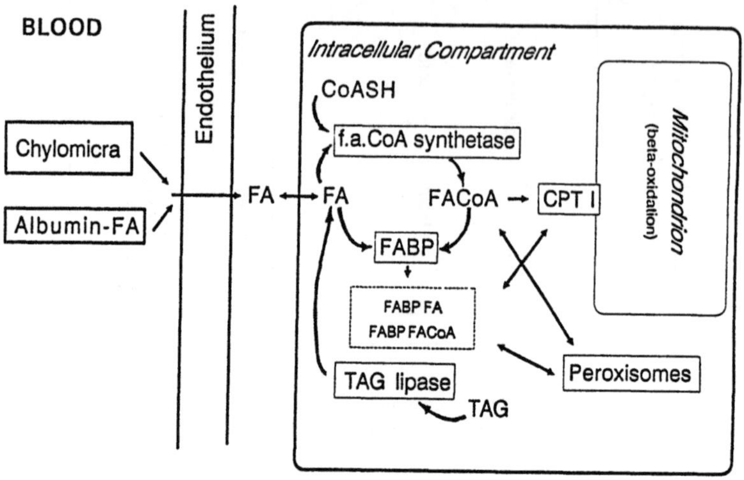

Fig. 3. Schematic representation of intracellular fatty-acid catabolism within a generalized somatic cell. FA Non-esterified fatty acid; FABP Intracellular fatty acid binding protein; FACoA fatty acyl CoA ester; TAG triacylglycerol; CPT I carnitine palmitoyltranseferase I

by an apparently carrier-mediated exchange across the inner membrane, which delivers free carnitine from the mitochondrion to the cytosol in a 1:1 stoichiometry for each acylcarnitine transported into the mitochondrion. Once inside the mitochondrial matrix, the acyl group is transferred from acylcarnitine back to CoASH by carnitine palmitoyltransferase II (CPT II), again forming an activated acyl CoA thioster which can now enter the ß-oxidation spiral of energy metabolism. Free carnitine is again returned to the cytosol in exchange for another acylcarnitine derivative.

Compelling evidence is available to indicate that CPT I is a regulated and flux-generating step in mitochondrial fatty acid oxidation by vertebrate tissues (Crabtree and Newsholme 1972b; Zammit and Newsholme 1979; Bremer and Osmundsen 1984). A frequently neglected consideration in assessing activity of this enzyme is that the transferase is capable of converting a wide range of long-chain, fatty acyl CoA thioesters to their acylcarnitine derivatives (Bieber and Fiol 1984). Recent evidence suggests that monounsaturated, long-chain fatty acids preferentially are mobilized for energy metabolism by fish tissues during starvation (rainbow trout, Jezierska et al. 1982; cod, Takima et al. 1985) and that their CoA thioesters may be preferred substrates for mitochondial CPT I (Sidell and Driedzic 1985). To ascertain whether a significant element of preference for monounsaturated fats may be accounted for in this critical metabolic step, we have examined the substrate selectivity of CPT from oxidative muncles of antarctic fish (Fig. 4).

At saturating substrate concentrations, carnitine palmitoyltransferase from two antarctic fish species, *Notothenia gibberifrons* and *Trematomus newnesi*, shows a significant preference for monounaturated palmitoleoyl CoA (16:1) compared to a variety of other fatty acyl CoA substrates (Fig. 4). Two notes of caution should be sounded here. First, crude preparations of enzyme were used in these studies, making distinction between catalyctic activities of CPT I (the rate-limiting step) and

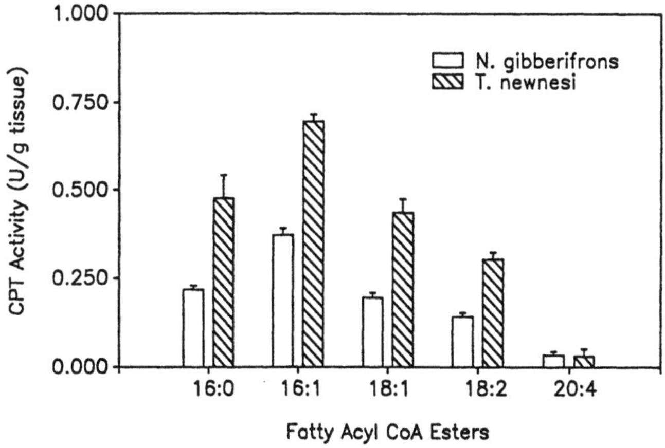

Fig. 4. Substrate selectivity of carnitine palmitoyltransferase in crude preparations from pectoral adductor muscles of antarctic fish species. All assays were conducted at saturating substrate concentration and assays temperature of 1 C. There is a significant difference ($P < 0.05$) between 16:0 and 16:1 CoAs as substrates in both species. Acyl CoA esters = 16:1 (palmitoyl), 18:1 (oleoyl), 18:2 (linoleoyl), 20:4 (arachidonyl). (Sidell et al., in prep.)

CPT II impossible. Second, assays have been performed under conditions of saturating substrate concentration and may miss important differences in activity which occur at more physiological concentrations of acyl CoA substrates near the enzyme's K_m. We presently are investigating these questions. Despite these cautionary notes, however, the available data permit us to conclude that although the specific preference between palmitoyl (16:0) and oleoyl (18:1) substrates displayed by the intact tissues is not mimicked by CPT, the general pattern of preferential usage of monoenoic fats over saturated fats remains consistent.

At this point in the development of our thesis, it again is worthwhile to point out that each of the studies mentioned above describes the behavior of metabolic systems in vitro and under conditions where the experimenter, not the animal, provides the alternative substrates. In short, the physiological relevance of the tissues' ability to oxidize monoenoic fatty acyl substrates is dependent upon these very same substrates being found in significant concentrations in vivo. To resolve this question, we recenty have completed quantitative analyses of the fatty acyl compositions of both triacylglycerol (TAG) and nonesterified fatty acid (FFA) pools of sera, adipose and oxidative skeletal muscle tissues of both *Notothenia gibberifrons* and *Trematomus newnesi* (Lund and Sidell 1991).

Each of the monoenoic fatty acids implicated as important substrates for energy metabolism of antarctic fish in our metabolic studies are prominent components of the fatty acyl pools of the animals under study (Table 2). Palmitoleic acid (16:1) accounts for 5-10% of the total acyl pools of both TAG and FFA in *N. gibberifrons*, while isomers of oleic acid (18:1) comprise 15-30% of the fatty acyl composition of tissues from this species (Table 2). These data lend further credence to the idea that the energy metabolism of oxidative tissues from antarctic fish is primarily reliant upon catabolism of monoenoic fatty acyl substrates. The very high corporeal lipid

Table 2. Relative percentages of fatty acids in TAG and FFA in serum and tissues of *Notothenia gibberifrons*

Fatty acid	Adipose TAG	Pectoral FFA	Pectoral TAG	Serum FFA	Serum TAG
	a	b	b	b	b
14:0	8.47 ± 1.19	2.68 ± 0.25	2.46 ± 0.26	1.96 ± 0.02	2.73 ± 0.5i
	n.s.	n.s.	n.s.	n.s.	n.s.
16:0	16.23 ± 0.80	18.31 ± 3.26	14.08 ± 1.23	16.07 ± 2.88	12.65 ± 1.47
	a	bc	ab	c	ab
16:1w7	9.40 ± 0.34	6.00 ± 0.53	7.54 ± 0.53	5.01 ± 0.68	7.79 ± 0.64
	b	d	cd	c	a
16:2	1.98 ± 0.23	0.51 ± 0.06	0.76 ± 0.12	1.02 ± 0.16	7.77 ± 1.07
	bc	a	ab	bc	c
18:0	2.72 ± 0.31	5.26 ± 1.16	4.23 ± 0.34	3.15 ± 0.27	2.39 ± 0.25
	b	b	a	b	a
18:1w9	13.12 ± 1.70	12.13 ± 1.48	19.82 ± 2.23	9.46 ± 0.91	20.18 ± 2.07
	ab	bc	a	bc	c
18:1w7	6.27 ± 0.40	4.65 ± 0.54	7.81 ± 0.36	5.29 ± 0.38	4.49 ± 0.65
	b	b	b	a	b
20:5w3	13.10 ± 2.17	15.39 ± 4.06	13.02 ± 1.63	26.26 ± 2.05	15.68 ± 2.33
	c	b	a	a	bc
22:6w3	4.56 ± 0.50	8.17 ± 1.46	12.02 ± 1.50	11.87 ± 1.18	7.38 ± 0.09
16:4	0.98 ± 0.47	0.72 ± 0.33	0.15 ± 0.06	50.63 ± 0.22	0.44 ± 0.07
18:2w6	1.03 ± 0.16	1.58 6 ± 0.45	0.88 ± 0.10	0.60 ± 0.14	0.89 ± 0.12
18:3w3	0.45 ± 0.09	0.30 ± 0.08	0.31 ± 0.03	0.34 ± 0.04	0.24 ± 0.05
18:4w3	2.64 ± 0.83	3.30 ± 1.25	1.42 ± 0.37	3.01 ± 0.65	2.64 ± 0.43
20:1w9 +11	2.29 ± 0.49	1.70 ± 0.27	3.05 ± 0.14	1.13 ± 0.24	1.72 ± 0.30
20:2w6	1.16 ± 0.73	2.64 ± 0.61	0.18 ± 0.06	3.30 ± 1.07	2.61 ± 1.47
20:4w6	0.59 ± 0.29	0.30 ± 0.10	0.45 ± 0.12	0.33 ± 0.30	0.39 ± 0.19
20:4w3	0.51 ± 0.14	0.56 ± 0.14	0.33 ± 0.09	0.67 ± 0.20	0.39 ± 0.10
22:1w11	0.16 ± 0.09	0.26 ± 0.15	0.53 ± 0.20	0.06 ± 0.04	0.39 ± 0.16
22:1w9	0.68 ± 0.20	0.1 ± 0.11	0.65 ± 0.16	0.40 ± 0.13	0.27 ± 0.12
21:5w3	0.52 ± 0.14	0.83 ± 0.23	0.62 ± 0.07	1.18 ± 0.13	0.81 ± 0.22
22:5w6	0.02 ± 0.01	0.06 ± 0.03	0.10 ± 0.03	trace	0.03 ± 0.02
22:5w3	0.93 ± 0.17	0.79 ± 0.21	1.46 ± 0.32	1.08 ± 0.25	1.03 ± 0.26
24:1w9	0.21 ± 0.06	0.22 ± 0.15	0.51 ± 0.10	0.25 ± 0.04	0.10 ± 0.05

All values are means plus or minus standard error of the mean (n=6). Results of the least significant difference multiple range tests ($p < 0.05$) appear over the mean and standard error values for fatty acids comprising greater than 5% of one or more lipid pools (Lund and Sidell 1991).

content of these species thus appears significant from both standpoints of buoyancy regulation (Eastman and DeVries 1982) and energy metabolism. Buoyancy and energy metabolism, however, may not be the only provinces of physiological significance for the lipid-rich nature of aerobic tissues in antarctic fish.

4 High Lipid Content May Enhance Oxygen Transfer Through Aerobic Tissues

Prompted by the observation that oxidative muscles of some temperate-zone fish display dramatic accumulation of intracellular lipid during cold acclimation (Egginton and Sidell 1989), our laboratory recently has advanced the hypothesis that high lipid content of tissues may aid movement of oxygen from capillary blood supply to the cellular mitochondrial population. A more detailed development of this model has been presented elsewhere (Sidell 1988; Egginton and Sidell 1989) and I will only describe the broadest conceptual elements of this hypothesis below.

The rate at which oxygen diffuses through any solvent system is dependent upon the oxygen diffusion constant, KO_2, for that solvent system in the following relationship:

$$dO_2/dt = KO_2 \cdot A \cdot [PO_2/X],$$

where A = the area across which diffusion occurs, PO_2 = the partial pressure gradient for oxygen over the diffusion path, and X = the diffusion path length. KO_2, in turn, is the multiplicative product of the diffusion coefficient for oxygen (DO_2) and the solubility of oxygen (α) in the solvent system. In the case of two juxtaposed solvent systems along any one-dimensional diffusion path, the KO_2s of each solvent contribute to describing overall flux in direct proportion to their prescence along the diffusion path.

At 25 °C, oxygen is 4.4 times more soluble in olive oil than in pure water and, unlike aqueous behavior, the solubility of oxygen in nonpolar solvents is affected little by temperature change (Battino *et al.* 1968). Glycerides of oleic acid are the primary constituents of olive oil, making it reasonably analogous to fish oils and a frequently cited model in considering biological permeabilities. Because the aqueous cytoplasm of muscle contains considerable dissolved solutes which will decrease gaseous solubility, a partition coefficient of 4.4 undoubtedly is conservative for estimating the relative solubility of oxygen in lipid versus cytoplasm. Less is known about the absolute values of diffusion coefficients for oxygen through lipids. Limited extant data suggest that, at the same temperature, DO_2 in long-chain hydrocarbons is roughly equivalent to that in water, despite large differences in macroscopic viscosities of the solvent systems (Windrem and Plachy 1980; Fischkoff and Vanderkooi 1975; Subczynski and Hyde 1984). With respect to comparison of DO_2s between intracellular lipid and cytoplasm, these estimates again appear conservative, because the two- to sixfold greater viscosity of cytoplasm than pure water should reduce DO_2 considerably (Sidell and Hazel 1987). On the basis of these combined solubility and diffusivity characteristics of oxygen in lipids, we have hypothesized two potential roles of intracellular lipid inclusions in the oxygen economy of oxidative tissues which may be of particular significance in antarctic fish:

1. The presence of lipid droplets may accelerate the rate of oxygen diffusion at any given PO_2 gradient between capillaries and mitochondria.
2. Intracellular lipid may function as a major oxygen store, damping out otherwise large variations in oxygen supply to mitochondria during changes in muscular activity.

Results from a recent quantitative ultrastructural study of oxidative muscles from two antarctic fish support these hypotheses (Londraville and Sidell 1990).

As part of our recent study, we employed stereological techniques to quantify morphological parameters estimating capacities for oxygen supply and demand in oxidative skeletal muscles from two phylogenetically related, yet ecotypically dissimilar antarctic fish. *Notothenia gibberifrons* is a sluggish demersal species and *Trematomus newnesi* is an active pelagic species. both are labriform swimmers and use their pectoral adductor muscles to power their primary mode of locomotion; pectoral adductors of both species are composed exclusively of slow oxidative muscle fibers.

Pectoral adductor muscles of both *N. gibberifrons* and *T. newnesi* display substantial mitochodrial volume densities (Table 3), a characteristic common to other antarctic fish species (e.g., see Johnston and Camm 1987). Consistent with their respective activity patterns, muscle of *T. newnesi* shows significantly greater mitochondrial density than that of the more sluggish *N. gibberifrons* (Table 3), suggesting a considerably greater potential demand for oxygen in the former species. Although muscle fibers of the two species are sized similarly, capillary bore size, capillary length density, and capillary:fiber ratio of *T. newnesi* are all significantly less than those of the more sluggish *N. gibberifrons*. As a consequence, the maximal radial distance that oxygen must diffuse to supply the tissue, conventionally referred to as Krogh's radius, is considerably greater for T. newnesi than *N. gibberifrons* (Table 3). We, thus, are presented with the apparently perplexing observation of greater potential oxygen demand, but lower capacity to supply oxygen in the former species.

Although the above picture appears paradoxical initially, further examination of ultrastructural differences between the muscles of these two species reveals a possible explanation. *T. newnesi* oxidative muscle fibers contain significantly more intracellular lipid droplets than those of *N. gibberifrons* (Table 3). Each of these lipid inclusions that occurs along any one-dimensional diffusion path between an oxygen source (capillary) and site of oxygen utilization (mitochondrion) should represent an area of low resistance to oxygen flux compared with the higher resistance of the aqueous segments of the diffusion path. We have used our quantitative ultrastructural data and the diffusion/solubility information mentioned above to construct a simple model estimating dO_2/dt to mitochondria residing at the

Table 3. Selected structural parameters related to capacities for oxygen supply and demand in oxidative muscle of two antarctic fish (Londraville and Sidell 1990)

Structural Parameters[a]	Species[b]	
	T. newnesi	*N. gibberifrons*
A(f) um^2	2153.4 ± 215.3	2102.8 ± 193.7
A(c) um^2	39.62 ± 5.17	52.39 ± 0.07**
C:F	0.62 ± 0.66	0.91 ± 0.07*
Jv (c,f) ($mm\ mm^{-3}$)	481.3 ± 49.0	696.3 ± 33.7*
Krogh's R (um)	26.3 ± 1.64	21.5 ± 0.51*
Vv (mit, f)	0.348 ± 0.012	0.249 ± 0.007***
Vv (lip,f)	0.026 ± 0.002	0.012 ± 0.004**

[a] A (f), mean cross-sectional area of fibers; C:F, numerical capillary-to-fiber ratio; A(c), mean cross-sectional area of individual capillaries; Vv, percentage of muscle fiber volume (volume density) occupied; (mit,f) mitochondria, and (lip,f) lipid.

[b] Values are means ± SEM, N=5 per species; * $P < 0.05$, ** $P < 0.01$, *** $P < 0.001$ for between-species comparisons.

end of Krogh's radius for muscle structures of both *T. newnesi* and *N. gibberifrons* (Londraville and Sidell 1990). Our results suggest that, by virtue of the species' greater intracellular lipid content, *T. newnesi* can support an equivalent delivery of oxygen to the periphery of its oxygenated tissue cylinder, despite significantly larger O_2 diffusion path length than that of *N. gibberifrons* (schematically represented in Fig. 5). Such an enhancement of transcellular oxygen movement and the potential of these intracellular lipid droplets as oxygen reservoirs could be of considerable adaptive significance to chronically cold-bodied antarctic fish species that are extremely lipid-rich, and either do not possess myoglobin (Channichthyid icefish) or express the pigment at only very low concentrations in their oxidative muscles (Fitch et al. 1984).

In summary, I have attempted to describe some insights that are emerging about the physiological significance of high lipid content long recognized as a characteristic of tissues from antarctic fish. There is little doubt that these considerable corporeal lipid stores have contributed to increasing the static lift of this swim-bladderless group of fish, permitting a number of species to exploit niches higher in the water column (Eastman and DeVries 1982). Our recent data further suggest that these fatty compounds represent an important metabolic fuel for aerobic respiration of these animals. Finally, regardless of other functions, the substantial lipid phase of muscles from antarctic fish has chemical properties that may increase O_2 flux through their cells.

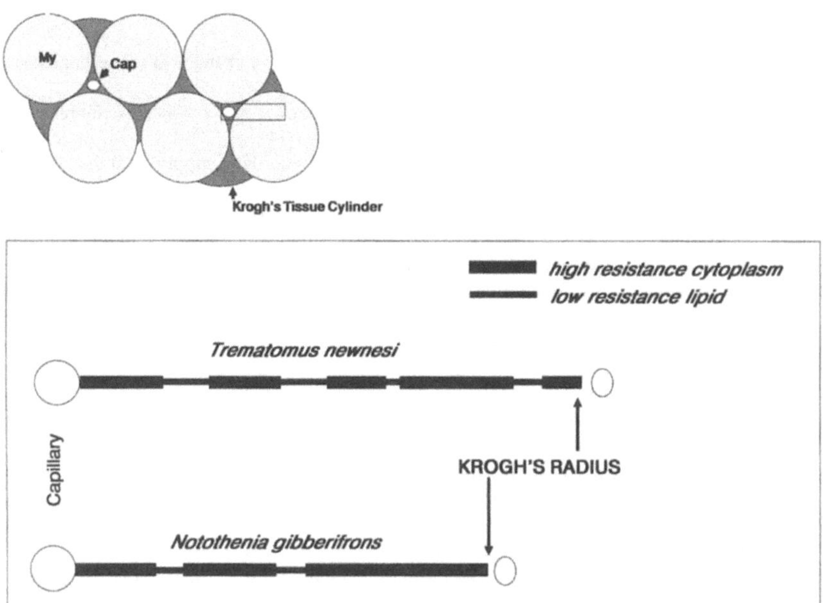

Fig. 5. Schematic diagram illustrating the effect of lipid on oxygen through skeletal muscle (see text for details). My Muscle fibers in cross-section; Cap capillary

Acknowledgments. Work contributing to this communication has been supported by U.S. National Science Foundation grants DPP 85-16137 and DPP 88-19469 to the author. I am especially grateful to the support staff of the U.S. Antarctic Program research station, Palmer Station, and the masters and crews of *R/V "Polar Duke"*, whose efforts have been essential to execution of our research. I also wish to thank my graduate students, Elizabeth Crockett and Richard Londraville, for insightful discussions and assistance in graphic production which added greatly to this contribution.

References

Battino R, Evans FD, Danforth WF (1968) The solubilities of seven gases in olive oil with reference to theories of transport through the cell membrane. J Am Oil Chem Soc 45:830–841

Bieber LL, Fiol CJ (1984) Kinetic analysis of multisubstrate carnitine palmitoyltransferase. Curr Top Cell Reg 24:111–117

Bremer J. Osmundsen H (1984) Fatty acid oxidation and its regulation. In: Numa S (ed) New comprehensive biochemistry, vol. 7, Fatty acid metabolism and its regulation. Elsevier, New York, pp 113–154

Clarke A (1983) Life in cold water: the physiological ecology of polar marine ectotherms. Oceanogr Mar Biol Annu Rev 43:22–-236

Crabtree B, Newsholme EA (1972a) The activities of phosphorylase, hexokinase, phosphofructokinase, lactate dehydrogenase and the glycerol 3-phosphate dehydrogenase in muscles from vertebrates and invertebrates. Biochem J 12:49–58

Crabtree B, Newsholme EA (1972b) The activities of lipases and carnitine palmitoylstransferase in muscles from vertebrates and invertebrates. Biochem J 130:697–05

Crockett EL, Sidell BD (1990) Some pathways of energy metabolism are cold adapted in antarctic fishes. Physiol Zool 63:472–488

DeVries AL (1983) Biological antifreeze agents. Annu Rev Physiol 45:245–260

Eastman JT, DeVries AL (1982) Buoyancy studies of notothenioid fishes in McMurdo Sound, Antarctica. Copeia 1982:385–393

Egginton S, Sidell BD (1989) Thermal acclimation induces adaptive changes in subcellular structure of fish skeletal muscle. Am J Physiol 256:R1–R9

Fischkoff S, Vanderkooi J (1975) Oxygen diffusion in biological and artificial membranes determined by the fluorochrome pyrene. J Gen Physiol 65:663–674

Fitch NA, Johnston IA, Wood RE (1984) Skeletal muscle capillary supply in a fish that lacks respiratory pigments. Respir Physiol 57:20–213

Hemmingsen EA, Douglas EL (1977) Respiratory and circulatory adaptations to the absence of hemoglobin in chaenichthyid fishes. In: Llano GA (ed) Adaptations within antarctic ecosystems. Smithsonian Inst, Washington, DC, pp 479–487

Hemmingsen EA, Douglas EL, Johansen K, Millard RW (1972) Aortic blood flow and cardiac output in the hemoglobin-free fish *Chaenocephalus aceratus*. Comp Biochem Physiol 43A:1045–1051

Holeton GF (1974) Metabolic cold adaptation of polar fish: fact or artefact? Physiol Zool 47:137–152

Jezierska B, Hazel JR, Gerking SD (1982) Lipid mobilization during starvation in the rainbow trout, *Salmo gairdneri*. J Fish Biol 21:681–692

Johnston IA, Camm JP (1987) Muscle structure and differentiation in pelalgic and demersal stages of the antarctic teleost *Notothenia neglecta*. Mar Biol 94:183–190

Lin Y, Dobbs GH, DeVries AL (1974) Oxygen consumption and lipid content in red and white muscles of antarctic fishes. J Exp Zool 189:379–386

Londraville RL, Sidell BD (1990) Ultrastructure of aerobic muscle in antarctic fishes may contribute to maintenance of diffusive fluxes. J Exp Biol 150:205–220

Lund EL, Sidell BD (1991) Neutral lipid compositions of antarctic fish tissues may reflect use of fatty acyl substrates by catabolic systems. Mar Biol (in press)

Murthy MSR, Pande SV (1987) Malonyl CoA binding site and overt carnitine palmitoyltransferase activity reside on the opposite sides of the outer mitochondrial membrane. Proc Natl Acad Sci USA 84:378–382

Ralph R, Everson I (1968) Respiration of antarctic fish. Comp Biochem Physiol 27:299–307

Scholander PF, Flagg W, Irving L (1953) Climatic adaptation in arctic and tropical poikilotherms. Physiol Zool 26:67–92

Sidell BD (1988) Diffusion and ultrastructural adaptive responses in ectotherms. In: Jones DP (ed) Microcompartmentation. CRC Press, Boca Raton, pp 71–92

Sidell BD, Driedzic WR (1985) Relationship or cardiac energy metabolism and cardiac work demand in fishes. In: Gilles R (ed) Circulation, respiration and metabolism. Springer , Berlin, Heidelberg, New York Tokyo, pp 103–120

Sidell BD, Hazel JR (1987) Temperature affects the diffusion of small molecules through cytosol of fish muscle. J Exp Biol 129:19—203

Sidell BD, Crockett EL, Driedzic WR (1988) Metabolic characteristics of muscle tissues from antarctic fishes. Antarc J US 23:138–140

Sidell BD, Crockett EL, Driedzic WR (in preparation) Monoenoic fatty acids are the preferred substrates for energy metabolism of antarctic fishes

Subczynski WK, Hyde JS (1984) Diffusion of oxygen in water and hydrocarbons using an electron spin resonance spin-label technique. Biophys J 45:743–748

Takima L, Love RM, Smith GL (1985) Selectivity in mobilization of stored fatty acids by maturing cod, *Gadus morhua L.* Comp Biochem Physiol 80B:713–718

Windrem DA, Plachy WZ (1980) The diffusion-solubility of oxygen in lipid bilayers. Biochim Biophys Acta 600:655–662

Wohlschlag DE (1960) Metabolism of an antarctic fish and the phenomenon of cold adaptation. Ecology 38:502–510

Wohlschlag DE (1964) Respiratory metabolism and ecological characteristics of some fishes in McMurdo Sound, Antarctica. In: Wells HW (ed) Biology of the antarctic seas. Antarctic res ser, vol 1. Am Geophys Union, Washington, DC, pp 33–62

Zammit VA, Newsholme EA (1979) Activities of enzymes of fat and ketone body metabolism in teleost and elasmobranch fish. Biochem J 184:313–322

Biochemical Mechanisms of Cold Adaptation and Stenothermality in Antarctic Fish

G.N. Somero

1 Introduction

Antarctic fish thrive at water temperatures near the freezing point of seawater, thermal conditions which would be lethal for virtually all other fish. The physiological and biochemical adaptations that underlie the abilities of antarctic fish to survive at these low temperatures are widespread, and represent the extremes of low-temperature adaptation found among the vertebrates. Certainly the most dramatic, and essential, of these adaptations are those that prevent the hypoosmotic antarctic fish from freezing. Freezing avoidance is due to the presence of freezing-point-depressing "antifreeze" molecules, either peptides or glycopeptides, in most of the extracellular fluids of these fish (reviewed in DeVries 1988; DeVries, this Vol.). These antifreezes protect the fish from ice crystal formation and/or crystal growth, even when the animals encounter ice in their habitats.

Although these adaptations for minimizing the dangers of ice formation in the body fluids of antarctic fish are critical for survival in the ice-laden antarctic waters, these key adaptations do not play a role in establishing two additional, and very important, aspects of the thermal biology of these fish: (1) the evolution of a metabolic apparatus that can function at high ("cold adapted") rates and in a closely regulated manner, at subzero temperatures; and (2) the development of an extreme degree of stenothermality of physiological function. This chapter will focus primarily on these two aspects of biochemical adaptation to the extremely low (and highly stable) temperatures encountered by antarctic fish. Emphasis will be on enzymatic and structural proteins; the vital role of membranes and diverse neurobiological processes in cold adaptation and stenothermality is considered by Macdonald and colleagues (Macdonald and Montgomery 1982; Macdonald et al. 1987, 1988; Macdonald, this Vol.).

The adaptive modifications of enzymatic and structural proteins which underlie high rates of controlled catalytic function and physiologically appropriate subunit assembly processes at subzero temperatures will be considered first. Profound evolutionary changes have occurred in diverse enzymatic and structural proteins of antarctic fish which allow the organisms to sustain the same types of closely regulated metabolic and locomotory activities found in other organisms, and often

Marine Biology Research Division, Scripps Institution of Oceanography University of California, San Diego, La Jolla, CA 920-93-0202, USA to express these activities at rates similar to those found in fish with substantially higher cell temperatures.

The second issue to be addressed stems from the observation that antarctic fish die of heat death at remarkably low temperatures. The upper lethal temperatures of three –1.9 °C-adapted notothenioid species from McMurdo Sound, *Pagothenia borchgrevinki*, *Trematomus bernacchii*, and *Trematomus hansoni*, were only approximately 6 °C (Somero and DeVries 1967) This is the lowest temperature of heat death known for any animal. Thus, cold adaptation in antarctic fish is paired with an extreme degree of stenothermality. The stenothermality of antarctic fish is also indicated by the low temperatures at which synthesis of heat shock proteins is induced (Maresca et al. 1988; this Vol.). The biochemical determinants of stenothermality in enzymatic systems will be considered in the context of thermal optima in enzyme function. This discussion will include an analysis of the physiologically appropriate criteria for defining "optimal" temperatures for enzymatic activity. This analysis will also consider the widely different temperatures at which optimal enzyme function, on the one hand, and protein structural integrity, on the other hand, are lost as temperature is increased.

2 Metabolic Cold Adaptation

The respiratory rates of antarctic fish generally are higher than the rates that are predicted to occur in fish living in –1.9 ° waters, i.e., antarctic fish exhibit metabolic cold adaptation (Wohlschlag 1964; Somero et al. 1968; Torres and Somero 1988a,b). Comparing metabolic rates of different species of fish from different environments is fraught with potential problems, however (Holeton 1974; DeVries and Eastman 1981; Dunn 1988). Artifacts from different levels of excitement of specimens can cloud results; however, this problem can be resolved with appropriate methodologies (cf. Macdonald et al. 1988). The most difficult problem derives from differences among species in their inherent levels of metabolic and locomotory activity. Thus, fish from any single environment differ enormously in metabolic rate as a function of lifestyle, e.g., pelagic vs demersal habit. Therefore, comparisons of fish from different environments must be made using species that are similar as possible in feeding strategy, locomotory ability, body size, dietary state, and other characteristics that influence metabolic rate.

One way to minimize the problems arising from "apples to oranges" artifacts in the study of metabolic cold adaptation is to compare confamilial or, ideally, congeneric species from different latitudes. By judicious choice of experimental subjects, the only factor affecting metabolic rate that will differ between the study species is habitat temperature. This objective was largely met in the comparisons of respiratory rates of mesopelagic fish from antarctic and Southern California waters by Torres and Somero (1988a,b). These comparisons showed clear evidence for metabolic cold adaptation in antarctic fish, some of which were congeners of the Southern California species (Fig. 1). Note that when the metabolic rates of the Southern California fish, which had been determined at either 10 °C (species occurring above 100 m) or 5 °C (deeper-occurring species), were adjusted to 0.5 °C, the temperature at which the measurements of respiration of the antarctic fish were

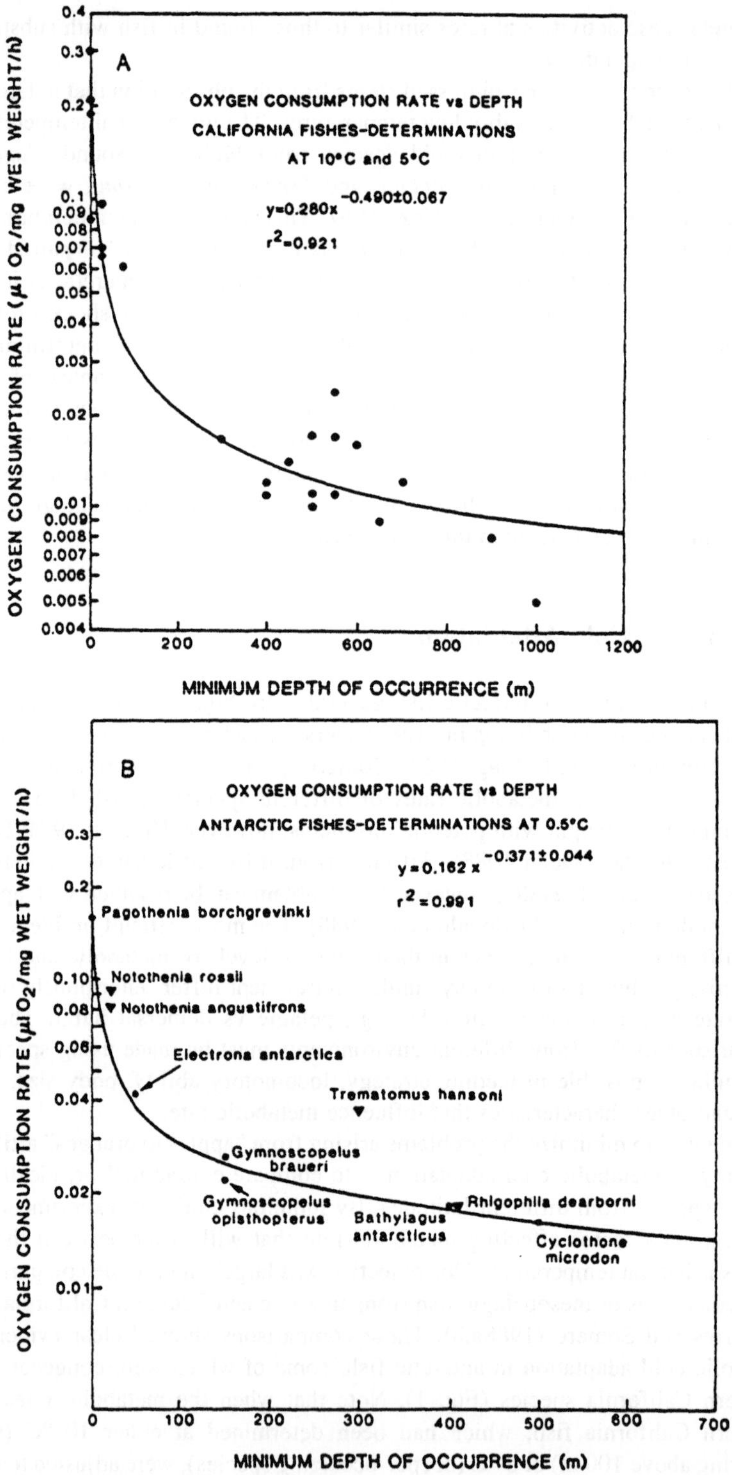
A
OXYGEN CONSUMPTION RATE vs DEPTH
CALIFORNIA FISHES-DETERMINATIONS
AT 10°C and 5°C
y=0.280x^-0.490±0.067
r^2=0.921
OXYGEN CONSUMPTION RATE (μl O2/mg WET WEIGHT/h)
MINIMUM DEPTH OF OCCURRENCE (m)
B
OXYGEN CONSUMPTION RATE vs DEPTH
ANTARCTIC FISHES-DETERMINATIONS AT 0.5°C
y=0.162 x^-0.371±0.044
r^2=0.991
Pagothenia borchgrevinki
Notothenia rossii
Notothenia angustifrons
Electrona antarctica
Trematomus hansoni
Gymnoscopelus braueri
Gymnoscopelus opisthopterus
Bathylagus antarcticus
Rhigophila dearborni
Cyclothone microdon
OXYGEN CONSUMPTION RATE (μl O2/mg WET WEIGHT/h)
MINIMUM DEPTH OF OCCURRENCE (m)

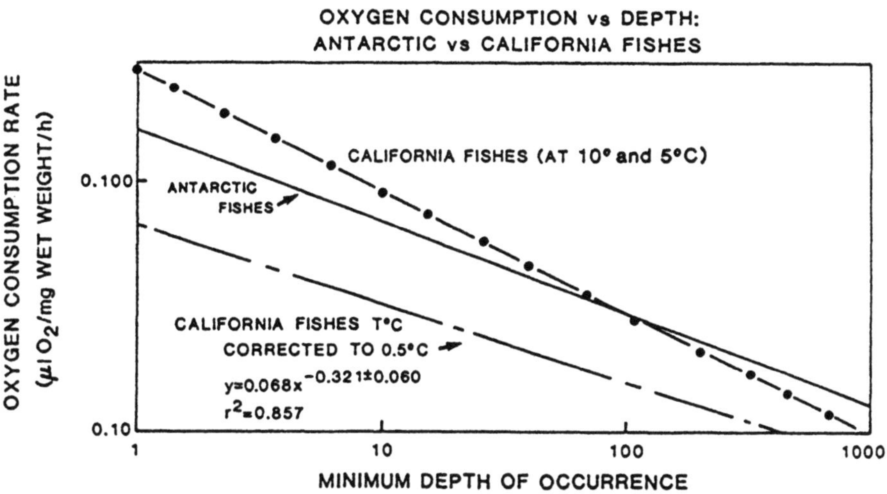

Fig. 2. The influence of temperature on oxygen consumption vs MDOC for mesopelagic fish from southern California and Antarctica. The line showing the predicted rates for the southern California fish at 0.5 °C was calculated using a Q_{10} of 2.0 (Torres and Somero 1988a)

made, antarctic fish from all depths had higher rates of oxygen consumption than the California species (Fig. 2).

What biochemical factors underlie these cold-adapted rates of oxygen consumption by the antarctic species? There appear to be two types of adaptations in enzymatic systems that facilitate high rates of catalysis at low temperatures (Hazel and Prosser 1974; Hochachka and Somero 1984). First, the intracellular concentrations of enzymes, especially enzymes associated with aerobic respiration, may be relatively high in cold-adapted or cold-acclimated organisms. This increase in number of catalytic sites offsets the reduced rate of activity per site due to Q_{10} effects. Second, the inherent catalytic activity per active site may be higher in enzymes from highly cold-adapted species (Low et al. 1973; Low and Somero 1976; Johnston and Walesby 1977; Somero and Siebenaller 1979; Yancey and Siebenaller 1987; Genicot et al. 1988). This second mechanism of metabolic cold adaptation has the advantage of achieving more units of metabolic activity by using fewer copies of faster-working enzymes, rather than higher numbers of copies of less efficient enzymes.

It is not clear what role each of these two mechanisms plays in metabolic cold adaptation in antarctic fish. Although all of the enzymes of antarctic fish examined to date have been found to have higher catalytic efficiencies than the homologous enzymes of warm-adapted species (Table 1), cold-adaptive changes in enzyme concentration probably are important as well. This conjecture is based on the

Fig. 1. Oxygen consumption rates of Southern California and antarctic mesopelagic fish. **A** Oxygen consumption versus minimal depth of occurrence (MDOC) of southern California mesopelagic fish, measured at either 10 °C (species with (MDOC) < 100 m) or 5 °C (Torres and Somero 1988a). **B** Oxygen consumption vs MDOC of antarctic mesopelagic fish, measured at 0.5 °C (Torres and Somero 1988a)

Table 1. Catalytic efficiencies (relative k_{cat} values) of enzymes from differently-adapted fishes and mammals[a]

	Relative k_{cat}
M_4-lactate dehydrogenases[b] (5 °C)	
Pagothenia borchgrevinki (-1.9 °C)	1.00
Sebastes pinninger (10-17 °C)	1.01
Platichthys stellatus (10-17 °C)	1.08
Sebastolobus alascanus (4-12 °C)	0.98
Thunnus thynnus (15-30 °C)	0.76
Pterois volitans (27-30 °C)	0.52
Rabbit (37 °C)	0.54
Mg^{2+} - Ca^{2+}-myofibrillar ATPase[c] (0 °C)	
Champsocephalus gunnari (-1 to 2 °C)	1.00
Notothenia neglecta (0.2 °C)	0.62
Cottus bubalis (3-12 °C)	0.45
Dascyllus carneus (18-26 °C)	0.05
Pomatocentrus uniocellatus (18-26 °C)	0.04
Trypsin[d] (5 °C)	
Paranothenia magellanica (~5 °C)	1.00
Salmo gairdneri (~4-20 °C)	0.57
Bovine (37 °C)	0.19

[a]The rates of activity of all homologues of each class of enzyme are expressed relative to the rate of the enzyme from the antarctic species having the highest k_{cat}. Body temperatures for each species are given in parenthesis after the species' name.
[b]Data from Somero and Siebenaller (1979) and Yancey and Siebenaller (1987).
[c]Data from Johnston and Walesby (1977).
[d]Data from Genicot et al. (1988).

finding that the catalytic efficiency differences between homologous enzymes of antarctic and temperate fish may be only about twofold, or slightly less (e.g., LDH and trypsin; Table 1), whereas the required increase in metabolic potential to offset the effects of a 10–20 °C decrease in temperature would be somewhat more than twofold. Further study will be necessary to determine how fully metabolic cold adaptation can be achieved by using "better" enzymes, rather than by producing more copies of enzymes. Measurements of tissue enzymatic activity alone cannot resolve this question, because such data do not indicate the relative importance of these two mechanisms of adaptation. Definitive study must involve the quantification of the number of enzyme molecules in a given mass of tissue and determination of their specific activities (k_{cat} values).

3 Thermal Optima and Stenothermality of Enzymatic Function

The potential for sustaining a high rate of metabolic function is not, in and of itself, sufficient to ensure that metabolism will occur in a physiologically adequate manner under a given thermal regime. The ability to closely regulate metabolic flux, in concert with the varying physiological demands of the organism, must also be sustained. Metabolic regulation is based on several levels of control, including transcriptional and translational regulation of the types and amounts of metabolic enzymes, covalent modification of enzymes, shifts in the distribution of enzymes within the cells ("compartmentation"), modulator-induced changes in activity, and changes in rate due to variations in substrate concentration. All of these levels of metabolic regulation can be affected by changes in temperature.

In the following discussion of thermal optima and thermal limits of controlled enzymatic function, consideration will be centered on certain kinetic properties of enzymes that play major roles in governing the second-by-second flux of substrates through metabolic pathways. Particular emphasis will be given to the initial step in the catalytic sequence, the binding of ligands (substrates and cofactors) to the enzyme. The binding ability ("affinity") of an enzyme for its substrate is often approximated by the apparent Michaelis-Menten constant (K_m) of the reaction. Binding events are highly important in controlling metabolic flux, and they are highly sensitive to perturbation by temperature (Hochachka and Somero 1984). It follows that metabolic integrity is dependent on the conservation of these kinetic properties within the relatively narrow ranges that are optimal for physiological function. We will see, too, that neither the highest nor the lowest value for binding parameters is apt to be optimal for physiological function.

Before examining data on three enzymes for which the temperature dependence of K_m has been determined for a wide variety of vertebrates, including antarctic fish, general criteria for defining "thermal optima" for biochemical functions will be considered. The basic question to be addressed is: for any temperature-dependent biochemical property of an organism, over what temperature range does this property attain values that are optimal for physiological function? In addressing this question, it will be important to keep in mind that what might appear as an "optimal" situation under in vitro conditions need not be "optimal" in a physiological sense. For example, enzymatic activity normally increases with temperature to temperatures well above the organism's upper lethal temperature. Thus, it is apparent that the temperature yielding the maximal rate of activity in vitro is not necessarily an index of what temperature is optimal for the organism.

Figure 3 presents a generalized model of how temperature might affect some biochemical property of the cell, for instance, an enzymatic kinetic property like a K_m value, or the rate of a metabolic process. Homologous systems, e.g., enzymes, from five species adapted to different temperatures are compared (see legend). For each of the five species, this biochemical trait is strongly dependent on temperature, and the trait shows its maximal values at temperatures much above the upper lethal temperatures of the species (above the intersection of the upper dashed horizontal line with the rate versus temperature curve for the species; see legend). However, when the five species' systems are examined at the animals' physiological temperatures (temperatures corresponding to the intersections between the rate curves and the horizontal lines, as shown by the darker portions of the five curves), the biochemical trait is found to be conserved within a relatively narrow range (the range bounded by the intersection of dashed lines with the ordinate). Thus, even though the magnitude of this biochemical property may increase up to temperatures well beyond the upper lethal temperature of the organism, under in vitro conditions, the values for the trait that appear to be optimal (as judged by their strong conserva-

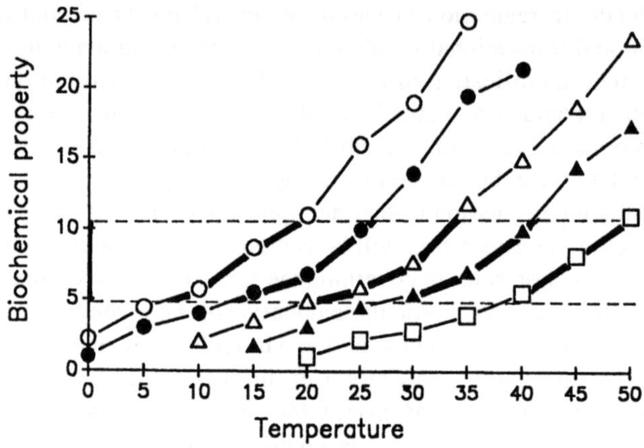

Fig. 3. A generalized model showing how biochemical properties of organisms, e.g., K_m values, are conserved within narrow ranges at different species, physiological temperatures. The five curves illustrate the temperature dependence of a biochemical property common to five different species having different ranges of body temperatures. The body temperature range for each species is encompassed by the intercepts on the abscissa that would result if vertical lines were drawn from the intersection of the *horizontal dashed lines* and the *lines* connecting each set of *symbols*. For example, the most cold-adapted species (*open circles*) has a body temperature range of approximately 6–20 °C; the most thermophilic species (*open squares*) has a body temperature range of approximately 38–48 °C. At physiological temperatures, the biochemical parameter is conserved within the range of approximately 5–10 (*darker sections* of curves; range encompassed by the intersections of the dashed *horizontal lines* and *Y-axis*)

tion at physiological temperatures) are of intermediate value. It is, then, incorrect, in a physiological context, to define thermal optima for enzymatic systems by emphasizing the temperatures at which one or more properties, e.g., catalytic activity or substrate binding ability, is the highest. The physiologically appropriate way to define thermal optima is, instead, in terms of the temperatures at which values for enzymatic properties are held within the range strongly conserved among all species.

Support for this conjecture is given by data showing how strongly K_m values are conserved among species under physiological conditions of temperature, hydrostatic pressure, osmotic concentration, and pH, factors that are highly perturbing of enzyme-ligand interactions (Hochachka and Somero 1984). The conservation of K_m values has been interpreted as a mechanism for retaining optimal regulatory ability under different temperatures, pressures, and solute conditions (Hochachka and Somero 1984). For example, by maintaining the K_m in the conserved range, the enzyme retains maximal responsiveness to changes in substrate concentration and to the influences of allosteric modulators.

Figure 4 presents data from comparisons of the effects of adaptation and measurement temperature on the K_m of pyruvate for the skeletal muscle type ($M_4 = A_4$) isozyme of lactate dehydrogenase (LDH) from seven vertebrates whose body temperatures span an almost 50 °C range, as indicated by the body temperature data given in the figure (Yancey and Somero 1978; Somero 1987; Coppes and

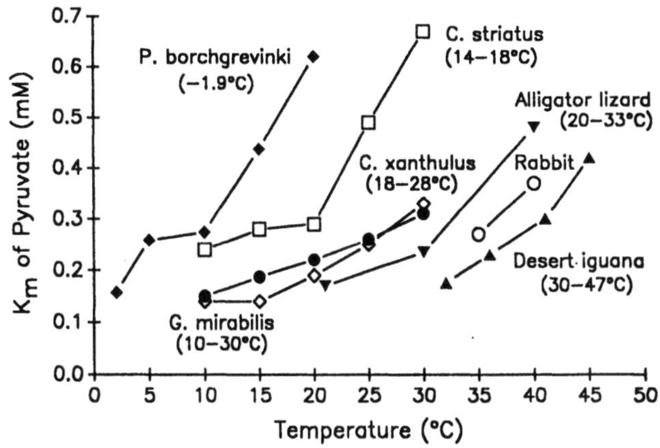

Fig. 4. The effect of temperature on the K_m of pyruvate for M_4-LDHs purified from vertebrates having different body temperatures (*numbers* in parentheses below species' names). Data from Yancey and Somero (1978); Somero (1987); Coppes and Somero (1990)

Somero 1990). The K_m of pyruvate is conserved within the range of approximately 0.15–0.35 mM pyruvate at the physiological temperatures of these species, which include an antarctic fish (*Pagothenia borchgrevinki*), several temperate-zone fish, a temperate-zone lizard, the rabbit, and a thermophilic lizard (desert iguana). Data for M_4-LDHs of approximately 30 additional species display the same conservation of K_m shown by the M_4-LDHs of these seven vertebrates (Yancey and Somero 1978; Graves and Somero 1982; Graves et al. 1983; Coppes and Somero 1990).

Although K_m is strongly conserved at physiological temperatures, K_m often rises sharply at temperatures just above the physiological range, especially for the LDHs of stenothermal species. These sharp increases in K_m of pyruvate are especially evident for the LDHs of *Cyanosion striatus*, with a maximal habitat temperature near 18 °C (compare the stenothermal *C. striatus* to its eurythermal congener, *C. xanthulus*), and for the antarctic species *P. borchgrevinki*. For the LDH of *P. borchgrevinki*, the K_m of pyruvate has risen above the physiologically conserved range (0.15–0.35 mM) by 10–15 °C. By the criterion of K_m conservation, therefore, the LDH of *P. borchgrevinki* is not adapted to function well at temperatures above approximately 10 °C, even though the enzyme's structure is stable at temperatures up to approximately 40 °C (see below).

Comparable patterns of K_m conservation at physiological temperatures, and rapid increases in K_m above physiological temperatures, have been found for other enzymes. Figure 5 shows the effect of temperature on the K_m of phosphoenolpyruvate (PEP) for pyruvate kinases (PKs) purified from skeletal muscle of five vertebrates (Low and Somero 1976). At physiological temperatures, the K_m of PEP is near 0.03–0.06 mM for all species. The PK of *P. borchgrevinki* exhibits this value between −2 and 0 °C. Above 0 °C, the K_m drops slightly, before rising sharply at higher measurement temperatures. As in the case of the K_m of pyruvate of LDH, at temperatures between 10 and 15 °C the K_m of PEP rises above the conserved range. It is interesting that the PK of the highly thermophilic desert fish *Cyprinodon*

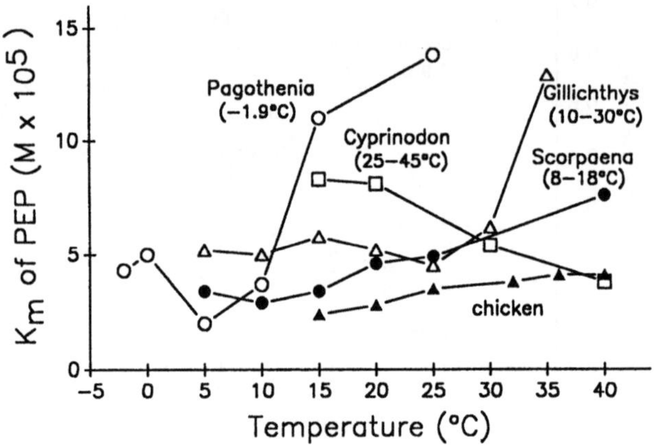

Fig. 5. The effect of temperature on the K_m of phosphoenolpyruvate (PEP) for muscle pyruvate kinases purified from vertebrates having different body temperatures (*numbers* in parentheses below species' names): Data from Low and Somero (1976)

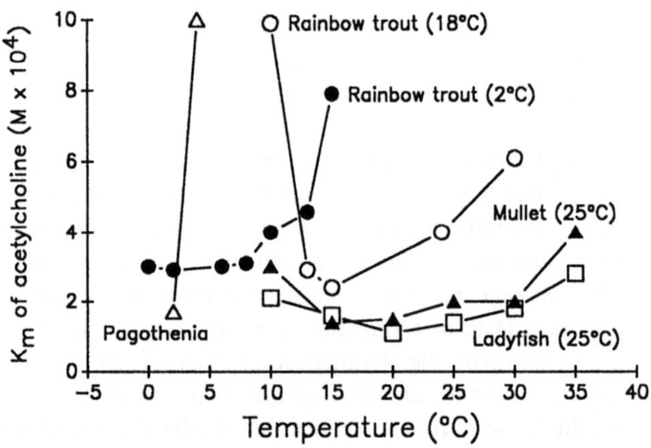

Fig. 6. The effect of temperature on the K_m of acetylcholine (ACh) for acetylcholinesterases of fish adapted or acclimated to different temperatures. Data from Baldwin and Hochachka (1970); Baldwin (1971)

macularius exhibits a decrease in K_m with rising temperature throughout the measurement range. This fish occurs in waters with temperatures up to approximately 40–45 °C.

Figure 6 shows the effect of temperature on the K_m of acetylcholine (ACh) for acetylcholinesterases (AChE) from four species of teleosts, including warm- and cold-acclimated populations of rainbow trout (*Salmo gairdneri*; Baldwin and Hochachka 1970; Baldwin 1971). The K_m of ACh is conserved within the range of approximately 0.1–0.3 mM at physiological temperatures for these species. The rise in K_m with increasing temperature for the enzyme from *P. borchgrevinki* is especially sharp in the case of AChE.

This strong thermal perturbation of an enzyme that is critical for synaptic transmission might underlie some of the neurological impairment that has been proposed to account, in part at least, for heat death of notothenioid fish at temperatures below 10 °C (cf. Somero and DeVries 1967). Indeed, Macdonald and Montgomery (1982) showed that the rate of spontaneous release of ACh in extraocular muscles of *P. borchgrevinki* is high (cold-adapted) at low temperatures and rises very dramatically with increasing temperature up to ~13 °C; spontaneous release then falls irreversibly near 15 °C (also see Macdonald et al. 1988). If massive release of ACh is occurring at the same time, i.e., as temperature is elevated to ~13 °C, while AChE is losing its ability to bind ACh, then large amounts of ACh could accumulate at synapses and disrupt synaptic transmission. It is not unreasonable to suggest that this coupled breakdown in control of transmitter release and transmitter hydrolysis could be a key factor in the heat death of *P. borchgrevinki* and other antarctic fish.

The similarities in thermal responses of these three enzymes serve as a basis for developing perspectives on thermal optima for protein function, and thermal limits for these functions, that differ from concepts based on maximal rates of catalysis and protein denaturation. As stressed earlier, the optimal temperatures for an enzyme, as defined physiologically, are the temperatures at which critical functional properties like K_m values lie within the range that is conserved at physiological temperatures of all species. Neither the temperature at which catalytic rate is maximal, nor the highest temperature at which structural stability is maintained, is a valid index of the physiologically optimal temperature of an enzyme.

Closely linked to this redefinition of thermal optimum are questions of thermal limits of controlled metabolic function and stenothermality of enzyme function. The thermal limits of controlled metabolic function fall within the range at which optimal (=conserved among species) values of K_m are maintained. The eurythermality (or stenothermality) of enzyme function is gauged by the breadth (or narrowness) of the temperature range over which optimal values for K_m are conserved for a given homologue of a particular enzyme. By this criterion, the enzymes of antarctic fish are relatively stenothermal in function. Compare, for example, the data for *P. borchgrevinki* in Figs. 4 and 5 with the data for the extremely eurythermal goby *Gillichthys mirabilis*.

The sharp upswings in K_m that are commonly seen at temperatures just above the physiological range may play some role in establishing the upper thermal tolerance of the whole organism, albeit this linkage remains to be established. The conjecture made above concerning the possibly lethal effect arising from the simultaneous loss of ACh binding by AChE and massive release of ACh at the synapse does illustrate, however, how a failure in substrate binding could have a highly detrimental effect on an organism. While disruption of ATP-generating pathways, i.e., of enzymes like LDH and PK, may be less threatening than disruption of synaptic function, even decreases of metabolic regulatory function by ATP generating enzymes could lead to significant metabolic impairment of the organism.

4 Protein Polymerization: Conservation of Subunit-Subunit Affinity

Much as enzymes must retain an optimal level of affinity for their substrates and cofactors to ensure proper metabolic regulation, the subunits of multimeric proteins must retain the correct affinities to ensure reversible self-assembly under the particular physical conditions of the environment. That is, the equilibrium constant of the monomer to polymer transformation must be closely conserved in all environments. This requirement, like the need for enzymes to retain correct K_m values, is a critical factor in protein evolution, because of the strong effects that temperature and hydrostatic pressure exert on subunit self-assembly equilibria (Swezey and Somero 1982, 1985).

Adaptation to temperature of subunit self-assembly processes has been noted for tubulins (Detrick et al. 1989 this Vol. and skeletal muscle actins (Swezey and Somero 1982, 1985). The conservation of actin self-assembly is indicated by the high similarity in the equilibrium constants which characterize the globular to filamentous transformation of actins from species adapted to widely different temperatures and hydrostatic pressures (Swezey and Somero 1982). The conservation of the equilibrium constant is achieved through adjustments in the enthalpy and entropy changes that accompany the self-assembly process. Actins from cold-adapted species like the antarctic fish *P. borchgrevinki* and *Gymnodraco acuticeps* polymerize with extremely small changes in enthalpy and entropy (Fig. 7). The low enthalpy change characterizing actin self-assembly in antarctic fish is interpreted as adaptive, in that a lower heat input is needed to drive self-assembly in this subzero environment. The low enthalpy and entropy changes have been proposed to reflect a reduced reliance on hydrophobic effects in actin self-assembly in cold-adapted species (Swezey and Somero 1982, 1985). The formation of hydrophobic interactions involves positive changes in both enthalpy and entropy. However,

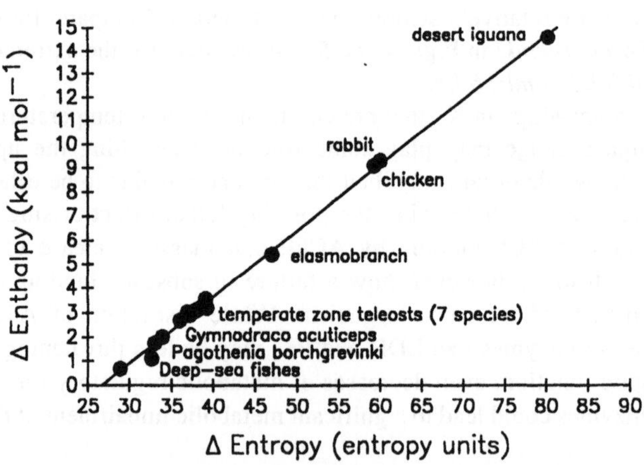

Fig. 7. Changes in enthalpy and entropy accompanying the polymerization of globular (G) actin to filamentous (F) actin, for skeletal muscle actins purified from vertebrates having different body temperatures. Data from Swezey and Somero (1982)

Detrick et al. (1989) reached a different conclusion about the importance of hydrophobic effects at low temperature in their analysis of the thermodynamics of tubulin polymerization.

5 Protein Structural Stability and Temperature Adaptation

It is well established that proteins of warm-adapted species are more resistant to thermal denaturation than the homologous proteins from cold-adapted species, at least when comparisons are restricted to organisms adapted to approximately 1 atm pressure (Hochachka and Somero 1984). Figure 8 shows how the thermal stability of globular (G) actin varies among species adapted to different temperatures and hydrostatic pressures (Swezey and Somero 1982). Except for the three deep-sea fish whose actins are extremely resistant to heat denaturation, the thermal stabilities of G-actins closely reflect the body temperatures of the organisms from which the actins were purified. The thermal resistance of the actins from the three deep-sea species, which are adapted to very low temperatures (2–5 °C), shows that adaptation temperature per se is not a complete determinant of a protein's thermal stability. High pressures, like high temperatures, appear to select for extremely strong protein structures (Siebenaller and Somero 1989). Deep-sea fish with low body temperatures may have proteins whose thermal stabilities exceed those of birds and mammals, and are only slightly less than those of highly thermophilic desert reptiles (Fig. 8).

Strong relationships between adaptation temperature and thermal stability have been demonstrated for pyruvate kinases (Fig. 9; Low and Somero 1976) and M_4-LDHs (Fig. 10; Donahue 1982) of vertebrates, including several species for which the effect of temperature on K_m of substrate (PEP or pyruvate, respectively) has been

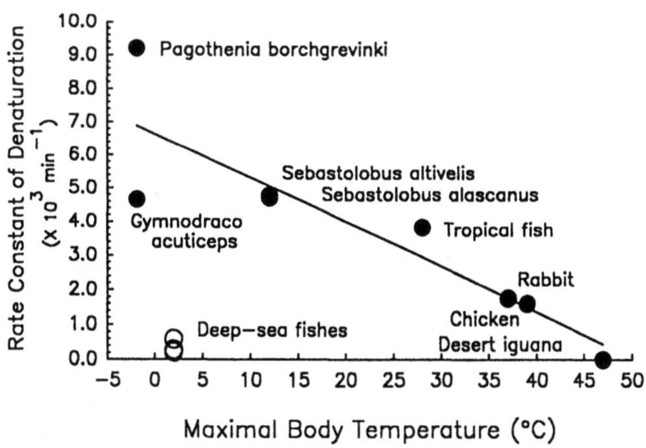

Fig. 8. Thermal denaturation of globular (G) actins purified from skeletal muscle of vertebrates adapted to different temperatures and pressures. Purified G-actin was incubated at 38 °C for varying time periods, and the rate of loss of native structure was quantified by measuring loss of G-actin's ability to inhibit DNAse-I (Swezey and Somero 1982)

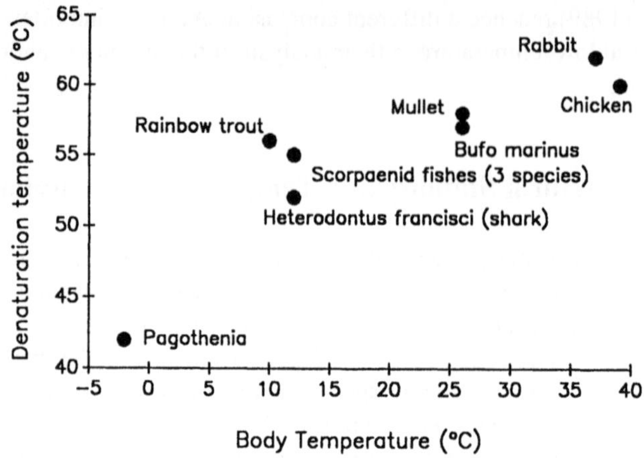

Fig. 9. Heat denaturation of pyruvate kinases purified from skeletal muscle of vertebrates having different adaptation temperatures. The denaturation temperature is the temperature at which loss of activity was first detected. Data from Low and Somero (1976)

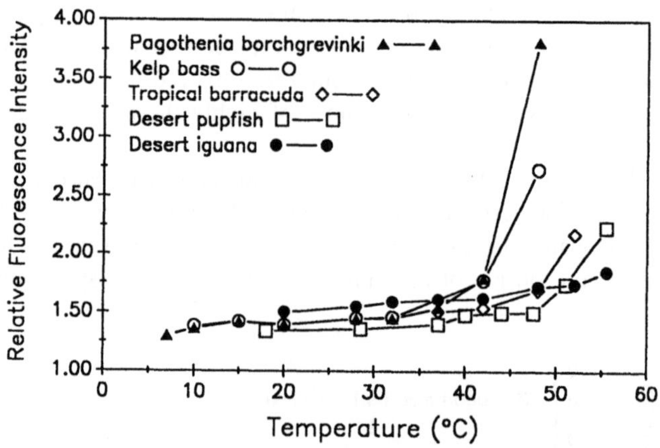

Fig. 10. Thermally induced unfolding of M_4-lactate dehydrogenases, as measured by quenching by acrylamide of fluorescence due to tryptophyl residues. Acrylamide concentration was 0.15 M. Access of the quenching agent, acrylamide, to tryptophyl residues is temperature-dependent, as indicated by the increase in the relative fluorescence intensity ratio at high temperatures. The relative fluorescence intensity at each temperature is the ratio of tryptophyl fluorescence in the absence of acrylamide to the fluorescence measured in the presence of 0.15 M acrylamide. Thus, a high relative fluorescence ratio is indicative of strong quenching by acrylamide (high access of acrylamide to tryptophyl residues); Donahue (1982)

measured. For PK and M_4-LDH of *P. borchgrevinki*, gross denaturation did not occur until temperatures near 40–45 °C, i.e., temperatures 35–40 °C above the fish's upper lethal temperature, and temperatures some 30 °C above the temperatures at which K_m values rose above the optimal (= physiologically conserved) range (see Figs. 4 and 5).

In the comparative studies of thermal perturbation of M_4-LDH structure, a quenching probe (acrylamide) was used to gauge the "openness" of the enzyme's structure. The more open the enzyme's structure, the more readily could the acrylamide penetrate the enzyme and quench the fluorescence due to tryptophyl residues. The ordinate in Figure 10 indicates relative fluorescence at each temperature: fluorescence in the absence of acrylamide divided by fluorescence in the presence of 0.15 M acrylamide. A high value of relative fluorescence is, then, an indication of an "open" protein structure. As shown by the data in Fig. 10, none of the M_4-LDH homologues showed an increase in structural "openness" until the measurement temperature was raised above 35 °C. At temperatures above 35 °C, the M_4-LDHs showed distinct differences based on body (adaptation) temperature. The enzyme of *P. borchgrevinki* unfolded at the lowest temperature, and the enzyme of the desert iguana was most stable. The point to be emphasized, however, is that these structural perturbation data give little hint that the effects of temperature on the K_m of pyruvate would differ so strikingly among species (Fig. 4).

The gross unfolding of protein structure, therefore, is a poor index of how temperature increases affect the functional properties of a protein. This is true for at least two reasons. First, thermal denaturation of proteins typically occurs many degrees above the upper lethal temperature of the organism. Second, loss of optimal protein function can occur at temperatures much lower than those at which protein denaturation commences. Therefore, although thermal stability of proteins does strongly correlate with adaptation temperature, as shown clearly by the data in Figs. 7–10, this correlation must be interpreted with caution, and in a context of loss of physiologically important functional properties, rather than in terms of gross unfolding of protein structure.

The basis for the correlation between protein denaturation temperature and adaptation temperature is not clearly established, although several hypotheses have been advanced to rationalize this relationship. One model discusses the linkage between protein flexibility and enzyme function. Because enzyme function involves alterations in protein conformation, e.g., during binding and release of ligands, the enzymes of cold-adapted species may benefit from relatively flexible structures at their low operating temperatures, a flexibility that an enzyme of a warm-adapted species would lack at the low body temperature of the cold-adapted species. The relatively flexible enzymes of cold-adapted species may owe their higher rate of catalysis to the relative energetic ease with which they can undergo the reversible conformational changes that are part of the catalytic process. Enzymes with high k_{cat} values generally have lower thermal stabilities than less efficient homologues of the enzyme (Table 1; Fig. 10; Hochachka and Somero 1984).

6 Summary

The proteins of cold–adapted antarctic fish differ from the homologous proteins of other vertebrates in ways that effect adaptation to low temperature, and may establish in part the extreme stenothermality of these fishes. Enzymes of antarctic fish are highly efficient as catalysts, as indexed by their high k_{cat} values. The thermal optima of enzymes from antarctic fish, as indexed by the temperatures at which the important kinetic parameter, the Michaelis-Menten constant (K_m) is conserved, are low. Additionally, the K_m of substrate typically increases sharply at temperatures only several degrees above 0 °C. These rapid increases in K_m at temperatures restrict the optimal functioning of these enzymes to low temperatures, and are proposed to be important in establishing the high degree of stenothermality noted for these species.

Acknowledgment. Portions of this work were supported by National Science Foundation grant DCB88-12180.

References

Baldwin J (1971) Adaptation of enzymes to temperature: acetylcholinesterases in the central nervous system of fishes. Comp Biochem Physiol 40:181–187

Baldwin J, Hochachka PW (1970) Functional significance of isoenzymes in thermal acclimatization: acetylcholinesterases from trout brain. Biochem J 116:883–887

Coppes Z, Somero GN (1990) Temperature adaptive differences between the M_4-lactate dehydrogenases of stenothermal and eurythermal sciaenid fishes. J Exp Zool 254:127–131

Detrick HW III, Johnson KA, Marchese-Ragona SP (1989) Polymerization of antarctic fish tubulins at low temperatures: energetic aspects. Biochemistry 28:10085–10093

DeVries AL (1988) The role of antifreeze glycopeptides and peptides in the freezing avoidance of antarctic fishes. Comp Biochem Physiol 90A:611–621

DeVries AL, Eastman JT (1981) Physiology and ecology of notothenioid fishes of the Ross Sea. J R Soc NZ 11:329–340

Donahue EV (1982) Lactate dehydrogenase: structural aspects of environmental adaptation. Doctoral dissertation, Univ California, San Diego

Dunn JF (1988) Muscle metabolism in antarctic fish. Comp Biochem Physiol 90B:539–545

Genicot S, Feller G, Gerday C (1988) Trypsin from antarctic fish (*Paranotothenia magellanica* Forster) as compared with trout (*Salmo gairdneri*) trypsin. Comp Biochem Physiol 90B:601-609

Graves JE, Somero GN (1982) Electrophoretic and functional enzymic evolution in four species of eastern Pacific barracudas from different thermal environments. Evolution 36:97–106

Graves JE, Rosenblatt RH, Somero GN (1983) Kinetic and electrophoretic differentiation of lactate dehydrogenases of teleost species-pairs from the Atlantic and Pacific Coasts of Panama. Evolution 37:30–37

Hazel JR, Prosser CL (1974) Molecular mechanisms of temperature compensation in poikilotherms. Physiol Rev 54:620–677

Hochachka PW, Somero GN (1984) Biochemical adaptation. Princeton Univ Press, Princeton, New Jersey

Holeton GF (1974) Metabolic cold adaptation of polar fish: fact or artefact? Physiol Zool 47:137–152

Johnston IA, Walesby NJ (1977) Molecular mechanisms of temperature adaptation in fish myofibrillar adenosine triphosphatases. J Comp Physiol 119:195–206

Low PS, Bada JL, Somero GN (1973) Temperature adaptation of enzymes: roles of the free energy, the enthalpy, and the entropy of activation. Proc Natl Acad USA 70:430–432

Low PS, Somero GN (1976) Adaptation of muscle pyruvate kinases to environmental temperature and pressure. J Exp Zool 198:1–12

Maresca B, Patriarca E, Goldenberg C, Sacco M (1988) Heat shock and cold adaptation in antarctic fishes: a molecular approach. Comp Biochem Physiol 90B:623–629

Macdonald JA, Montgomery JC (1982) Thermal limits of neuromuscular function in an antarctic fish. J Comp Physiol 147:237–250

Macdonald JA, Montgomery JC, Wells RMG (1988) The physiology of McMurdo Sound fishes: current New Zealand research. Comp Biochem Physiol 90B:567–578

Siebenaller JF, Somero GN (1989) Biochemical adaptation to the deep sea. CRC Crit Rev Aquat Sci 1:1–25

Somero GN (1987) Protein adaptation and biogeography: threshold effects on molecular evolution. Trends Ecol Evol 1:124–127

Somero GN, DeVries AL (1967) Temperature tolerance of some antarctic fishes. Science 156:257–258

Somero GN, Giese AC, Wohlshlag DE (1968) Cold adaptation of the antarctic fish *Trematomus bernacchii*. Comp Biochem Physiol 26:223–233

Somero GN, Siebenaller JF (1979) Inefficient lactate dehydrogenases of deep-sea fishes. Nature (Lond) 282:100–102

Swezey RR, Somero GN (1982) Polymerization thermodynamics and structural stabilities of skeletal muscle actins from vertebrates adapted to different temperatures and hydrostatic pressures. Biochemistry 21:4496–4502

Swezey RR, Somero GN (1985) Pressure effects on actin self-assembly:interspecific differences in the equilibrium and kinetics of the G to F transformation. Biochemistry 24:852–860

Torres JJ, Somero GN (1988a) Vertical distribution and metabolism in antarctic mesopelagic fishes. Comp Biochem Physiol 90B:521–528

Torres JJ, Somero GN (1988b) Metabolism, enzymic activities, and cold adaptation in antarctic mesopelagic fishes. Mar Biol 98:169–180

Wohlschlag DE (1964) Respiratory metabolism and ecological characteristics of some fishes in McMurdo Sound, Antarctica. In: Lee MO (ed) Biology of the antarctic seas, vol 1. Am Geophys Union, Washington, DC, pp 33–62

Yancey PH, Siebenaller JF (1987) Coenzyme binding ability of homologs of M_4-lactate dehydrogenase in temperature adaptation. Biochim Biophys Acta 924:483–491

Yancey PH, Somero GN (1978) Temperature dependence of intracellular pH: its role in the conservation of pyruvate apparent K_m values of vertebrate lactate dehydrogenases. J Comp Physiol 125:129–134

Polymerization of Microtubule Proteins from Antarctic Fish

H.W. Detrich, III

1 Introduction

Microtubules are a major component of the cytoskeleton of most eukaryotic cells. They participate in many fundamental processes, including mitosis, nerve growth and regeneration, the intracellular transport of organelles, and the determination of cell shape. The formation of microtubules from their subunit proteins, tubulin αβ dimers and microtubule-associated proteins (MAPs), is an entropically driven process that is favored by high temperatures (Correia and Williams 1983). For example, the microtubule proteins of mammals and birds polymerize to yield microtubules at temperatures near 37 °C, but these polymers are "cold-labile"; they disassemble to their subunits at low temperatures (0–4 °C). How, then, do the microtubules of cold-living poikilotherms (e.g., the fish of the antarctic marine ecosystem) assemble and function at body temperatures as low as –1.9 °C?

One adaptive strategy that conserves the efficient polymerization of a cytoskeletal polymer under widely divergent temperatures has been described by Swezey and Somero (1982). They examined the polymerization energetics of skeletal muscle actins from 14 vertebrate species that have average body temperatures from –1.9 to 39 °C. Their results indicate that the standard enthalpy and standard entropy of actin polymerization *increases* in a regular fashion as average body temperature increases. Based on these observations, Swezey and Somero (1982) suggested that cold-adapted poikilotherms depend less on endothermic hydrophobic interactions, and more on exothermic polar bonds, for the stabilization of their actin filaments than do homeotherms. One may ask whether the polymerization of the microtubules of cold-living poikilotherms is based on a similar adaptive strategy.

The goal of our studies is to determine the molecular features that enable the microtubule proteins of poikilotherms to polymerize efficiently at low temperatures. We have chosen the coastal fish of the Antarctic Peninsula as the model system for our work. Near Palmer Station these cold-adapted osteichthyans, which diverged from temperate fish approximately 40 million years ago as the Southern Ocean cooled (DeWitt 1971), live at an annual mean bottom temperature of –1.0 °C (seasonal range –0.1 °C in January to –1.9 °C in August; Lowry 1975). Thus, antarctic fish are well suited to studies of macromolecular adaptation to low temperatures.

Department of Biology, Northeastern University, Boston, Massachusetts 02115, USA

At the molecular level, the modifications that favor microtubule assembly at low temperatures may reside in the tubulin subunits, in microtubule-stabilizing factors such as the MAPs, or in both. We (Detrich et al. 1989) and others (Williams et al. 1985) have shown that pure, MAP-free brain tubulins from antarctic fish polymerize efficiently at temperatures within the physiological range (–1.9 to +2 °C). Evidently, the major functional adaptations of the microtubule proteins of antarctic fish are present in their tubulins. In this chapter we review our studies of the polymerization energetics of antarctic fish tubulins (Detrich et al. 1989) and of the role of antarctic fish MAPs in microtubule assembly (Detrich et al. 1990). We conclude that maintenance of efficient microtubule assembly at low temperatures results from increases in the proportion or the strength (or both) of the hydrophobic interactions at sites of tubulin-tubulin contact. Thus, the adaptive strategy that favors tubulin polymerization at low temperatures differs from that of the actins.

2 Polymerization of Tubulins from Antarctic Fish

2.1 Purification of Tubulins

At the outset of our studies, we developed a method (Detrich and Overton 1986; Detrich et al. 1989) for the rapid purification of tubulin from the brain tissues of several antarctic fish (e.g., *Notothenia gibberifrons*, *N. coriiceps neglecta*, and *Chaenocephalus aceratus*). The procedure employs DEAE ion-exchange chromatography to produce a fraction enriched in tubulin from a high-speed supernatant extract of brain, followed by one cycle of microtubule assembly in vitro to obtain tubulin free of residual contaminating proteins. Figure 1 presents a sodium dodecyl sulfate (SDS) urea-polyacrylamide gel that contains tubulin from an antarctic cod, *N. gibberifrons*. The purified protein contains the α and β tubulins and is free of MAPs. Approximately 12 mg of tubulin can be obtained from 10 g of fish brain tissue in a period of 6–7 h. Furthermore, the overall yield is tenfold greater than that produced by a method involving two cycles of temperature-dependent microtubule assembly/disassembly and phosphocellulose chromatography (Williams et al. 1985).

2.2 Reversible Polymerization

When warmed from 0 to 21.1 °C in an appropriately buffered solution containing GTP, *N. gibberifrons* tubulin (0.5 mg/ml) polymerized as judged by turbidimetry (Fig. 2). Following a brief lag, solution turbidity increased toward a plateau value. Numerous filaments were seen in this solution by dark-field light microscopy (not shown), and the filaments were found to be microtubules when similar solutions were observed by negative-stain electron microscopy (Detrich et al. 1989). Subsequent cooling of this solution (Fig. 2, see arrows specifying times of the temperature changes) produced progressive decreases in turbidity, which approached a new plateau value at each temperature. Furthermore, the final turbidity obtained after cooling to 0.4 °C was almost identical to that observed at zero time. These results

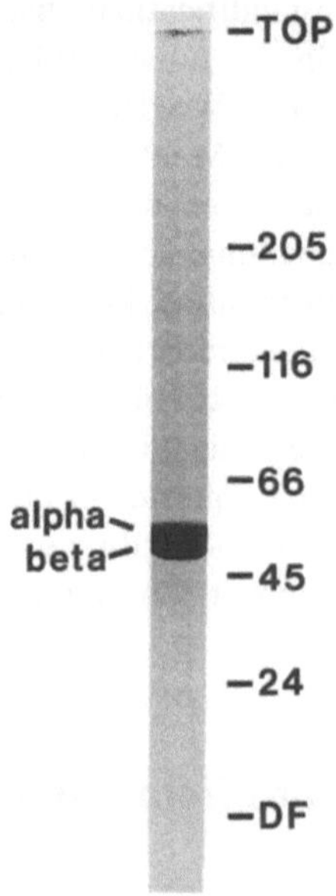

Fig. 1. Electrophoretic analysis of brain tubulin from *Notothenia gibberifrons*. Tubulin was purified from brain tissue by DEAE ion-exchange chromatography and one round of microtubule polymerization as described previously (Detrich and Overton 1986; Detrich et al. 1989). The purified protein was electrophoresed on an SDS urea-polyacrylamide gradient gel (Kim et al. 1979), and the gel was stained with Coomassie Brilliant Blue R-250 (Detrich and Overton 1986; Detrich et al. 1989). Electrophoretic migration was from top to bottom. The molecular weights of standards run on the same gel are given in thousands on the *vertical axis*, and the positions of the tubulin chains (α and β), the dye front *(DF)*, and the *top* of the gel are indicated

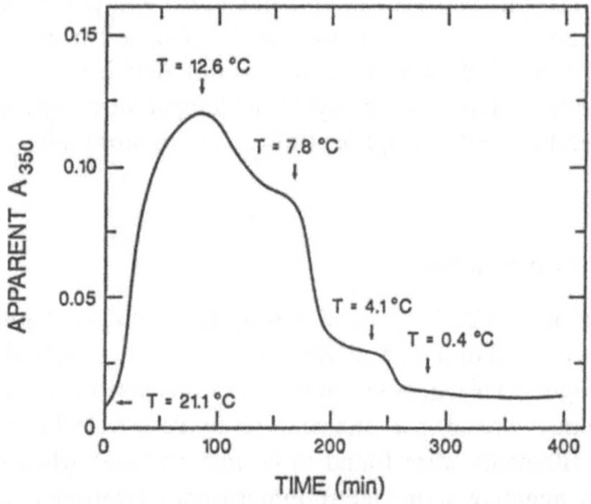

Fig. 2. Reversible polymerization of *N. gibberifrons* brain tubulin. A sample of purified tubulin (0.5 mg/ml) in PME buffer [0.1 M piperazinediethanesulfonic acid-NaOH (Pipes-NaOH), 1 mM $MgSO_4$, 1 mM ethylene glycol bis(β-aminoethyl ether)-*N, N, N', N'*- tetraacetic aid (EGTA), pH 6.82 at 20 °C] containing 1 mM guanosine 5'-triphosphate (GTP) was warmed from 0 to 21.1 °C at zero time, and the turbidity generated by the sample was observed spectrophotometrically (apparent A_{350}). At intervals (see *arrows*), the temperature of the solution was reduced to the values indicated. (Reprinted with permission from Detrich et al. 1989. Copyright 1989 American Chemical Society)

indicate that tubulin from *N. gibberifrons* polymerizes reversibly in response to changes in temperature.

2.3 Temperature Dependence of Microtubule Assembly

At a given temperature, the assembly of microtubules in vitro requires a minimal, or "critical", concentration of tubulin. Below the critical concentration, polymerization does not occur, whereas above it the extent of assembly varies linearly with the total tubulin concentration (Gaskin et al. 1974; Johnson and Borisy 1975). To determine the temperature dependence of the assembly of antarctic fish microtubules, we measured the critical concentrations for polymerization of tubulins from two fish species, *N. gibberifrons* and *N. coriiceps neglecta,* at eight temperatures between 0 and 18 °C by a quantitative sedimentation assay (see Johnson and Borisy 1975 and Detrich et al. 1989 for details). This method gives the steady-state distribution of assembled microtubule protein in terms of two operationally defined species: monomer (the polymerizing unit) and polymer (microtubules). Figure 3A shows measurements of the critical concentrations for *N. gibberifrons* tubulin at four temperatures. The values, determined as the *y* intercepts of the linear regression lines through the appropriate supernatant (monomer) data points (see Detrich et al. 1989), decreased from 0.87 mg/ml at 0 °C to 0.02 mg/ml at 18 °C. Clearly, the tubulins of antarctic fish are able to form microtubules efficiently at low temperatures. By contrast, critical concentrations for pure mammalian brain tubulins at like temperatures and under identical solution conditions are estimated to be two orders of magnitude larger (Williams et al. 1985).

Figure 3B presents a van't Hoff plot of the critical concentration data. In constructing this plot, we have assumed that the apparent association constant for addition of a tubulin dimer to the end of a microtubule is equal to the reciprocal of the critical concentration (Gaskin et al. 1974; Lee and Timasheff 1977). The standard enthalpy change, ΔH°, for microtubule elongation, calculated from the slope of the linear regression line, is +26.9 kcal/mol. The corresponding standard entropy change, ΔS°, is +123 eu. Thus, the polymerization of antarctic fish tubulins is entropically driven,

2.4 Salt Dependence of Microtubule Assembly

Noncovalent protein association reactions are mediated largely by four molecular interactions: (1) hydrogen bonds; (2) van der Waals bonds; (3) electrostatic (ionic) bonds; and (4) hydrophobic interactions. Only the latter two make positive contributions to the enthalpy and entropy changes of subunit association reactions (Ross and Subramanian 1981). Thus, we hypothesize that the entropy-driven polymerization of antarctic fish tubulins reflects a reliance on ionic bonds and/or hydrophobic interactions at sites of intersubunit contact. To estimate the relative contributions of these two interactions, we measured the critical concentration for polymerization of *N. gibberifrons* tubulin at 10 °C as a function of the concentration of a structure-making salt, NaCl (Detrich et al. 1989). High concentrations of NaCl strengthen hydrophobic interactions and weaken electrostatic bonds. As shown in Fig. 4, the critical concentration rose modestly, from 0.041 to 0.34 mg/ml, as the salt concentration was increased from 0 to 0.4 M. In contrast, NaCl at concentrations

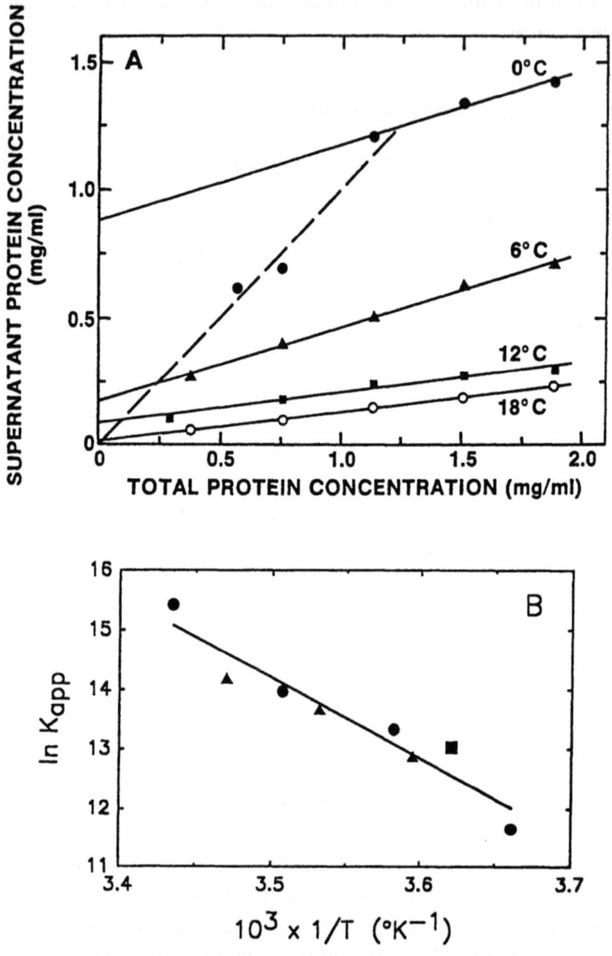

Fig. 3. Thermodynamics of the assembly of antarctic fish microtubules in vitro. **A** Temperature dependence of the critical concentration for polymerization of *N. gibberifrons* tubulin. Critical concentrations were measured at temperatures between 0 and 18 °C by means of a quantitative sedimentation assay (Johnson and Borisy 1975) as described previously (Detrich et al. 1989). Briefly, solutions of microtubules (in 0.1 M Pipes-NaOH, 1 mM $MgSO_4$, 1 mM GTP, 5 or 8 mM $CaCl_2$, 6.2 or 10 mM EGTA, and, in some cases, 1 mM dithioerythritol, pH 6.82 at 20 ° C) at different total protein concentrations were incubated at the desired experimental temperature for an interval sufficient to attain a new steady-state level of polymerization. The samples were then centrifuged isothermally, the supernatants were recovered to yield the monomer fractions, and their protein concentrations were measured (Detrich et al. 1989). Concentrations of the supernatant fractions are plotted as a function of the total tubulin concentration for determinations at 0 °C *(closed circles)*, at 6 °C (*triangles*), at 12 °C (*squares*), and at 18 °C (*open circles*). For each temperature, the critical concentration was estimated as the *y* intercept of the linear regression line through the monomer data points (neglecting data points obtained at total protein concentrations that did not yield detectable polymer pellets). **B** van't Hoff analysis of the apparent equilibrium constant for microtubule elongation. The natural logarithm of the apparent association constant for microtubule elongation (i.e., the natural logarithm of the reciprocal of the critical concentration) is shown as a function of the reciprocal of the absolute temperature. Three tubulin preparations from two antarctic fish were used in these studies: *circles, N. gibberifrons* tubulin, preparation 1; *triangles, N. gibberifrons* tubulin, preparation 2; *square, N. coriiceps neglecta* tubulin. The best-fitting straight line through the data points was determined by linear-regression analysis. (Reprinted with permission from Detrich et al. 1989. Copyright 1989 American Chemical Society)

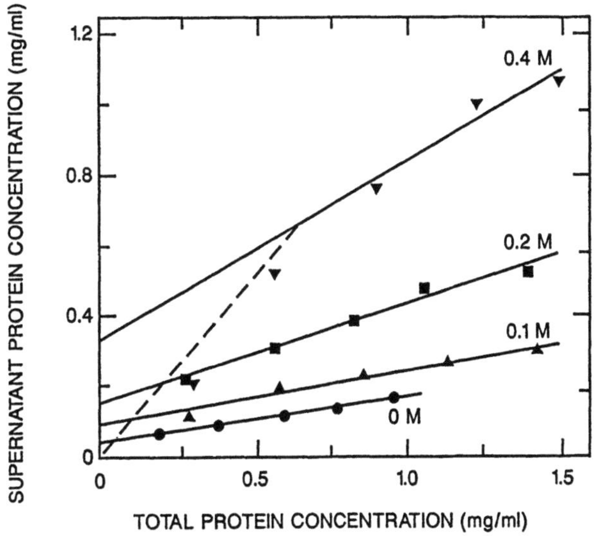

Fig. 4. Salt dependence of the critical concentration for polymerization of *N. gibberifrons* tubulin. Critical concentrations at 10 °C were measured by means of the quantitative sedimentation assay (Detrich et al. 1989; see also the legend to Fig. 3) in a buffer (0.1 M Pipes-NaOH, 2 mM $MgSO_4$, 1 mM GTP, 8 mM $CaCl_2$, 10 mM EGTA, and 1 mM dithiothreitol, pH 6.82 at 20 °C) supplemented with concentrations of NaCl from 0 to 0.4 M. Concentrations of the supernatant (monomer) fractions are plotted as a function of the total tubulin concentration for determinations at 0 M NaCl (*circles*), at 0.1 M NaCl (*triangles*), at 0.2 M NaCl (*squares*), and at 0.4 M NaCl (*inverted triangles*). For each salt concentration, the critical concentration was estimated as the *y* intercept of the linear regression line through the monomer data points (neglecting data points obtained at total protein concentrations that did not yield detectable polymer pellets)

exceeding 0.25 M suppresses completely the polymerization of tubulins or total microtubule proteins from mammals (Olmsted and Borisy 1975; Himes et al. 1977). Thus, we suggest that hydrophobic interactions make larger contributions to the polymerization of antarctic fish tubulins than they do to the assembly of mammalian tubulins.

3 Role of MAPs in Microtubule Assembly

The results presented above indicate that much of the "cold-stability" of these fish microtubules is a property of their tubulins. Nevertheless, one may ask whether these poikilotherms also possess MAPs that make additional contributions to the energetics of microtubule formation at their low body temperatures. Below we consider the effects of MAPs from antarctic fish and from mammals on the reversible polymerization of fish tubulin at low temperatures.

3.1 Purification of MAPs from Antarctic Fish

We have employed a modification of the taxol-dependent procedure of Vallee (1982) to isolate microtubules and MAPs from the brain tissues of *N. coriiceps neglecta, N. gibberifrons, C. aceratus,* and other antarctic fish. As shown in Fig. 5, microtubules prepared from *C. aceratus* (lane A) contain tubulin and a complement of presumptive MAPs. Sedimentation of the associated proteins did not occur in the absence of taxol, nor did the proteins pellet from brain extracts that had been depleted of tubulin prior to incubation with taxol (Detrich et al. 1990). Therefore, these proteins were associated specifically with the taxol-induced microtubules. Nearly quantitative displacement of the MAPs was obtained when the taxol-stabilized microtubules were exposed to elevated concentrations of NaCl; MAPs were recovered in the supernatant fraction (lane B) and tubulin in the microtubule pellet (lane C) following centrifugal resolution of the NaCl-treated sample. The yield of microtubule protein obtained by this method, ~1 mg/g of brain tissue, was identical to that reported by Vallee (1982) for calf brain cerebral cortex. Of the total microtubule protein, ~15% was recovered in the MAP fraction.

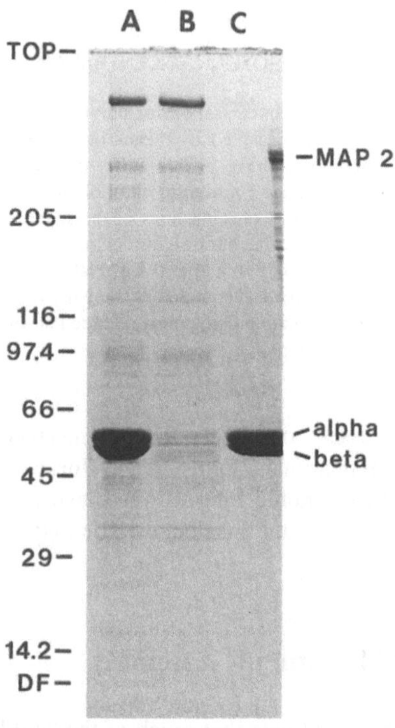

Fig. 5. Taxol-dependent purification of microtubule proteins (tubulin and MAPs) from brain tissues of the icefish, *Chaenocephalus aceratus.* Microtubules and MAPs were isolated by a modification (Detrich et al. 1990) of the method of Vallee (1982). Samples from different steps of the purification were electrophoresed on an SDS-urea polyacrylamide gradient gel (Kim et al. 1979), and the gel was stained with Coomassie Brilliant Blue R-250. Samples *A*, microtubules from taxol-dependent assembly; *B* and *C*, MAP-containing supernatant and tubulin-containing pellet, respectively, obtained by exposure of sample *A* to elevated ionic strength [i.e., PME buffer (see legend to Fig. 2) containing 1 mM GTP, 30 μM taxol, and 0.36 M NaCl] followed by centrifugal resolution. Electrophoretic migration was from *top* to *bottom*. The molecular weights of standards are given in thousands on the *vertical axis*, and the positions of the tubulin chains (α and β), bovine MAP 2 ($M_{ra} \simeq 275\,000$), the dye front (*DF*), and the *top* of the gel are indicated

The protein compositions of MAP fractions from each of the antarctic fish were quite similar. Prominent among the MAPs is a protein of unusually high apparent molecular weight ($M_{ra} \simeq$ 410 000–430 000, depending upon the fish species), and there are clusters of proteins in the molecular weight ranges 220 000–270 000, 140 000–155 000, 85 000–95 000, 40 000–45 000, and 32 000–34 000 (Detrich et al. 1990; for *C. aceratus* MAPs, see Fig. 5, lane B). Some of these proteins are probably functional homologs of the high molecular weight MAPs 1 and 2 ($M_{ra} \simeq$ 270 000–350 000), the tau proteins ($M_{ra} \simeq$ 55 000-62 000), and the low molecular weight MAPs ($M_{ra} \simeq$ 30 000-35 000) of vertebrate homeotherms (reviewed by Olmsted 1986 and Matus 1988).

3.2 Assembly of Microtubules by the MAPs and Tubulins of Antarctic Fish

The brain MAPs of mammals promote the assembly of microtubules at 30–37 °C when added to solutions of pure mammalian tubulin (Sloboda et al. 1976; Murphy et al. 1977). Do the MAPs of antarctic fish perform a similar function at low temperatures? To address this question, we have examined microtubule assembly systems reconstituted from pure fish tubulins and taxol-purified fish MAPs.

When solutions containing tubulin and MAPs from *N. gibberifrons* were released from calcium-mediated assembly inhibition by addition of excess EGTA, the MAPs induced the formation of microtubules at 2 °C (Fig. 6). Both the initial rate and the final extent of assembly increased with increasing concentrations of the MAPs (curves 3–4). The MAP-induced microtubules depolymerized (i.e., sample

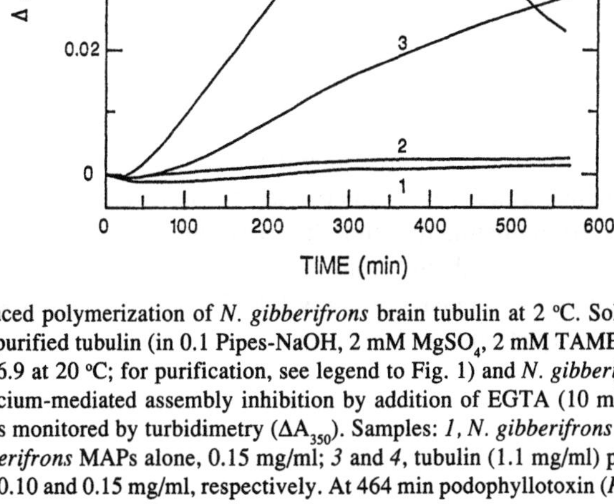

Fig. 6. MAP-induced polymerization of *N. gibberifrons* brain tubulin at 2 °C. Solutions containing DEAE/assembly-purified tubulin (in 0.1 Pipes-NaOH, 2 mM $MgSO_4$, 2 mM TAME, 1 mM DTT, and 8 mM $CaCl_2$, pH 6.9 at 20 °C; for purification, see legend to Fig. 1) and *N. gibberifrons* MAPs were released from calcium-mediated assembly inhibition by addition of EGTA (10 mM), and polymerization at 2 °C was monitored by turbidimetry (ΔA_{350}). Samples: *1, N. gibberifrons* tubulin alone, 1.1 mg/ml; *2, N. gibberifrons* MAPs alone, 0.15 mg/ml; *3* and *4*, tubulin (1.1 mg/ml) plus MAPs at final concentrations of 0.10 and 0.15 mg/ml, respectively. At 464 min podophyllotoxin (*PLN*) was added to sample *4* to a final concentration of 50 μM (*arrow*)

turbidity returned to baseline levels) upon addition of podophyllotoxin to 50 μM (Fig. 6) or $CaCl_2$ to 10 mM (not shown). A sample of tubulin alone (at the same subcritical concentration used for the MAP/tubulin reconstitutions) did not polymerize (curve 1), nor did the MAPs alone develop appreciable turbidity (curve 2). Evidently, some component or components of the MAP preparations increase both the rate and the extent of assembly of microtubules from antarctic fish tubulin at physiological temperatures.

Similar results were obtained for solutions of *N. gibberifrons* tubulin and *C. aceratus* MAPs at 0 °C (Detrich et al. 1990). Numerous filamentous elements were observed when these solutions were examined by dark-field light microscopy at 0 °C, and the filaments were found to be microtubules when examined by negative-stain electron microscopy.

Microtubules from the experiment of Fig. 6 (samples 3, 4) and from other experiments (Detrich et al. 1990) were collected by centrifugation and subjected to electrophoresis. Analysis of the Coomassie-stained gels showed that small quantities of at least seven proteins with apparent molecular weights between 60 000 and 430 000 co-sedimented with the microtubules. These proteins did not sediment from samples containing MAPs alone and must therefore have been attached specifically to microtubules. We conclude that the co-sedimenting proteins are bona fide MAPs that promote the formation of microtubules at low temperatures.

3.3 Formation of Microtubules by Heterologous Combinations of Tubulins and MAPs

To determine if the antarctic fish MAPs contain significant adaptations that contribute to microtubule formation at low temperatures, we examined the polymerization of heterologous combinations of tubulins and MAPs from these fish and from mammals. Addition of bovine MAPs to subcritical concentrations of *N. gibberifrons* tubulin stimulated the assembly of microtubules at 2 °C (Detrich et al. 1990). Compared on a weight basis, both bovine and fish MAPs induced comparable extents of microtubule formation at low temperatures. Bovine proteins corresponding to MAP 2, a MAP 1 species, and several proteins with lower molecular weights (possibly including some tau proteins) were incorporated into the hybrid microtubules. Therefore, mammalian MAPs, like the MAPs of antarctic fish, are able to promote microtubule assembly at temperatures near 0 °C. By contrast, reciprocal combinations containing bovine tubulin and antarctic fish MAPs polymerized only at the elevated temperatures (33 °C) necessary for assembly of homologous MAP/tubulin systems from mammals, and this assembly was cold-reversible. Thus, MAPs from cold-adapted fish and from the mammal function similarly at both low and high temperatures. It appears unlikely that the MAPs of antarctic fish possess major functional adaptations that are absent in other MAPs.

4 Discussion

The cytoplasmic microtubules of cold-adapted antarctic fish assemble efficiently from their tubulin and MAP subunits in a thermal environment (–1.9 to +2 °C) that promotes the depolymerization of microtubules from temperate poikilotherms and from homeotherms. Williams et al. (1985) have proposed that cold-living poikilotherms overcame the destabilizing effects of low temperatures on microtubules by the evolution of modified tubulins that possess large association constants. The results reviewed here provide strong support for this interpretation. We find that pure tubulins from the brains of antarctic fish polymerize readily at temperatures as low as 0 °C. The MAPs of these poikilotherms make additional contributions to the energetics of microtubule assembly at physiological temperatures, but mammalian MAPs are equally effective in this regard. Thus, with respect to polymerization at cold temperatures, the major locus of functional adaptation appears to be the tubulin dimer.

4.1 Functional Adaptation of Antarctic Fish Tubulins

4.1.1 Energetics of Polymerization

The positive enthalpy and entropy changes ($\Delta H° = +26.9$ kcal/mol, $\Delta S° = +123$ eu) for the polymerization of antarctic fish tubulins indicate that the reaction is driven strongly by an increase in system entropy (see also Williams et al. 1985). These values are considerably greater than the enthalpies and entropies reported for the polymerization of pure tubulins from temperate poikilotherms or from homeotherms (see Detrich et al. 1989 for a complete discussion). For example, Robinson and Engelborghs (1982) reported that the van't Hoff plot for the assembly of porcine brain tubulin between 10 and 35 °C was linear and gave $\Delta H° = +6.3$ kcal/mol, $\Delta S° = +44.4$ eu. Estimates for $\Delta H°$ and $\Delta S°$ for polymerization of egg tubulin from a temperate sea urchin (~11-16 kcal/mol and ~61–79 eu, respectively; derived from the data of Detrich and Wilson 1983, Suprenant and Rebhun 1983, and Detrich et al. 1985) fall between the values for the antarctic fish and the mammal. Although the comparisons are limited and the buffer conditions used in these studies were not identical, it would appear that the entropic control over microtubule assembly increases with decreasing average body temperature.

The interspecific differences in polymerization thermodynamics have an important adaptive consequence: they conserve the critical concentration for microtubule assembly within a narrow range of tubulin concentrations. For example, the critical concentrations for polymerization of tubulins from cold-adapted antarctic fish, from temperate invertebrates (e.g., sea urchins, clams), and from mammals, measured in vitro at physiologically appropriate temperatures, fall within the range 0.36–2.5 mg/ml (Herzog and Weber 1977; Suprenant and Rebhun 1983, 1984; Detrich et al. 1985, 1989; Williams et al. 1985). Thus, each of these organisms is able to assemble microtubules effectively within the normal limits of its body temperature.

4.1.2 Molecular Interactions

The enhanced polymerization capacity of the antarctic fish tubulins must result from changes in the molecular interactions between tubulin dimers in the microtubule. Polymer-stabilizing interactions may be increased in strength, destabilizing interactions may be reduced, or changes of both kinds may be involved. Before discussing the nature and location of the adaptations of the fish tubulins, we review briefly the structural organization of the tubulin subunits of mammals.

Each of the monomers of the native tubulin αβ heterodimer appears to be folded to give a two-lobed structure, composed of larger amino-terminal and smaller carboxy-terminal globular domains (approximately two-thirds and one-third of each subunit, respectively) linked by an exposed connecting region (Mandelkow et al. 1985; Sackett and Wolff 1986). In addition, a short, extended, highly acidic carboxy-terminal "tail" terminates each chain (Sackett and Wolff 1986). The major *inter*dimer contacts within the microtubule appear to be formed by the carboxy-terminal domain of α tubulin and the amino-terminal domain of the β chain (Kirchner and Mandelkow 1985). Conversely, the *intra*dimer interactions should be formed by the amino-terminal domain of α tubulin and the carboxy-terminal domain of the β chain (Kirchner and Mandelkow 1985; Serrano and Avila 1985). The highly charged carboxy-terminal tails of both subunits project into the solvent and make few contacts with other tubulin domains, both in the free dimer and in the microtubule (Sackett and Wolff 1986; Breitling and Little 1986).

Two lines of evidence suggest that the functional adaptation of the antarctic fish tubulins is based partly on a greater reliance on hydrophobic contacts between tubulin dimers. First, the polymerization energetics demonstrate that the antarctic fish tubulins rely to a greater extent on entropy-generating interactions (i.e., hydrophobic interactions and/or electrostatic bonds) than do other tubulins. Second, the small perturbation of the critical concentration (hence, the association constant for microtubule elongation) by elevated ionic strength indicates that, with respect to other tubulins, the greater entropic control over the polymerization of the fish tubulin is due to an increase in the relative proportion of hydrophobic interactions to ionic bonds. Together, these results suggest that antarctic fish have mitigated the destabilizing influence of low temperatures on entropically driven reactions through the evolution of tubulins that form increased numbers of, and/or qualitatively stronger, hydrophobic interactions between the domains involved in interdimer contact. We predict that such changes would be located primarily in the carboxy-terminal (small) globular domain of the α chain, in the amino-terminal (large) globular domain of β tubulin, or possibly in both regions. The structural divergence of the α tubulins of antarctic fish from those of other organisms (Detrich et al. 1987) favors the first possibility.

The acidic carboxyl termini of mammalian tubulins hinder microtubule assembly in vitro, probably because they generate electrostatic repulsion between dimers (Serrano et al. 1984; Bhattacharyya et al. 1985; Sackett et al. 1985). Thus, reduction of the acidic character of the carboxy-terminal tails of the antarctic fish tubulins would also be expected to strengthen the interaction between tubulin dimers. Previous work from this laboratory suggests that the antarctic fish tubulins may have fewer negative charges in their tails than do tubulins from other organisms (Detrich and Overton 1986, 1988). In addition, the carboxy-terminal tail of the α chain of native *N. coriiceps neglecta* tubulin resists cleavage by subtilisin under conditions conducive to carboxy-terminal proteolysis of bovine α tubulin (Detrich et al. 1987). Thus, carboxy-terminal charge reduction may also contribute to the enhanced polymerization efficiency of the antarctic fish tubulins.

4.1.3 Temperature Adaptation of Tubulins and Actins: Alternative Strategies

In contrast to the trend we report for pure tubulins, Swezey and Somero (1982) found that the enthalpy and entropy changes for actin assembly *decrease* with decreasing average body temperature. The polymerization of actins from antarctic and deep-sea fish, the organisms with the lowest body temperatures, was characterized by the smallest values for ΔH° and ΔS°. It would appear that cold-adapted organisms are less dependent on hydrophobic interactions for the maintenance of their actin filaments than are organisms with higher body temperatures. Thus, there are at least two adaptive strategies that conserve cytoskeletal polymer assembly at low temperatures: (1) quantitative and/or qualitative modification of the bond types at subunit contact surfaces to provide an entropy of association sufficient to overcome unfavorable enthalpy changes (e.g., tubulin); or (2) mitigation of destabilizing enthalpy changes by greater reliance on bond types that make negative contributions to the overall enthalpy of polymerization (e.g., actin).

4.2 MAPs and Cold-Stable Microtubules

4.2.1 Function and Structure of Antarctic Fish MAPs

Stimulation by the fish MAPs of microtubule assembly from fish tubulin follows the same qualitative pattern at 0 °C that is found at 37 °C in the classic mammalian case (Sloboda et al. 1976; Murphy et al. 1977). Both the initial rate and the final extent of assembly are enhanced, and the degree of enhancement increases with the concentration of added MAPs. Assembly is reversed by the addition of podo–phyllotoxin or calcium. It therefore seems likely that the antarctic fish MAPs enhance microtubule assembly through the same mechanisms that operate for mammalian MAPs: promotion of microtubule nucleation and enhancement of microtubule elongation by stabilization of formed polymer (Murphy et al. 1977). This conclusion is strengthened by the observed interchangeability of the MAPs with respect to their assembly-enhancing function: bovine MAPs stimulate polymerization of fish tubulin, and fish MAPs stimulate polymerization of bovine tubulin. Thus, bovine and fish MAPs may have similar binding domains and may occupy similar sites on the microtubule lattice.

The MAP fraction from each of the antarctic fish contains a protein of unusually high apparent molecular weight ($M_{ra} \simeq$ 415 000-430 000, depending on the species). MAPs of comparable size have not been found in the brain tissues of mammals or of birds. However, Strömberg et al. (1989) recently described an apparent MAP of similar molecular weight ($M_{ra} \simeq$ 400 000) that is present in a cold-labile brain microtubule fraction from the Atlantic cod, *Gadus morhua*. We suggest that these large MAPs from the austral and boreal fish may be structurally and functionally homologous.

4.2.2 Comparison of Antarctic Fish MAPs and Mammalian STOPs

In addition to their cold-labile microtubules, many mammalian cells contain a subpopulation of microtubules that resist depolymerization at temperatures near

0 °C (Jones et al. 1980). One possible explanation for the cold stability of such microtubules invokes the presence of proteins, termed STOPs, that cap or block the ends of the polymers (Job et al. 1982; Margolis et al. 1986). When exposed at the ends of microtubules, these protein caps are thought to prevent the endwise depolymerization of the polymer that is normally produced by low temperature, by antimitotic drugs (e.g., podophyllotoxin), or by other agents. By contrast, microtubule polymerization supported by the antarctic fish MAPs is completely reversible. Thus, the MAPs of antarctic fish do not function as capping proteins that inhibit subunit exchange at microtubule ends.

5 Concluding Remarks

Our results argue strongly that the tubulins of antarctic fish have acquired, through mutation and natural selection, modifications that maintain efficient polymerization at low temperatures. Many, if not most, of these adaptations are likely to reside in their structurally divergent α chains (Detrich and Overton 1986, 1988; Detrich et al. 1987). MAPs do not appear to contribute to the temperature adaptation of the microtubules of antarctic fish. Thus, our current efforts are directed toward determination of the primary structures and posttranslational modifications of the α and β tubulins of *N. coriiceps neglecta*.

Acknowledgments. This paper is based on work supported by National Science Foundation Grants DPP-8317724 and DPP-8614788. I am deeply indebted to Drs. K.A. Johnson, S.P. Marchese-Ragona, R.D. Sloboda, and R.C. Williams, Jr., and to Ms. B.W. Neighbors for their contributions to this work and for many valuable discussions and suggestions. I gratefully aknowledge the logistical support provided to the project by the staff of the Division of Polar Programs of the National Science Foundation, by the personnel of ITT Antarctic Services, Inc., and by the captains and crews of *R/V Polar Duke*.

References

Bhattacharyya B, Sackett DL, Wolff J (1985) Tubulin, hybrid dimers, and tubulin S: stepwise charge reduction and polymerization. J Biol Chem 260:10208–10216

Breitling F, Little M (1986) Carboxy-terminal regions on the surface of tubulin and microtubules: epitope locations of YOL1/34, DM1A and DM1B. J Mol Biol 189:367–370

Correia JJ, Williams RC Jr (1983) Mechanisms of assembly and disassembly of microtubules. Annu Rev Biophys Bioeng 12:211–235

Detrich HW III, Overton SA (1986) Heterogeneity and structure of brain tubulins from cold-adapted antarctic fishes: comparison to brain tubulins from a temperate fish and a mammal. J Biol Chem 261:10922–10930

Detrich HW III, Overton SA (1988) Antarctic fish tubulins: heterogeneity, structure, amino acid compositions, and charge. Comp Biochem Physiol 90B:593–600

Detrich HW III, Wilson L (1983) Purification, characterization, and assembly properties of tubulin from unfertilized eggs of the sea urchin *Strongylocentrotus purpuratus*. Biochemistry 22:2453–2462

Detrich HW III, Jordan MA, Wilson L, Williams RC Jr (1985) Mechanism of microtubule assembly: changes in polymer structure and organization during assembly of sea urchin egg tubulin. J Biol Chem 260:9479–9490

Detrich HW III, Prasad V, Ludueña RF (1987) Cold-stable microtubules from antarctic fishes contain unique α tubulins. J Biol Chem 262:8360–8366

Detrich HW III, Johnson KA, Marchese-Ragona SP (1989) Polymerization of antarctic fish tubulins at low temperatures: energetic aspects. Biochemistry 28:10085–10093
Detrich HW III, Neighbors BW, Sloboda RD, Williams RC Jr (1990) Microtubule-associated proteins from antarctic fishes. Cell Motil Cytoskel 17:174–186
DeWitt HH (1971) Coastal and deep-water benthic fishes of the Antarctic. In: Bushnell VC (ed) Antarctic map folio series, Folio 15. Am Geogr Soc, New York, pp 1–10
Gaskin F, Cantor CR, Shelanski ML (1974) Turbidimetric studies of the in vitro assembly and disassembly of porcine neurotubules. J Mol Biol 89:737–758
Herzog W, Weber K (1977) In vitro assembly of pure tubulin into microtubules in the absence of microtubule-associated proteins and glycerol. Proc Natl Acad Sci USA 74:1860–1864
Himes RH, Burton PR, Gaito JM (1977) Dimethyl sulfoxide-induced self-assembly of tubulin lacking associated proteins. J Biol Chem 252:6222–6228
Job D, Rauch CT, Fischer EH, Margolis RL (1982) Recycling of cold-stable microtubules: evidence that cold stability is due to substoichiometric polymer blocks. Biochemistry 21:509–515
Johnson KA, Borisy GG (1975) The equilibrium assembly of microtubules in vitro. In: Inoue' S, Stephens RE (eds) Molecules and cell movement. Raven, New York, pp 119-139
Jones DH, Gray EG, Barron J (1980) Cold stable microtubules in brain studied in fractions and slices. J Neurocytol 9:493–504
Kim H, Binder LI, Rosenbaum JL (1979) The periodic association of MAP_2 with brain microtubules in vitro. J Cell Biol 80:266–276
Kirchner K, Mandelkow E-M (1985) Tubulin domains responsible for assembly of dimers and protofilaments. EMBO J 4:2397–2402
Lee JC, Timasheff SN (1977) In vitro reconstitution of calf brain microtubules: effects of solution variables. Biochemistry 16:1754–1764
Lowry JK (1975) Soft bottom macrobenthic community of Arthur Harbor, Antarctica. In: Pawson DL (ed) Biology of the antarctic seas, vol 5. Antarctic Res Ser 23. Am Geophys Union, Washington, pp 1–19
Mandelkow E-M, Herrmann M, Ruhl U (1985) Tubulin domains probed by limited proteolysis and subunit-specific antibodies. J Mol Biol 185:311–327
Margolis RL, Rauch CT, Job D (1986) Purification and assay of a 145-kDa protein ($STOP_{145}$) with microtubule-stabilizing and motility behavior. Proc Natl Acad Sci USA 83:639–643
Matus A (1988) Microtubule-associated proteins: their potential role in determining neuronal morphology. Ann Rev Neurosci 11:29–44
Murphy DB, Johnson KA, Borisy GG (1977) Role of tubulin-associated proteins in microtubule nucleation and elongation. J Mol Biol 117:33–52
Olmsted JB (1986) Microtubule-associated proteins. Ann Rev Cell Biol 2:421–457
Olmsted JB, Borisy GG (1975) Ionic and nucleotide requirements for microtubule polymerization in vitro. Biochemistry 14:2996–3005
Robinson J, Engelborghs Y (1982) Tubulin polymerization in dimethyl sulfoxide. J Biol Chem 257:5367–5371
Ross PD, Subramanian S (1981) Thermodynamics of protein association reactions: forces contributing to stability. Biochemistry 20:3096–3102
Sackett DL, Wolff J (1986) Proteolysis of tubulin and the substructure of the tubulin dimer. J Biol Chem 261:9070–9076
Sackett DL, Bhattacharyya B, Wolff J (1985) Tubulin subunit carboxyl termini determine polymerization efficiency. J Biol Chem 260:43–45
Serrano L, Avila J (1985) The interaction between subunits in the tubulin dimer. Biochem J 230:551–556
Serrano L, de la Torre J, Maccioni RB, Avila J (1984) Involvement of the carboxy-terminal domain of tubulin in the regulation of its assembly. Proc Natl Acad Sci USA 81:5989–5993
Sloboda RD, Dentler WL, Rosenbaum JL (1976) Microtubule-associated proteins and the stimulation of tubulin assembly in vitro. Biochemistry 15:4497–4505
Strömberg E, Serrano L, Avila J, Wallin W (1989) Unusual properties of a cold-labile fraction of Atlantic cod (*Gadus morhua*) brain microtubules. Biochem Cell Biol 67:791–800
Suprenant KA, Rebhun LI (1983) Assembly of unfertilized sea urchin egg tubulin at physiological temperatures. J Biol Chem 258:4518–4525
Suprenant KA, Rebhun LI (1984) Purification and characterization of oocyte cytoplasmic tubulin and meiotic spindle tubulin of the surf clam, *Spisula solidissima*. J Cell Biol 98:253–266

Swezey RR, Somero GN (1982) Polymerization thermodynamics and structural stabilities of skeletal muscle actins from vertebrates adapted to different temperatures and hydrostatic pressures. Biochemistry 21:4496–4503
Vallee RB (1982) A taxol-dependent procedure for the isolation of microtubules and microtubule-associated proteins (MAPs). J Cell Biol 92:435–442
Williams RC Jr, Correia JJ, DeVries AL (1985) Formation of microtubules at low temperatures by tubulin from antarctic fish. Biochemistry 24:2790–2798

The Biochemistry of Oxygen Transport in Red-Blooded Antarctic Fish

G. di Prisco, R. D'Avino, C. Caruso, M. Tamburini, L. Camardella, B. Rutigliano, V. Carratore, and M. Romano

1 Introduction

The temperature of the coastal antarctic waters, in which fish from temperate waters would be unable to survive, is close to the constant value of −1.87 °C, the equilibrium temperature of the ice-seawater mixture. During the increasing geographic and climatic isolation south of the Antarctic Convergence, initiated approximately 65 Ma, the physiology of antarctic fish became gradually adjusted to tolerate the progressive cooling of the environment. Because of the development of cold adaptation, the Antarctic Ocean is now the ideal habitat for the fish fauna, by virtue of the evolutionary response at different levels of life organization (organ, cell, molecule) to the many constraints of this habitat. The Convergence became a natural barrier to migration in both directions, thus representing a key factor for fish isolation and evolution.

The progressively evolving environmental conditions dictated a very severe selection. So far, 203 antarctic teleost species have been identified (Andriashev 1987); 109 of these belong to the suborder Notothenioidei, 97% endemic, which probably appeared in the Lower Tertiary. (The first teleosts appeared in the Triassic, over 200 Ma; their evolution continued through the Jurassic and Cretaceous. The first fossil records of most percoid families can be traced to the Eocene.) Notothenioidei comprise six families (Table 1). The high endemism and the modest diversification in families and species are most likely due to the long isolation and increasing harshness of the climatic conditions during the cooling process south of the Convergence.

In the process of cold adaptation, the evolutionary trends of antarctic fish include unique specializations. One is the modification of the hematological characteristics, which clearly differentiate fish below the Convergence from fish of temperate or tropical climates. The blood of antarctic fish contains fewer erythrocytes and less hemoglobin (Everson and Ralph 1968; Hureau et al. 1977; Wells et al. 1980). At the extreme of such evolution, the colourless blood of the 16 species of the family Channichthyidae totally lacks hemoglobin (Ruud 1954). Since the subzero seawater temperature would greatly increase the viscosity of a corpusculate fluid such as blood, the decrease in the number of erythrocytes and in the amount of hemoglobin, counteracting the temperature-induced viscosity increase, greatly

Institute of Protein Biochemistry and Enzymology, C.N.R., Via Marconi 10, 80125 Naples, Italy

Table 1. The families of the suborder Notothenioidei

Family	No. of species
Bovichthyidae	1
Nototheniidae	52
Bathydraconidae	16
Harpagiferidae	5
Artedidraconidae	19
Channichthyidae	16

facilitates the cardiac work and brings the energy demand to levels which the organism is able to tolerate. Hence, this modification is related to cold adaptation.

The search for correlation between fish hematology and the exceptionally extreme conditions of the antarctic environment raises a number of intriguing questions and justifies a study on the biochemistry of oxygen transport, with special attention on the relationship between the molecular structure and the oxygen-binding properties of hemoglobin. The long-range aim of this study is to gain some insight into the molecular basis of cold adaptation, as well as into the history of fish evolution during the isolation that followed the separation and drift of Antarctica from Gondwana.

In reviewing these investigations, which is the main purpose of this chapter, it has also become clear that some of the questions which have originated in the course of this study have a general bearing, not merely related to the antarctic habitat.

2 Experimental

Twenty-six species of six families (Nototheniidae, Bathydraconidae, Harpagiferidae, Artedidraconidae, Rajidae and Zoarcidae; the first four belong to the suborder Notothenioidei) have been analyzed in the studies of our group so far. Specimens were captured in two geographically distinct areas: (1) the Antarctic Peninsula (Dallman Bay and Low Island), by bottom trawling; and (2) the Ross Sea, by the use of gill nets, set at least 100 m deep on the sea-floor. Fish were kept in aquaria supplied with running, aerated seawater, at Palmer Station (64° 46' S, 64° 03' W) and at Terra Nova Bay Station (74° 42' S, 164° 07' E), respectively. During visits to McMurdo Station, samples of the blood of four species was kindly supplied by A.L. DeVries.

Blood was drawn from the caudal vein of unanaesthetized fish by means of heparinized syringes. The separation and lysis of washed erythrocytes was followed by electrophoretic analysis on cellulose acetate of each hemolysate, by ion-exchange chromatography purification and (in ten cases) by subsequent crystallization of the hemoglobins (di Prisco 1986a, 1988; D'Avino and di Prisco 1988, 1989; di Prisco and D'Avino 1989; di Prisco et al. 1990). The hemoglobins of 19 and 15 species, respectively, were characterized structurally and functionally.

3 Hemoglobin and Molecular Adaptation of the Blood

Several perspectives and indications arising from the amount of evidence gathered on various aspects of oxygen transport are related to the adaptation of antarctic fish to environmental conditions and to evolution strategies. On the other hand, some have a more general bearing, which is not strictly limited to considerations based on geographic isolation. Such an additional source of information is also very valuable, considering the yet limited degree of knowledge available on the molecular basis of oxygen transport in fish of all latitudes.

Both points of view will be taken into account in the considerations which follow, on the oxygen-transport system in red-blooded antarctic fish.

3.1 Adaptations in Response to Environmental Changes

3.1.1 Blood Viscosity

The potentially negative physiological effects due to the exponential increase in blood viscosity, when temperatures drop below 0 °C, have been counterbalanced, as previous studies have pointed out (Hemmingsen and Douglas 1970, 1977; Macdonald et al. 1987; Wells et al. 1990), by the reduction, or elimination, of erythrocytes and hemoglobin (Ruud 1954; Everson and Ralph 1968; Hureau et al. 1977; Wells et al. 1980). Thus, an evolutionary pathway generating fish with colourless blood has eventually diverged from that of fish having "less red" blood, posing interesting and yet unanswered questions about gene regulation and expression.

3.1.2 Hemoglobin Multiplicity

Another feature, which follows a common trend in endemic antarctic families and seems to be related to the habitat, is the hemoglobin multiplicity. Characteristically, the blood of most fish contains multiple hemoglobins (Riggs 1970), that in active and fast swimmers, such as trout (Binotti et al. 1971), often show functional differences in oxygen binding, thus enabling oxygen transport to tissues even under conditions of extreme acidosis. In some instances, however, all components are functionally indistinguishable from one another and have pH-dependent oxygen affinities; carp, a bottom feeder, is one example (Tan et al. 1972; Gillen and Riggs 1972). In contrast, the electrophoretic examination of the hemolysates of 22 species (mostly sedentary bottom feeders) of the largely endemic suborder Notothenioidei (Table 2) reveals a single major hemoglobin (Hb 1), often accompanied by a second, minor component (Hb 2, about 5% of total), which is invariably functionally indistinguishable from Hb 1 (di Prisco and D'Avino 1989; di Prisco et al. 1990). Multiplicity and functional differences indicate the need to adapt to environmental changes; in view of the relatively constant physicochemical features of the antarctic marine environment, it is not surprising that antarctic fish do not have more than one hemoglobin. Within this suborder, the nototheniid *Trematomus newnesi* is so far the only exception. It has two, functionally distinct, major hemo-

Table 2. Hemoglobin components in the blood of antarctic fish

Family	Species[a]	Hemoglobin components
Nototheniidae	** *Notothenia coriiceps negl.*	Hb 1 (95%), Hb 2 (5%)
	** *N. rossii*	Hb 1 (95%), Hb 2 (5%)
	** *N. gibberifrons*	Hb 1 (90%), Hb 2 (10%)
	Nototheniops nudifrons	Hb 1 (95%), Hb 2 (5%)
	N. larseni	Hb 1 (95%), Hb 2 (5%)
	* *Pagothenia hansoni*	Hb 1 (95%), Hb 2 (5%)
	** *P. bernacchii*	Hb 1 (97%) (Hb 2 ?)
	* *P. borchgrevinki*	Hb 1, Hb 2 (Hb 3 ?)
	** *Trematomus newnesi*	Hb C (25%), Hb 1 (70%) Hb 2 (5%)
	* *T. nicolai*	Hb 1 (95%), Hb 2 (5%)
	** *T. centronotus*	Hb 1 (95%), Hb 2 (5%)
	** *T. loennbergi*	Hb 1 (95%), Hb 2 (5%)
	* *T. eulepidotus*	Hb 1 (95%), Hb 2 (5%)
	* *Dissostichus mawsoni*	Hb 1 (97%), (Hb 2 ?)
	* *Aethotaxis mitopteryx*	One Hb (99%)
Bathydraconidae	* *Parachaenichthys charcoti*	One Hb (99%)
	** *Gymnodraco acuticeps*	One Hb (99%)
	* *Cygnodraco mawsoni*	Hb 1 (97%), Hb 2 (3%)
	* *Bathydraco marri*	One Hb (99%)
Harpagiferidae	*Harpagifer antarcticus*	One Hb (99%)
Artedidraconidae	*Artedidraco skottsbergi*	One Hb (99%)
	Histiodraco velifer	One Hb (99%)
Rajidae	*Raja georgiana*	One Hb (99%)
Zoarcidae	*Lycenchelys nigripalatum*	Hb 1, Hb 2, Hb 3, Hb 4
	** *Lycodichthys dearborni*	Hb 1, Hb 2, Hb 3, Hb 4
	** *Austrolycicthys brachyceph.*	Hb 1, Hb 2, Hb 3, Hb 4 (Hb 5 ?)

[a] * Obtained in purified form;
** purified Hb components, obtained in crystalline form.

globins (Hb 1 and Hb C, 65–70% and 20–25% of total, respectively) and one minor (Hb 2) component; traces of Hb C, less than 1% of total, are present in all the species. This fish, however, is not a bottom feeder; it is cryopelagic, very active and feeds underneath the pack and fast ice. In comparison with other Notothenioidei, the behaviour and life style of *T. newnesi* are much more similar to those of trout, which also has hemoglobins with different functional properties.

The general trend of Notothenioidei is not followed by Zoarcoidei, found at all latitudes. Three species of the family Zoarcidae of this nonendemic suborder, similar to fish of temperate waters, show a higher multiplicity, with four to five functionally distinct, major components (di Prisco and D'Avino 1989; di Prisco et al. 1990), suggesting separate evolutionary pathways following suborder divergence. In support of this hypothesis, Notothenioidei have identical "antifreeze" glycoproteins (AFGPs) for freezing resistance, which in the antarctic Zoarcidae is instead provided by peptides with a totally different molecular structure (DeVries 1980). Thus, it is likely that family diversification in Notothenioidei took place after the cooling event which led to appearance of AFGPs.

3.1.3 Reduced Physiological Role of Hemoglobin in Routine Oxygen Transport

Both the concentration reduction and the limited multiplicity of hemoglobin strongly suggest that the role of this oxygen carrier tends to become less critical in a stable, cold environment. It is beyond doubt that Channichthyidae are not disadvantaged by their lack of hemoglobin.

Anthony (1961) had shown that carbon monoxide poisoning of the blood of goldfish has no effect on survival at low temperatures. In *Pagothenia bernacchii*, the induced reduction of the hematocrit to less than 1–2% had no apparent deleterious effect (Wells et al. 1990). Moreover, in vivo complete conversion of hemoglobin to the carbon monoxide derivative in the blood of *P. bernacchii* has no discernible effect on the vital functions (di Prisco et al., in preparation). The results are strongly suggestive of the existence, in red-blooded antarctic fish, of an alternative (to be used for emergencies), or parallel, oxygen-transport system: like Channichthyidae, red-blooded antarctic fish can carry routinely needed oxygen dissolved in plasma. This conclusion is extremely interesting (and further work along these lines is essential) since possible similarities in the physiology of oxygen transport may represent an important link between the red-blooded and the hemoglobinless species.

The hemoglobinless, scaleless Channichthyidae have a high proportion of cutaneous oxygen uptake (Hemmingsen and Douglas 1970). The zoarcid *Lycodichthys dearborni*, also scaleless, under resting conditions takes over 35% of the oxygen through the skin (Wells 1986); lower, but significant proportions of cutaneous respiration have been demonstrated in other red-blooded species, despite the presence of scales. Thus, cutaneous oxygen uptake is probably an important respiratory strategy also for red-blooded antarctic fish; interestingly, south of the Convergence, scaleless species are very common.

3.1.4 Oxygen-Binding Properties

Studies on the oxygen transport in the blood of several antarctic species have shown that the oxygen affinity (a property which controls both the binding of oxygen at the exchange surface and its release to the tissues) of hemoglobin is quite low (Grigg 1967; Raymond and DeVries 1976; Qvist et al. 1977; Wells and Jokumsen 1982; Tetens et al. 1984; di Prisco 1986b; di Prisco et al. 1986, 1988), as indicated by the values of $P_{1/2}$ (the partial pressure of oxygen required to achieve half saturation of the hemoglobin). This characteristic is presumably related to the high concentrations of dissolved oxygen in the cold antarctic waters. The hemoglobin of the more active species appears to have the lowest oxygen affinity.

The binding and release of oxygen to tissues may be regulated by several factors, which include pH, physiological effectors and temperature. If the affinity of hemoglobin for oxygen decreases in the lower physiological range of pH, we observe the alkaline Bohr effect; if the extent of the decrease is such that hemoglobin can no longer become saturated even when exposed to pressures of several atmospheres of pure oxygen, and the cooperativity of oxygen binding, linked with the allosteric transition between the T and R conformations (Monod et al. 1965) is lost, we are dealing with an "exaggerated" Bohr effect, often observed in fish hemoglobins: the Root effect (Root 1931; Brittain 1987). The presence of the latter implies the presence of the Bohr effect, but not vice versa. The presence or absence of these effects dictate how easily hemoglobin may bind sufficient oxygen at the gills or adequately release it to the tissues.

We have studied this regulation in the hemoglobins of several species, by analyzing intact erythrocytes, hemolysates (previously "stripped" in order to remove endogenous organic phosphates), and purified hemoglobin fractions (D'Avino and di Prisco 1989; di Prisco 1986b, 1988; di Prisco and D'Avino 1989; di Prisco et al. 1986, 1988, 1990, in prep.; Caruso et al., in prep.; Tamburrini et al., in prep., by means of tonometry and spectrophotometric techniques (Giardina and Amiconi 1981). Bohr and Root effects are displayed by the hemoglobins of all Notothenioidei (13 species) under examination (Table 3),with the exception of two of the three components of the nototheniid *T. newnesi* (the only notothenioid species found to have two major hemoglobins) and of the single hemoglobin of the bathydraconid *Gymnodraco acuticeps* (in agreement with Wells and Jokumsen 1982). To our knowledge, the latter species is the only example of a fish having a single hemoglobin with pH-independent oxygen binding. This peculiarity may be related to the postulated reduced role of hemoglobin in such an extreme environment; if so, it probably exists in other antarctic species.

Whenever the alkaline Bohr effect is displayed, subunit cooperativity also disappears at pH lower than 7.0, concomitantly with the decrease in oxygen affinity.

Root-effect hemoglobins have been associated with the presence of at least one of two anatomical structures which require the release of considerable amounts of oxygen: (1) the swim bladder and (2) the choroid rete (Riggs 1970; Wittenberg and Wittenberg 1974). The two species mentioned above lack a swim bladder (all antarctic teleosts do), but are also among those (which, interestingly enough, include the hemoglobinless Channichthyidae) also lacking the choroid rete (Eastman 1988), in line with having hemoglobins without Root effect.

When present in purified hemoglobins and "stripped" hemolysates, the decrease in oxygen affinity, typical of the Root effect, becomes dramatic at pH lower than 6.5–7.0 in the absence, and 7.5 in the presence, of endogenous effectors (ATP). In other experiments, the higher pH (7.3–8.3 and 7.9–8.3, in the absence and presence, respectively, of ATP) may be the reason why a Root effect was not detected in the "stripped" hemolysates of three notothenioid species (Wells and Jokumsen 1982).

Among the purified four to five components found in two species of Zoarcidae (suborder Zoarcoidei), only the hemoglobin with higher anodal electrophoretic mobility shows the Root effect, whereas the oxygen affinity of the less acidic

Table 3. Regulation by pH and physiological effectors of the oxygen binding of hemoglobin components in antarctic fish

Family	Species	Bohr and Root effects; effect of organophosphates
Nototheniidae	*N. coriiceps neglecta*	Present (Hb 1, Hb 2)
	N. rossii	Present (Hb 1, Hb 2)
	N. gibberifrons	Present (Hb 1, Hb 2)
	P. hansoni	Present (Hb 1, Hb 2)
	P. bernacchii	Present (Hb 1)
	T. newnesi	Present (Hb C)
		Absent (Hb 1, Hb 2)
	T. centronotus	Present (Hb 1, Hb 2)
	T. eulepidotus	Present (Hb 1)
	D. mawsoni	Present (Hb 1)
Bathydraconidae	*P. charcoti*	Present
	G. acuticeps	Absent
	C. mawsoni	Present (Hb 1, Hb 2)
Artedidraconidae		
	H. velifer	Present
Zoarcidae		
	L. dearborni	Absent[a] (Hb 1, Hb 2, Hb 3)
		Present (Hb 4)
	A. brachycephalus	Absent[a] (Hb 1, Hb 2, Hb 3)
		Present (Hb 4)

[a]The Bohr effect was not measured.

components is almost independent on protons and organic phosphates (di Prisco and D'Avino 1989; di Prisco et al. 1990).

Investigations on the effect of temperature on oxygen binding and release are presently being pursued. In fact, the thermodynamic parameters of these processes are likely to provide useful indications on the modulation of the hemoglobin function and its physiological meaning in extreme environments. At pH 7, the enthalpy change ΔH for oxygenation in *Notothenia coriiceps neglecta* (di Prisco et al. 1988; D'Avino and di Prisco 1989), *G. acuticeps* and *T. newnesi* (Tamburrini et al., in prep.; di Prisco et al., in prep.) appears unusually low. Relatively low apparent ΔH values for oxygenation have also been reported in *Dissostichus mawsoni*, a midwater predator (Qvist et al. 1977; Wells and Jokumsen 1982), and *Pagothenia borchgrevinki*, a cryopelagic species (Tetens et al. 1984).

In *P. bernacchii* (Tamburrini et al., in prep.), preliminary experiments indicate that changes in temperature have a negligible effect on the oxygen affinity of the main hemoglobin; surprisingly, the overall ΔH value of oxygen binding is positive (+2 kcal/mole of oxygen, after subtracting the heat contribution of oxygen solubilization). In other words, during the oxygenation-deoxygenation cycle, hemoglobin does not require any significant amount of energy. This energy-saving mechanism (never found before in poikilothermic fish) may well be an example of molecular adaptation to extreme life conditions under strong evolutionary pressure. Rossi-Fanelli and Antonini (1960) have shown that the temperature dependence of oxygenation of tuna hemoglobin in the range of pH 6.5-8.7 is close to zero; they

Table 4. Complete amino acid sequences of antarctic fish hemoglobins (total number of sequenced α and β chains, 17; total number of sequenced hemoglobins, 10)

Family	Species[a]	Sequenced hemoglobins
Nototheniidae	*N. coriiceps neglecta**	Hb 1, Hb 2
	P. bernacchii	Hb 1
	*T. newnesi***	Hb C, Hb 1, Hb 2
	A. mitopteryx	The single Hb
Bathydraconidae	*G. acuticeps*	The single Hb
	*C. mawsoni****	Hb 1, Hb 2
	Bathydraco marri	The β chain of the single Hb

[a] * Hb 1 and Hb 2 have the β chain in common;
** Hb C and Hb 1 have the α chain in common, Hb 1 and Hb 2 have the ß chain in common;
*** Hb 1 and Hb 2 have the α chain in common.

speculate that this feature may reflect a molecular adaptation allowing the fish to swim in waters of widely different temperatures without changing the blood oxygen transport properties. It is worth mentioning that tuna is not entirely poikilothermic and maintains a partial regulation of body temperature, which may be as much as 14 °C above that of the water (Carey and Teal 1966).

3.1.5 Molecular Structure

The structural characterization of antarctic fish hemoglobin includes the determination of the molecular weight of the native proteins and of the globin chains (separated by HPLC or FPLC), the globin electrophoretic mobilities, amino acid compositions and amino acids sequences.

The complete sequence of the 16 chains of 10 hemoglobins of 6 species and of the β chain of the hemoglobin of a seventh species (Table 4) has so far been elucidated (D'Avino et al. 1989a, 1989b, 1990; di Prisco et al. 1990; Caruso et al., in prep.; Tamburrini et al., in prep.; di Prisco et al., in prep.); others are under investigation. When present, the minor component has one of the two globin chains in common with the major one (D'Avino and di Prisco 1988, 1989; di Prisco et al., in prep.; Caruso et al., in prep.). Fig.1 and Fig.2 report the alignment of sequences of the α and ß chains of antarctic fish hemoglobins with the known sequences of hemoglobins of non-antarctic species and of human hemoglobin. The degree of sequence identity among antarctic species and among the latter ones and species from temperate waters has been calculated from the alignment (Tables 5 and 6).

The elucidation of the amino acid sequence of many hemoglobins provides confirmation of the remarkable conservation in the structure of the hemoglobin molecule during evolution. As expected, the molecular domains of special importance in the quaternary structure (assembly and spatial arrangement of the two pairs of chains), including the amino acid residues known to be invariant in all vertebrates (Dickerson and Geis 1983), are essentially unchanged. Most of the replacements in the sequences appear to be localized in domains where the structural requirements for the biological function are less stringent.

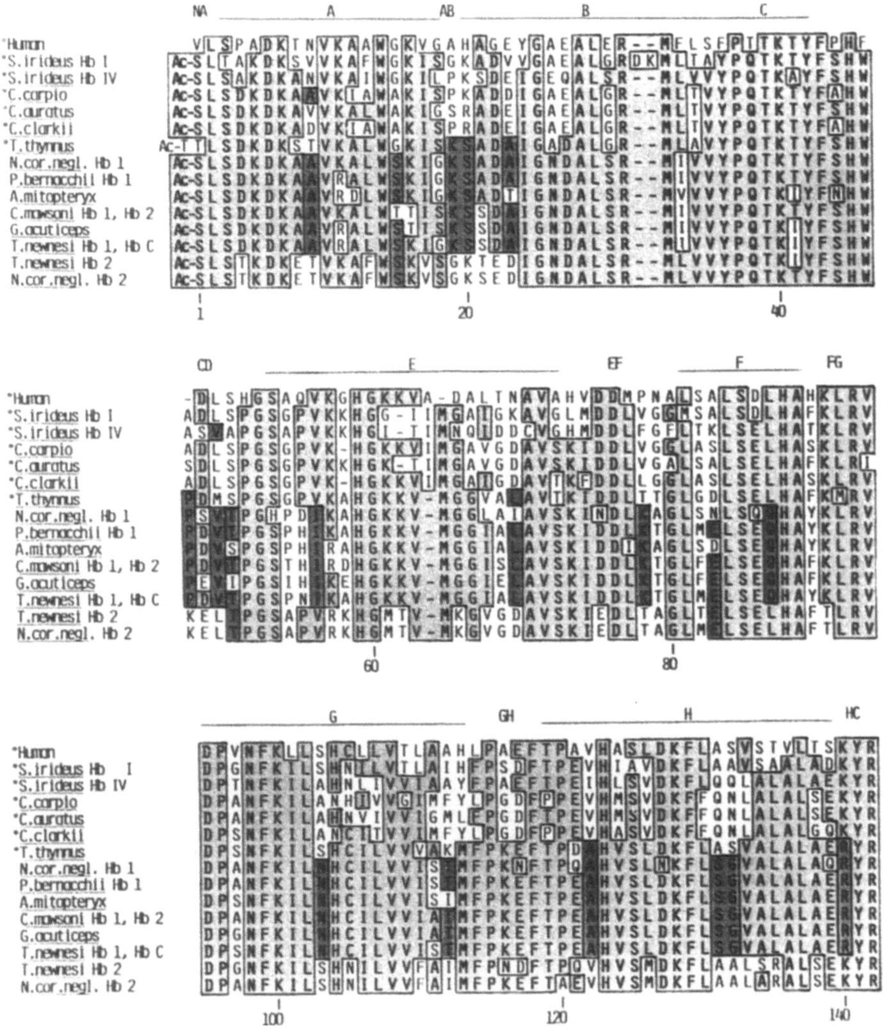

Fig. 1. Alignment of the amino acid sequence of the α chains of antarctic fish (*N. coriiceps neglecta, T. newnesi, P. bernacchii, A. mitopteryx, C. mawsoni, G. acuticeps*) hemoglobins with α chain sequences (denoted by *asterisk*) of nonantarctic fish (*Cyprinus carpio, Carassius auratus, Catostomus clarkii, Salmo irideus, Thunnus thynnus*) and human hemoglobins. The one-letter amino acid notation is used. Residue identities in at least eight sequences are *lightly shaded*. In the *nonshaded* areas, identities in at least five of the eight antarctic sequences are also shown (*darkly shaded*). The helical (*A,B,C,E,F,G,H*) and nonhelical (*NA, AB, CD, EF, FG, GH, HC*) regions, as established for mammalian hemoglobins (Perutz 1969), are indicated above the human sequence. In α chains, the helical region D is lacking

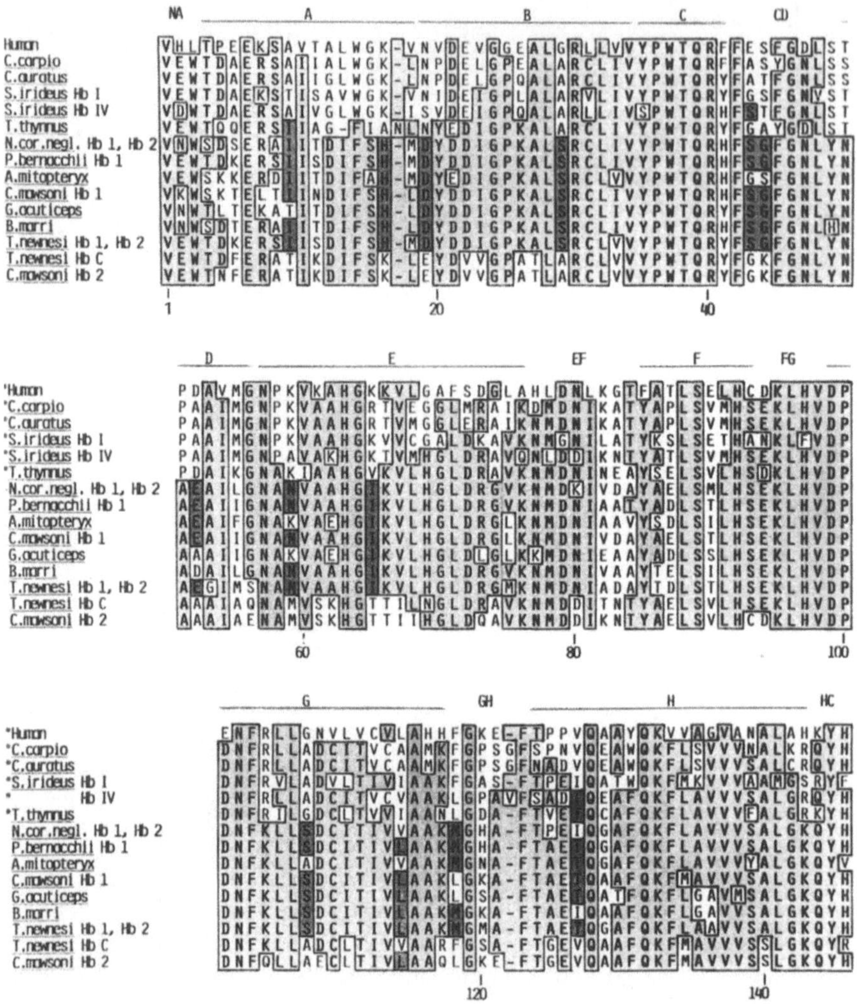

Fig. 2. Alignment of the amino acid sequence of the β chains of antarctic fish (*N. coriiceps neglecta, T. newnesi, P. bernacchii, A. mitopteryx, C. mawsoni, G. acuticeps, B. marri*) hemoglobins with β-chain sequences (denoted by *asterisk*) of nonantarctic fish (*Cyprinus carpio, Carassius auratus, Salmo irideus, Thunnus thynnus*) and human hemoglobins. Residue identities in at least eight sequences are *lightly shaded.* In the *nonshaded* areas, identities in at least five of the nine antarctic sequences are also shown (*darkly shaded*). The helical (*A, B, C, D, E, F, G, H*) and nonhelical (*NA, CD, EF, FG, GH, HC*) regions, as established for mammalian hemoglobins (Perutz 1969), are indicated above the human sequence

Table 5. Sequence identity (%) in α chains of fish hemoglobins

Species[a]	*C. auratus**	*C. clarkii**	*T. thynnus**	*C. carpio**	*S. irid.** Hb IV	*S. irid.** Hb I	*T. newnesi* Hb 2	*N.cor. negl.* Hb 2	*G. acut.*	*C. mawsoni* Hb 1,Hb 2	*A. mitopt.*	*P. bern.* Hb 1	*T. newnesi* Hb 1,Hb C
N. cor.negl. Hb 1	61	58	73	59	57	55	61	63	82	83	83	89	87
T. newnesi Hb 1, Hb C	63	58	76	58	62	52	63	66	92	90	90	97	
P. bernacchii Hb 1	65	61	78	64	62	57	65	70	91	91	90		
A. mitopteryx	62	61	75	62	59	55	62	65	84	84			
C. mawsoni Hb 1, Hb 2	63	59	77	60	62	53	64	69	93				
G. acuticeps	61	57	75	58	62	53	65	67					
N. cor.negl. Hb 2	65	58	68	63	63	62	93						
T. newnesi Hb 2	64	59	67	61	58	62							
*S. irideus** Hb I	67	65	65	66	60								
*S. irideus** Hb IV	68	63	60	63									
*C. carpio**	80	91	66										
*T. thynnus**	67	66											
*C. clarkii**	77												

[a]*Nonantarctic species.

Table 6. Sequence identity (%) in ß chains of fish hemoglobins

Species[a]	*C. auratus**	*T. thynnus**	*T. carpio**	*S. irid** Hb IV	*S. irid.** Hb I	*C. mawsoni** Hb 2	*T. newnesi* Hb C	*B. marri*	*G. acut.*	*C. mawsoni* Hb 1	*A. mitopt.*	*P. bern.* Hb 1	*T. newnesi* Hb 1, Hb 2
N. cor.negl. Hb 1, Hb 2	58	66	57	63	53	65	70	90	80	88	82	90	86
T. newnesi Hb 1, Hb 2	60	66	57	62	53	64	68	85	80	84	83	93	
P. bernacchii Hb 1	64	68	61	66	58	66	70	87	83	87	86		
A. mitopteryx	60	67	58	60	54	62	67	81	77	80			
C. mawsoni Hb 1	58	67	56	62	53	67	70	87	85				
G. acuticeps	59	63	56	59	55	65	67	85					
B. marri	60	66	57	61	54	65	68						
T. newnesi Hb C	61	61	57	62	57	89							
C. mawsoni Hb 2	58	60	55	60	54								
*S. irideus** Hb I	64	51	64	59									
*S. irideus** Hb IV	78	61	73										
*C. carpio**	91	60											
*T. thynnus**	62												

[a]*Nonantarctic species.

The degree of sequence identity of the major components, among not only different species, but also different families high, appears unusually (about 90%), in contrast to the identity (about 60%) found among non-antarctic families (in fact, *Carassius auratus* and *Cyprinus carpio*, with 80 and 91%, belong to the same family). This evidence emphasizes the effect of the long exposure of the entire fish fauna to the same evolutionary pressure building up under the Convergence. Conversely, the degree of identity among antarctic teleosts and the other fish appears lower than that among the nonarctic species themselves, with the exception of the nonpoikilothermic *Thunnus thynnus*. The latter, however, belongs to the same order, Perciformes (or, according to some textbooks, to a closely related one, Thunniformes) and is therefore phylogenetically less remote from the notothenioids.

It is of interest to note that the minor Hb 2 components (when present) and Hb C are very similar to one another, but show a highly reduced degree of identity with the major hemoglobin (of course, these considerations refer only to the chain which is not in common with Hb 1; see Table 4). In other words, the sequences of the main hemoglobins of different species or families are more similar to one another than Hb 1 and Hb 2 of the same species. This peculiarity also occurs among non-antarctic fish, although to a lesser extent: Hb IV of trout has 73% identity in the β chain with carp hemoglobin and only 59% with Hb I of the trout itself. However, trout Hb IV, unlike Hb I, is functionally similar to carp hemoglobin, whereas in antarctic species, Hb 1 and Hb 2 are functionally indistinguishable.

It is conceivable that the minor components Hb 2 may simply be evolutionary remnants, without physiological significance. This conclusion is further supported by the analysis of the sequence identity data. In 37 and 25 positions of the α and β chain respectively, Hb 2 and Hb C components (the two bottom sequences of Figs. 1 and 2) are identical with each other, and differ from all other antarctic hemoglobins. In 34 of the 37 α positions, and in 20 of the 25 β positions, residue identity is found in at least five of the other antarctic six α and seven β sequences, respectively. Close examination of Figs. 1 and 2 also reveals that, in the majority of a class of positions (not included in the lightly shaded areas because of insufficient degree of identity, and indicated by darkly shaded areas), only residues of Hb 2 and Hb C are identical to corresponding non-antarctic ones. It is also worth to recall that large amounts of Hb C have been found in *T. newnesi* only, and are indicative of the physiological relevance of this component (found only in traces in all other notothenioids) in this one species, where it differs functionally from Hb 1 and Hb 2. From this point of view, Hb C cannot be assimilated to Hb 2.

The non-cold-adapted nototheniid *Paranotothenia angustata*, commonly found in both the New Zealand and Patagonian regions, is similar to the antarctic nototheniids with respect to hemoglobin multiplicity, amino acid composition (di Prisco and D'Avino 1989; di Prisco et al. 1990) and oxygen-binding properties. From the composition, the percentage difference in the amino acid sequence of the globin chains can be estimated, suggesting a high degree of identity with the antarctic notothenioids. *P. angustata* probably established itself in the southern waters of New Zealand when the Campbell Plateau was still relatively close to West Antarctica in the Lower Tertiary (Andersen 1984). It is worth mentioning that at the end of the Miocene (5 Ma) and during the Pliocene, the Convergence may have moved northward up to 39° S, the present latitude of northern New Zealand (Kennett 1968), facilitating the migration of some fish species.

4 The Structure and Function of Antarctic Fish Hemoglobin and its Meaning Beyond the Correlation With the Environment

The new findings and the accordingly expanded knowledge of the molecular structure and biological function of antarctic fish hemoglobin have produced a number of implications of more general significance, not strictly limited to the antarctic environment. Figures 1 and 2 and Table 7 provide useful material for the considerations which follow.

4.1 Root/Bohr Effects and the Primary Structure

X-ray crystallography and model building have suggested that several amino acid residues have a key role in the functional properties (Arnone 1972; Barra et al. 1981; Perutz and Brunori 1982). Root and Bohr effects appear to be linked to the presence of several polar residues in the β chain: Lys EF6, Ser F9, Glu FG1, Arg H21, His HC3. In the deoxy T-structure, Ser F9 donates a hydrogen bond to the unbound oxygen atom of C-terminal His HC3 and accepts a hydrogen bond from the -NH- bond of the same residue, thereby stabilizing the C-terminal salt bridges (respectively, imidazole and -COOH of His HC3 with Glu FG1 and, in the α chain, Lys C5) in this structure. In trout Hb I, which has neither effect, all five polar residues have been replaced by apolar ones, and Lys HC1 (which in mammals forms a salt bridge with His HC3) by Arg, still able to form the salt bridge with the C-terminus and stabilize the quaternary R-conformation. On the other hand, in Root-effect trout Hb IV, in HC1 Lys is substituted by Gln, which cannot form the latter salt bridge and provide such stabilization; the strong interaction between His HC3 and Glu FG1 in the transition from the R- to the T-state (Perutz and Brunori 1982) is energetically favoured (in trout Hb I, in FG1 Glu is replaced by Asn). As an overall result, in Root-effect hemoglobins the T-state is stabilized and the R-state destabilized.

Table 7. Functionally important amino acid residues (Perutz and Brunori 1982) in the β chains of some antarctic and nonantarctic fish hemoglobins

Position	*S.irid.* Hb I	*S.irid.* Hb IV	*N. cor.negl.* Hb 1, Hb 2	*C. mawsoni* Hb 1	*C. mawsoni* Hb 2	*G. acut*	*T. newnesi* Hb 1, Hb 2	*T. newnesi* Hb C	*P. bern.* Hb 1
NA2	Glu	Asp	Asn	Lys	Glu	Asn	Glu	Glu	Glu
EF6	Leu	Lys	Val	Val	Lys	Glu	Ala	Thr	Ala
F9	Ala	Ser	Ser	Ser	Cys	Ser	Ser	Ser	Ser
FG1	Asn	Glu	Glu	Glu	Asp	Glu	Glu	Glu	Glu
H21	Ser	Arg	Lys	Lys	Lys	Lys	Lys	Lys	Lys
HC1	Arg	Gln	Gln	Gln	Gln	Gln	Gln	Gln	Gln
HC3	Phe	His	His	His	His	His	His	Arg	His
Bohr effect	–	+	+	+	+	–	–	+	+
Root effect	–	+	+	+	+	–	–	+	+
Effector	–	ATP	ATP	ATP	ATP	–	–	ATP	ATP

The first of these five residues (Lys EF6) is replaced in all antarctic β chains (except in *Cygnodraco mawsoni* Hb 2, the others are present, or conservatively replaced (again, Ser F9 and Glu FG1 are replaced in *C. mawsoni* Hb 2 only), also in hemoglobins which lack the Root and Bohr effects (Table 7). In summary, the antarctic sequences elucidated so far do not follow the predictions of the stereochemical model (see trout Hb IV) in the following cases: *(1)* Lys EF6 is replaced by an apolar or negatively charged residue in the sequences of all hemoglobins (except the minor component Hb 2 of *C. mawsoni*), regardless of the presence of Bohr and Root effects; *(2)* in F9, Ser is surprisingly replaced by Cys in *C. mawsoni* Bohr- and Root-effect Hb 2; on the other hand, Ser F9 is found in *G. acuticeps* hemoglobin and in *T. newnesi* Hb 1 and Hb 2, all lacking the effects: *(3)* in H21, Arg is conservatively replaced by Lys, also in the latter three hemoglobins; *(4)* in HC3, His is replaced by Arg in *T. newnesi* Hb C; *(5)* Gln HCl, unable to stabilize the R-conformation, is found in all sequences, including those of the hemoglobins which lack the effects.

4.2 Organophosphate Compounds

ATP and, to a lesser extent, GTP are among the allosteric effectors used by teleosts to regulate oxygen binding. The organophosphate binding site in the β chain (Arnone 1972) has been suggested to include Lys EF6, Asp NA2 and Arg H21 in trout Root-effect Hb IV (Perutz and Brunori 1982). In antarctic fish, the replacement of Lys EF6 has already been discussed, as well as that (conservative) of Arg H21; in NA2, the negatively charged residue is found in four hemoglobins (one of which lacks ATP regulation), but is replaced by Asn in two hemoglobins (one showing and the other lacking ATP regulation) and by Lys in an ATP-regulated hemoglobin (Table 7).

4.3 X-ray Diffraction Analysis

In proteins, the primary structure is responsible for the three-dimensional configuration of the molecule. If suitable crystals are available, in addition to the sequence, the structural configuration can be described and represented graphically through X-ray diffraction analysis of the crystal. Collaborative work with the groups of M.F. Perutz and L. Mazzarella on X-ray diffraction analysis of the crystals of the major components of the nototheniids *P. bernacchii* and *T. newnesi* (the former with pH-regulated oxygen binding, the latter without) is attempting to correlate the amino acid sequence with the three-dimensional structure in two functionally distinct hemoglobins.

5 Concluding Remarks

Some of the apparent contradictions arising during the development of this study may find an explanation in postulating the possible presence, in the erythrocytes of antarctic fish with hemoglobins found to lack the Root and Bohr effects in vitro, of some (unidentified) endogenous factor, playing the role of organophosphates and able to elicit these effects in vivo. This hypothesis is in fact supported by oxygen-saturation curves of *T. newnesi* erythrocytes, which suggest that the Root effect (no longer found after purification in two of the three hemoglobins, accounting for almost 80% of total) is indeed fully operative in the intact cells. This possibility implies original and interesting mechanisms of biological control and further work is of course needed.

The advances in sequencing hemoglobins of fish of the Antarctic Ocean suggest, however, that caution must be exerted in linking the various aspects of the hemoglobin functional behaviour to a single or a few amino acid residues. Replacements in other domains may induce changes in the molecular configuration and these may act in combination with, or substitute for, the role that specific residues most certainly have in many cases. Site-directed mutagenesis and model building are likely to become the tools of choice to shed light on this structure-function interplay. It is pertinent here to note that the present knowledge of antarctic fish hemoglobin sequences is not yet sufficiently parallelled by knowledge of sequences in nonantarctic species; it is needless to stress how urgent it is to promptly fill this gap, also in view of the above-mentioned considerations.

Estimates of blood oxygen capacity have indicated that the oxygen-carrying capacity of the red-blooded species is nine- to ten-fold higher than that of Channichthyidae (Holeton 1970), which however can pump three to four times as much blood; it has also been suggested that the channichthyid *Chaenocephalus aceratus* uses 63% of the blood oxygen-carrying capacity when resting, as compared with 21–33% of some red-blooded species. It has been calculated that the resting oxygen consumption of red-blooded and hemoglobinless fish is similar. However, the presence of hemoglobin and of its Root and Bohr effects enhances the rate of diffusion of oxygen by raising the blood-to-tissue oxygen tension gradient at sites where the pH becomes lower; the hemoglobinless fish probably need a high blood-to-tissue PO_2 gradient as a compensating factor necessary to keep moving oxygen from the blood to the tissues (Holeton 1970).

The molecular machinery of the Root effect is fully operative in many of the species under examination. Although in antarctic fish the swim bladder is lacking, there is a good correlation between the presence of Root-effect hemoglobins and of the other oxygen-secreting structure, the choroid rete. However, in these fish the eye is a physiologically active structure only in relation to depth and seasonal light. Thus, the functional relevance of the Root effect in producing a maximal but localized oxygen delivery, at the level of specific organs or in particular physiological conditions, is still relatively unclear. For instance, the release of small amounts of oxygen into other hollow organs, such as the intestine, or fat tissue, could dramatically affect the buoyancy (di Prisco et al. 1988).

Of course, these considerations acquire special interest in view of the possibility of a reduced role of hemoglobin (which, however, may become crucial under special circumstances) in routine oxygen transport. A number of intriguing and stimulating questions is raised by the existence in the same habitat of a suborder (Notothenioidei), comprising fish species that have completely eliminated hemoglobin and erythrocytes, as well as red-blooded fish (albeit "less red"), able to use the same oxygen transport system of Channichthyidae when their oxygen carrier is temporarily incapacitated or no longer available.

In conclusion, the biochemistry of oxygen transport in antarctic fish appears to offer a wealth of interesting indications, related both to the physiological adaptation of the organism to the climatic conditions of the antarctic habitat and, more in general, to the remarkable flexibility that hemoglobin exhibits in fish physiology. For this reason, basic information on the biochemistry of the hemoglobins from these organisms, supported by studies on the physiology of respiration and on genetics, will hopefully provide valuable insights into the evolutionary history and trends of antarctic fish; in addition, at the same time, it will help to improve the general understanding of the extremely complex structure-function relationship in this ancient and versatile protein.

Acknowledgements. This work is in the framework of the Italian National Programme for Antarctic Research. It was partially supported by Grant DPP 82-18356 from the Division of Polar Programs, National Science Foundation, Washington, D.C., USA.

References

Andersen NC (1984) Genera and subfamilies of the family Nototheniidae (Pisces, Perciformes) from the Antarctic and Subantarctic. Steenstrupia 10:1–34

Andriashev AP (1987) A general review of the antarctic bottom fish fauna. In: Kullander SO, Fernholm B (eds) Proc V Congr Europ Ichthyol, Stockholm 1985. Swed Mus Nat Hist, pp 357–372

Anthony EH (1961) Survival of goldfish in the presence of carbon monoxide. J Exp Biol 38:109-129

Arnone A (1972) X-ray diffraction study of binding of 2,3-diphosphoglycerate to human deoxyhaemoglobin. Nature (Lond) 237:146–149

Barra D, Bossa F, Brunori M (1981) Structure of binding sites for heterotropic effectors in fish hemoglobins. Nature (Lond) 293:587–588

Binotti I, Giovenco S, Giardina B, Antonini E, Brunori M, Wyman J (1971) Studies on the functional properties of fish hemoglobins. II. The oxygen equilibrium of the isolated hemoglobin components from trout blood. Arch Biochem Biophys 142:274-280

Brittain T (1987) The Root effect. Comp Biochem Physiol 86B:473-481

Carey F, Teal J (1966) Heat conservation in tuna fish muscle. Proc Natl Acad Sci US 56:1464–1469

D'Avino R, di Prisco G (1988) Antarctic fish hemoglobin: an outline of the molecular structure and oxygen binding properties. 1. Molecular structure. Comp Biochem Physiol 90B:579–584

D'Avino R, di Prisco G (1989) Hemoglobin from the antarctic fish *Notothenia coriiceps neglecta.* 1. Purification and characterization. Eur J Biochem 179:699–705

D'Avino R, Caruso C, Romano M, Camardella L, Rutigliano B, di Prisco G (1989a) Hemoglobin from the antarctic fish *Notothenia coriiceps neglecta.* 2. Amino acid sequence of the α chain of Hb 1. Eur J Biochem 179:707–713

D'Avino R, Caruso C, Schininà ME, Rutigliano B, Romano M, Camardella L, Bossa F, Barra D, di Prisco G (1989b) The amino acid sequence of the α- and β-chains of the two hemoglobins of the antarctic fish *Notothenia coriiceps neglecta.* FEBS Lett 250:53–56

D'Avino R, Caruso C, Schininà ME, Rutigliano B, Romano M, Camardella L, Bossa F, Barra D, di Prisco G (1990) Hemoglobin from the antarctic fish *Notothenia coriiceps neglecta.* Amino acid sequence of the beta chain. Comp Biochem Physiol 96B:367–373

DeVries AL (1980) Biological antifreezes and survival in freezing environments. In: Gilles R (ed) Animals and environmental fitness. Pergamon Press, Oxford, pp 583–607

Dickerson RE, Geis I (1983) Hemoglobin: structure, function, evolution and pathology. Benjamin/Cummings , Menlo Park, CA

di Prisco G (1986a) Antarctic fishes and cold adaptation. Proc 1st Symp on Mar Biochem, Ital Biochem Soc, Grasso, Bologna, pp 51–68

di Prisco G (1986b) Functional properties of hemoglobin from antarctic fishes. Antarct J US 21(5):215–216 di Prisco G (1988) A study of hemoglobin in antarctic fishes: Purification and characterisation of hemoglobins from four species. Comp Biochem Physiol 90B:631–637

di Prisco G, D'Avino R (1989) Molecular adaptation of the blood of antarctic teleosts to environmental conditions. Antarct Sci 1:119–124

di Prisco G, D'Avino R, Condò S, Giardina B, Brunori M (1986) Regulatory effects on oxygen binding in antarctic fish hemoglobin. Proc 1st Symp on Mar Biochem, Ital Biochem Soc, Grasso, Bologna, pp 113–115

di Prisco G, Giardina B, D'Avino R, Condò SG, Bellelli A, Brunori M (1988) Antarctic fish hemoglobin: an outline of the molecular structure and oxygen binding properties - II. Oxygen binding properties. Comp Biochem Physiol 90B:585–591

di Prisco G, D'Avino R, Camardella L, Caruso C, Romano M, Rutigliano B (1990) Structure and function of hemoglobin in antarctic fishes and evolutionary implications. Polar Biol 10:269–274

Eastman JT (1988) Ocular morphology in antarctic notothenioid fishes. J Morphol 196:283–306

Everson I, Ralph R (1968) Blood analyses of some antarctic fish. Bull Br Antarct Surv 15:59–62

Giardina B, Amiconi G (1981) Measurement of binding of gaseous and nongaseous ligands to hemoglobins by conventional spectrophotometric procedures. Methods Enzymol 76:417–427

Gillen RG, Riggs A (1972) Structure and function of the hemoglobins from the carp, *Cyprinus carpio*. J Biol Chem 247:6039–6046

Grigg GC (1967) Some respiratory properties of the blood of four species of antarctic fishes. Comp Biochem Physiol 23:139–148

Hemmingsen EA, Douglas EL (1970) Respiratory characteristics of the hemoglobin-free fish *Chaenocephalus aceratus*. Comp Biochem Physiol 33:733–744

Hemmingsen EA, Douglas EL (1977) Respiratory and circulatory adaptations to the absence of hemoglobin in Chaenichthyid fishes. In: Llano GA (ed) Adaptations within antarctic ecosystems. Smithsonian Inst, Washington, DC, pp 479–487

Holeton GF (1970) Oxygen uptake and circulation by a hemoglobinless antarctic fish (*Chaenocephalus aceratus* Lönnberg) compared with three red-blooded antarctic fish. Comp Biochem Physiol 34:457–471

Hureau J-C, Petit D, Fine JM, Marneux M (1977) New cytological, biochemical and physiological data on the colorless blood of the Channichthyidae (Pisces, Teleosteans, Perciformes). In: Llano GA (ed) Adaptations within antarctic ecosystems. Smithsonian Inst, Washington, DC, pp 459–477

Kennett JP (1968) Paleo-oceanographic aspects of the foraminiferal zonation in the Upper Miocene-Lower Pliocene of New Zealand. G Geol Ser 2, 35:143–156

Macdonald JA, Montgomery JC, Wells RMG (1987) Comparative physiology of antarctic fishes. Adv Mar Biol 24:321–388

Monod J, Wyman J, Changeux JP (1965) On the nature of allosteric transition: a plausible model. J Mol Biol 12:88–118

Perutz MF (1969) The haemoglobin molecule. Proc R Soc London, Ser B 173:113-140

Perutz MF, Brunori M (1982) Stereochemistry of cooperative effects in fish and amphibian hemoglobins. Nature (Lond) 299:421-426

Qvist J, Weber RE, DeVries AL, Zapol WM (1977) pH and hemoglobin oxygen affinity in blood from the antarctic cod *Dissostichus mawsoni*. J Exp Biol 67:77–88

Raymond JA, DeVries AL (1976) Some respiratory characteristic of the blood of four antarctic fishes. J Exp Zool 196:393–396

Riggs A (1970) Properties of fish hemoglobins. In: Hoar WS, Randall DJ (eds) Fish physiology, vol 4. Academic Press, New York, pp 209–252

Root RW (1931) The respiratory function of blood in marine organisms. Biol Bull Mar Biol Lab, Woods Hole 61:427–456

Rossi-Fanelli A, Antonini E (1960) Oxygen equilibrium of haemoglobin from *Thunnus thynnus*. Nature (Lond) 186:895–896

Ruud JT (1954) Vertebrates without erythrocytes and blood pigment. Nature (Lond) 173:848–850

Tan AL, De Young A, Noble RW (1972) The pH dependence of the affinity, kinetics, and cooperativity of ligand binding to carp hemoglobin, *Cyprinus carpio*. J Biol Chem 247:2493–2498

Tetens V, Wells RMG, DeVries AL (1984) Antarctic fish blood: respiratory properties and the effects of thermal acclimation. J Exp Biol 109:265–279

Wells RMG (1986) Cutaneous oxygen uptake in the antarctic icequab, *Rigophila dearborni* (Pisces; Zoarcidae). Polar Biol 5:175–179

Wells RMG, Jokumsen A (1982) Oxygen binding properties of hemoglobins from antarctic fishes. Comp Biochem Physiol 71B:469–473

Wells RMG, Ashby MD, Duncan SJ, Macdonald JA (1980) Comparative studies of the erythrocytes and haemoglobins in nototheniid fishes from Antarctica. J Fish Biol 17:517–527

Wells RMG, Macdonald JA, di Prisco G (1990) Thin-blooded antarctic fishes: a rheological comparison of the haemoglobin-free icefishes, *Chionodraco kathleenae* and *Cryodraco antarcticus*, with a red-blooded nototheniid, *Pagothenia bernacchii*. J Fish Biol 36:595–609

Wittenberg JB, Wittenberg DK (1974) The choroid rete mirabile. I. Oxygen secretion and structure: comparison with the swim bladder rete mirabile. Biol Bull 145:116–136

Subject Index

MIX
Papier aus verantwortungsvollen Quellen
Paper from responsible sources
FSC® C105338

If you have any concerns about our products,
you can contact us on
ProductSafety@springernature.com

In case Publisher is established outside the EU,
the EU authorized representative is:
Springer Nature Customer Service Center GmbH
Europaplatz 3, 69115 Heidelberg, Germany

Printed by Libri Plureos GmbH
in Hamburg, Germany